Skylab

America's Space Station

Springer
London
Berlin
Heidelberg
New York
Barcelona
Hong Kong
Milan
Paris
Santa Clara
Singapore
Tokyo

David J. Shayler

Skylab

America's Space Station

Springer

Published in association with

Praxis Publishing
Chichester, UK

David J. Shayler
Astronautical Historian
Astro Info Service
Halesowen
West Midlands
UK

SPRINGER–PRAXIS BOOKS IN ASTRONOMY AND SPACE SCIENCES
SUBJECT *ADVISORY EDITOR*: John Mason B.Sc., Ph.D.

ISBN 1-85233-407-X Springer-Verlag Berlin Heidelberg New York

British Library Cataloguing-in-Publication Data
Shayler, David J.
Skylab: America's space station. – (Springer–Praxis series in astronomy and space sciences)
1. Space stations 2. Astronautics – United States
I. Title
629.4′42′0973

ISBN 1-85233-407-X

Library of Congress Cataloging-in-Publication Data
Skylab: America's space station/David J. Shayler.
p. cm. – (Springer–Praxis series in astronomy and space sciences)
Includes bibliographical references.
ISBN 1-85233-407-X (alk. paper)
1. Skylab Program. I. Title. II. Series.

TL789.8.U6 S567 2001
629.44′2–dc21 2001020621

Printed by MPG Books Ltd, Bodmin, Cornwall, UK

Copy editing and graphics processing: R.A. Marriott
Cover design: Jim Wilkie
Typesetting: BookEns Ltd, Royston, Herts., UK

Printed on acid-free paper supplied by Precision Publishing Papers Ltd, UK

Dedicated to the memory of
Charles 'Pete' Conrad (1930–1999), Commander, Skylab 2

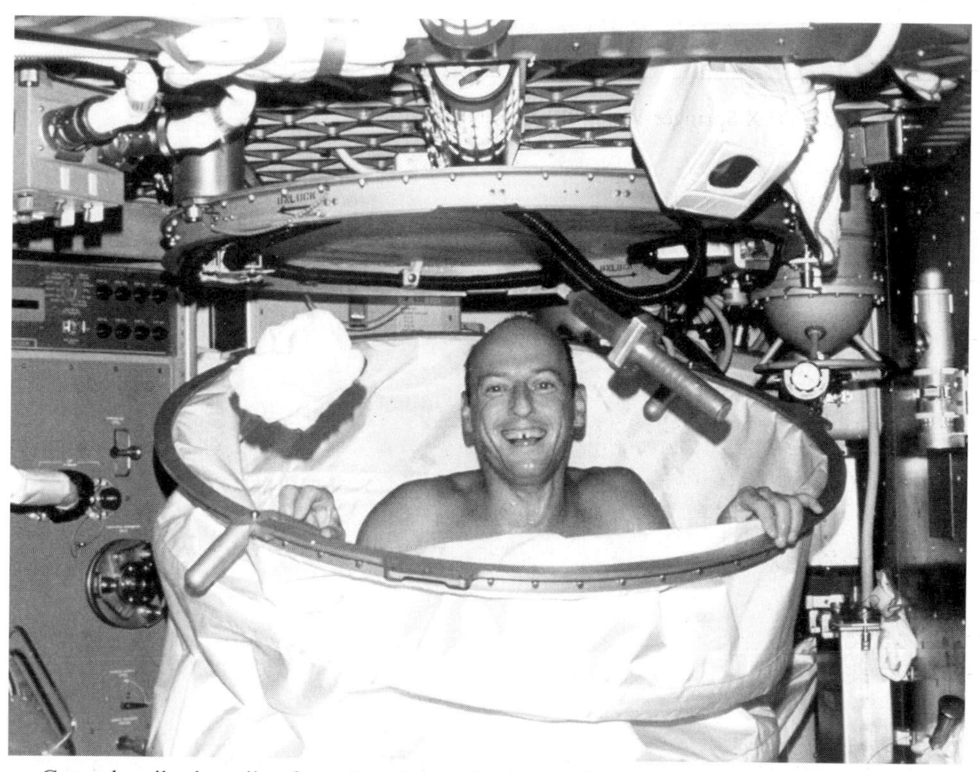

Conrad smiles happily after a hot shower in the wardroom of Skylab during his fourth and final mission in space

'If people say 'space' to me, I don't think about flying to the Moon.
I think about flying in the Skylab; it was a tremendous vehicle.
It was a great pleasure to work in that thing, and to operate it.'

Table of contents

Foreword

In 1958 my parents, my future wife and I stood out in the back yard of our home in Kenmore, New York, and watched mankind's first satellite – the Soviet Sputnik – pass overhead. Back then, I had never heard the word 'astronaut'; but just fifteen years later my parents stood out in that same back yard and watched me pass overhead. In that short period of time, America assembled a first-class space team, went to the Moon, and put a space station into orbit. The Mercury, Gemini, Apollo and Skylab programmes rapidly succeeded each other, and all of them were completed within little more than a decade.

Some considered Skylab to be the legacy of this Herculean effort to go to the Moon. Unused Apollo hardware, resulting from the cancellation of the last three lunar flights, became available to those who had a relatively low-cost application in mind, and at the time the sentiment of some was that 'it might as well be used for something'. Indeed, Skylab was originally called the Apollo Extension System, and then the Apollo Applications Program. Others saw the availability of Apollo hardware as an opportunity to initiate two essential steps in mankind's impending expansion into space. First, Skylab would begin the habitation and utilisation of this new world in earnest, just as settlers have always followed explorers. Second, as our first laboratory in near-Earth orbit, Skylab would not only conduct experiments across a wide range of scientific disciplines, but would also begin the indispensable preparation of man and systems for future long-duration manned missions to reach farther and to explore other bodies within our solar system.

But it was not an easy path to tread. The successful utilisation of Skylab required a much better understanding of the physical, physiological and operational challenges of long-duration spaceflight than was available at the time of the launch of Skylab 1. Space sickness became a more prominent factor than in earlier programmes because of the larger spacecraft volumes available for crew motion and the higher-intensity work-loads required in the early days of each mission. Counteracting the advantage of the lack of stress on the body lay one of the major problems of spaceflight: with no gravity to work against, muscles weaken; and without sufficient bodily exercise, bones slowly loose calcium and also weaken – just as they do in Earth-bound bed-ridden patients.

But perhaps the most elusive factor is that human productivity is maximized in

Scientist-astronaut and Skylab 4 Science Pilot Ed Gibson works at the Apollo Telescope Mount console in the Multiple Docking Adapter of Skylab during his record-breaking 84-day mission.

space, in much the same way as it is on Earth. Moving a human a few hundred miles above the Earth's surface does not reduce the value of achievements through intellectual challenge, individual initiative and sense of accomplishment. Yet the mode of operation carried over from short-duration flights (all actions, large or small, rigidly controlled by predefined timelines and checklists, thereby reducing most experiments to 'procedurements') does just that. Incorporating even the basics of human productivity into the early design and planning of operations and experiments presents a challenge that was inadequately addressed before Skylab. Will one of the lessons learned from Skylab again be that often we do not learn? Even now, this question remains unanswered.

Despite the problems encountered, anyone would have found life on Skylab an enjoyable experience. Within days, microgravity becomes an accepted and pleasurable way of life in which all motions are controlled by fingertip forces, and there is no real 'up' or 'down' – only six directions of 'sideways'. Within days you realise that space is neither foreign nor hostile; rather, it becomes familiar, and then simply comfortable, and you begin to think of your space station as a stable, three-bedroom home – possibly perched on a moving mountain 270 miles high. In fact, it feels so solid and so secure that it does not really feel like flying at all until you finally leave it in your re-entry vehicle. Then it is just like leaving your home on Earth, settling into a sports car, and getting back on the road again. Earth becomes like the face of an old friend. You come to recognise our planet's features – not by the outlines of land masses, but by colors and patterns and textures that seem to take on a life of their own. See a red wind-swept desert, and know you are over north Africa; see lines of clouds over a dense jungle, and you know you are over Brazil; or see plankton blooming at the edge of a current mixing with another current near a coastline, and you know you are over the Falkland current off the east coast of South America.

But what of our future in space? In some respects, the prospects for America have never been brighter. We still have within our people a spirit and will that wants nothing more than the ever-deeper exploration and profitable use of space. In America we have physical facilities second to none; we have our enthusiastic youth – equally motivated but far better trained than those young engineers who took us to the Moon and into Skylab; and we have our greybeards – engineers and managers with knowledge and wisdom accrued from decades of experience.

With the International Space Station we expect to develop new materials which are important in engineering and biology; and we must also extend our knowledge of how our bodies function without gravity, which is essential for a better understanding of how they function on Earth and how they will function on future long-term missions far from Earth. Next, we might see a return to the Moon – not only for exploration, but also to prepare us for the next stage. By using our experience of long-duration missions and lunar surface operations, we could then voyage to Mars – and when we land there, the importance of the planet may extend far beyond exploration. Mars was once a much hotter and wetter place, and it had a dense atmosphere and flowing rivers; but why is it now so barren? Once we understand, we will also know much more about the future of our own planet.

But what of exploration beyond the solar system? In addition to Earth orbit – the home of the Hubble Space Telescope – one of the best locations for carrying out astronomical observations would be the far side of the Moon. Instruments set up there could one day be capable of directly imaging planets around other stars – and if we were to find a blue planet with an oxygen atmosphere, the pull would be irresistible. There would be a crash programme for a 'warp drive,' and eventually – and inevitably – a mission that would fire our imagination far more than would any episode of *Star Trek*. But the distances are immense. On Skylab 4 we travelled the distance that light travels in just three minutes; yet light takes more than four years to reach a Centauri, the star nearest to our solar system. Even considering the total distance travelled by Skylab and during all of mankind's space missions, it is obvious that when it comes to real space travel we have barely put our toe out the front door.

Skylab has helped to place us at the forefront of this quest, and has been instrumental in guiding us towards potential achievements, which are much more profound than any of us can, at the moment, fully appreciate. Yet I believe that a deep enough scratch into the tough hide of even the most cynical or reticent space scientist or engineer will reveal a 'trekkie' – someone who realises that spaceprobes and their data are interesting, and sometimes even exciting, but that ultimately it is we who have to go there ourselves, to explore and experience new vistas before they can truly become a real part of our own world. And one day, astronauts and cosmonauts – driven by this innate understanding and by the pride, spirit and capabilities of all humans – will finally live up to their names and become those who do, in fact, travel among the stars.

Edward G. Gibson, PhD
Science Pilot, Skylab 4

Author's preface

Astronaut Pete Conrad was especially proud of his personal contribution to the mission that saved America's space station Skylab from total loss following launch damage on 14 May 1973. As Commander of the first manned mission, Conrad had gained more satisfaction in completing this mission – his fourth – than he had from commanding the second manned lunar landing mission four years earlier, or from either of his other two missions during the Gemini programme.

After joining NASA in September 1962, Conrad spent seven years training for a flight to the Moon, and his command of Apollo 12 became one of many high points in his distinguished career. Apollo 12 was the second mission to land on the Moon, and Conrad became the third man to walk on the lunar surface. It was a highly successful mission that thankfully followed the flight plan. Two long-time friends – Dick Gordon and Alan Bean – joined him on the flight, making the crew one of the very best, and he enjoyed the mission tremendously. Later, as further Moon-flights were being cancelled and as crew seats became scarce, Conrad took over the Skylab Branch Office within the Astronaut Office, and set himself the new goal of becoming the first Commander of an American space station.

On 14 May 1973 he watched the last Saturn V to be launched take the empty Skylab Orbital Workshop into orbit. Close by was the smaller Saturn 1B that would take Conrad's crew up to the station the very next day. However, all the work that he had put into training for his Skylab mission over the previous three years suddenly seemed to have been in vain. Seconds after the Saturn V left the pad it sustained significant structural damage during the ascent to orbit. Instead of launching to the station the following day, the next two weeks were spent incorporating contingency plans to save the crippled space station.

Conrad and his crew eventually launched, however, and successfully completed their planned 28-day mission. The mission – Skylab 2 – set the standards for the second crew, who flew a highly successful mission of 59 days. The third and final crew built upon and exceeded the achievements of their colleagues, and completed the programme with an unprecedented 84-day voyage around the Earth – an American manned spaceflight record that would endure for the next 21 years.

A team of twenty-two astronauts trained for Skylab, and nine of these lived on the station during three manned flights completed between May 1973 and February 1974. Much was achieved, and a great deal was learned about the Sun, the solar system, the space environment, the Earth, and the effects of extended-duration spaceflight on the human body. There could have been other Skylabs; but there were not, and many of the Skylab astronauts have often lamented that their achievements and contributions to human spaceflight have been overshadowed by the Apollo lunar missions, from which evolved the concept of the space station.

Skylab was developed from design studies in the 1960s, called the Apollo Applications Programme (AAP). These studies were evaluated for a range of alternative applications using hardware developed to put the first men on the Moon, including the development of a space station.

By 1971 – the tenth anniversary of the first American and Russian manned spaceflights – the longest period that American astronauts had remained in space was fourteen days, in the cramped Gemini spacecraft in 1965. The Russians had recently orbited the world's first space station, Salyut, and the three Soyuz 11 cosmonauts had spent 23 days on board; but the mission ended tragically with the loss of the crew during the recovery phase after leaving the station.

Skylab offered the Americans their first chance to gain experience of extended-duration flight on larger space stations then being proposed, with crews of up to 50 or even 100 astronauts on board, to be constructed with the help of a new vehicle called the Space Shuttle. The concept is only now being realised with the construction of the International Space Station, nearly thirty years after the three pioneering Skylab missions.

In the intervening three decades, the Russian Salyut space stations accrued an impressive record of achievements, only to be superseded by the highly successful Mir programme, during which cosmonauts regularly completed missions of several months. For America, the final decision to build a large space station came in 1984, after many years of discussion and planning, and a decade after the last Skylab crew returned home. It would be a further fifteen years before the first elements of the ISS were placed in orbit – and then only with the co-operation and partnership of another fifteen nations, including Russia.

While the Russians gradually developed and improved their space-station experience, the American Space Shuttle flew more than ninety highly successful missions, and completed a range of scientific, military, commercial and technological objectives that offered (but could not deliver) cheaper and more frequent access to space. In the 1980s and 1990s, America's equivalent of a space station was the European-developed space laboratory, Spacelab. It was housed in the payload bay of the Shuttle, and was very reliable, but the maximum duration of a mission was up to eighteen days.

The first phase in the creation of the ISS consisted of a series of American Shuttle missions to the Russian Mir complex, and the flights of cosmonauts on the Shuttle and astronauts on Mir from 1993 to 1998. These flights were designed to provide valuable experience in mission planning and flight operations that would be crucial for the construction and operation of the larger station.

Undoubtedly, the Russians have contributed a vast resource of experience and knowledge to the ISS, and an understanding of living and working in space for periods of several months. But the Americans, too, have an archive of data on long-duration spaceflight, not only from the seven astronauts who lived on Mir between 1995 and 1998, but from nine astronauts more than twenty years earlier.

I have always been surprised at the lack of published works on Skylab over the years, apart fom NASA publications, magazine articles, scientific papers and popular reviews of the missions. One of the first accounts of life onboard Skylab appeared in a series of articles by Henry S.F. Cooper Jnr in *New Yorker*. These were later expanded into the book *A House in Space* (1976), which followed 'a typical Skylab day, focusing on the third and last of the crews. They stayed up the longest and also proved the most independent, causing some unprogrammed differences with the scientists and engineers controlling the mission from the ground.' Many of the 'unprogrammed differences' were, however, part of the habitation evaluations that were planned as part of the Skylab flight programme before the astronauts left the ground. Unfortunately, these differences – along with the media misinterpretation of a 'strike' by the third crew – received more attention than the mission itself.

Since the last crew came home more than 25 years ago, no commercial publisher has dedicated a book to Skylab, and of the nine Skylab astronauts, only Bill Pogue has published personal accounts of his experiences, although the others have contributed comments, articles, interviews, conference papers, industry reports and countless public speeches on living and working on the station. Ed Gibson also produced a text-book on solar physics before flying on Skylab, and, like Pogue, has published a number of novels since his mission.

Important lessons were learned onboard Skylab during 1973 and 1974, some of which were not adopted during the Shuttle programme and are only now being re-examined for the ISS. It remains to be seen whether *all* of the lessons from the Skylab flights of Conrad and his colleagues have been learned.

In writing this book I wanted to provide the reader with an insight into the real nature of Skylab – how it evolved, how it was operated, the work and activities of the astronauts, and what they discovered. I did not want to compile a 'history', as this can be found in the official NASA publication *Living and Working in Space* (NASA SP4208, 1983), by W. David Compton and Charles D. Benson. Neither did I want to compile a diary of day-to-day events or a collection of scientific findings or results, as there exist countless articles and documents that specialise in those areas. Some of these are listed in the Bibliography at the end of this book, and are highly recommended for further reading and research.

What I desired for this book was an introduction to Skylab for those who are too young to remember the programme, or have recollections of a programme that emerged from the shadow of Apollo to become perhaps one of the most important contributions to pioneering work on the habitation of humans in space. Space-station habitability became more familiar during the period of Mir, and on the ISS it will become almost routine; but the skills of living and working away from Earth for several weeks were first learned on Skylab.

The format of this book follows the evolution of the Skylab Orbital Workshop, through the preparations to fly and support the missions, to the training and activities of the crews, and the lost opportunities of cancelled follow-on missions. I have also included a discussion on what followed Skylab, what lessons were learned that have found their way into the follow-on programmes, and what has not been incorporated. All of these issues are important for the much larger and more complex ISS programme.

Skylab is not merely a story of hardware, long missions and space science, but also of explorers and pioneers, and of lost opportunities. The development of the Mir programme would not have been possible without the earlier series of Salyut space stations. Had Skylab been followed by Skylab B as planned, America would probably have had its own permanent space station in orbit by about fifteen years ago. This could have been a major programme for the early Space Shuttle missions instead of trying to 'sell' the Shuttle as a commercial launch vehicle – a task for which it was clearly not designed.

When the Shuttle was planned to dock to Mir (where NASA astronauts were to spend many months), the old Skylab files were dusted off and reviewed; and Russian cosmonauts revealed that they also read the Skylab reports as part of their preparations for their own long-duration missions.

Skylab, then, was not the world's first space station, as that distinction belongs to Salyut. Neither can it be called America's first space station, as the ISS is an international effort. Skylab remains as being the *one* and *only* American space station to reach Earth orbit.

The Apollo missions are recognised as a milestone in human exploration. Skylab, with Salyut and Mir, should be remembered as a significant landmark in pushing the limits of human endurance in space. The valuable resources that Skylab provided are only now being re-evaluated as American astronauts again prepare for spaceflights lasting several months.

It is hoped that this book will assist in providing a wider understanding of Skylab's work and the importance of its achievements. It will undoubtedly stir memories and generate renewed interest; and it might perhaps inspire others to write books to document the full story of Skylab – America's space station.

David J. Shayler
West Midlands, England

Spring 2001

www.astroinfoservice.co.uk

Acknowledgements

Skylab was the first real spacecraft I saw flying in space. Admittedly, it was only a pinprick of light in the night sky, and I was standing on the south-west coast of England in 1973. I knew Skylab would be flying close by, over northern Spain and France, and I hoped that the sky would be clear enough to catch a glimpse. Sure enough, at the right time, a very weak 'star' travelled serenely across the sky. It was a brief, strained glance, but it was the right time and place, and I knew that it must be the station. I have always thought of Skylab with that memory of seeing it with my own eyes. Until then, the only other item of 'real' space hardware I had seen was the Apollo 10 Command Module on tour in the UK – but Skylab was really 'up' in space, and with astronauts on board!

For me, Skylab was as special as Apollo; but the media apparently did not agree, as the TV and news coverage was far less than during the Apollo missions. Because of this I have tried, over the years, to learn as much as possible about the programme. I have consulted NASA archives in Houston, and have talked to several former astronauts who were part of the programme; but the research for this book really started that summer evening in 1973, and has continued for almost three decades, during which time some outstanding people have offered assistance, shared their experiences and, with a glint in the eye, recalled treasured memories of a personally rewarding programme.

From the NASA History Office in Washington DC, Lee Saegesser, initially, and more recently Roger Launius, have offered continued support and interest for my studies of NASA human spaceflight programmes, offering valuable comments and access to official NASA publications and archives, as well as research direction.

In Houston, the History Office staff of Janet Kovacevich, Joey (Pellarin) Kuhlman, David Portree, and more recently Glenn Swanson, have provided repeated access to the wealth of stored archives. This is an extensive repository of information for those willing to sift through box after box of memos, correspondence, reports and notations. I was even allowed to commandeer a photocopier to save me from writing hundreds of notes. At Rice University, Joan Ferry and her staff enthustically provided access to the Skylab collection when it was located there, and to the archives of former astronaut Curt Michel, who worked in the CB AAP office in the mid- to late-1960s.

The continued support of a network of fellow space authors, journalists, historians, sleuths and enthusiasts that have provided a wealth of data over many years is once again gratefully acknowledged. Without their frequent and generous help, my research would be much more difficult, and the results would be incomplete. Special thanks are given to Colin Burgess, Mike Cassutt, Phil Clark, Rex Hall, Brian Harvey, Anders Hansonn, Bart Hendrickx, Neville Kidger, Andy Salmon and Bert Vis.

Special thanks are also extended to former Skylab astronauts for their comments, suggestions, insights and unselfish supply of information to detail this account. This is, after all, their story. I am indebted to Ed Gibson for his Foreword and for an insight into operating the ATM; to Jerry Carr, on details of living and working onboard Skylab; and to Bill Pogue, for explaining the Earth resources experiments and operations. 'Dr Bill' Thornton was instrumental in detailing information on the medical experiments and the SMEAT test; Don Lind provided a very personal view of training as a rescue crewman; and in several interviews a decade ago, (the late) Karl Henize offered an insight into the astronomical research carried out on Skylab. Others who have, over several years, provided encouragement or information by interview, correspondence or documentation, include Vance Brand, Walt Cunningham, Bruce McCandless, Story Musgrave and Rusty Schweickart. Paul Weitz offered assistance on SL-2 details, and Jack Lousma on SL-3 details, while Dick Truly and Bob Crippen took time out from their busy schedules to answer my e-mails concerning the CapCom and support assignments.

I am indebted to the staff of the Still Photo Library in Houston, especially Debbie Dodds, Lisa Vasquez, Jody Russell, Mary Wilkinson and Mike Gentry; the sound department of Diane Ormsbee and Pete Nubile; the JSC PAO staff of Iva 'Scotty' Scott, Barbara Schwartz, Jeff Carr, James Hartsfield; Eileen Hawley and Sherri Jules; Librarians Margaret Persinger and Kay Grinter at KSC; and Chuck Shaw, Flight Director at JSC, for information on mission control operations. Over the past fifteen years these people have continued to support my research in this and other projects. Without their dedication, access to the resources of NASA would have been impossible.

Thanks are again due to my brother Mike for the initial editing of the original text in such a professional way, which greatly enhanced the result. I must also thank David Hardy for allowing me to include his excellent conceptual illustration of a Soyuz–Skylab docking mission (first published in *Challenge of the Stars*, Patrick Moore, 1972). As with previous projects in the Springer–Praxis series, I am indebted to Clive Horwood, Chairman of Praxis, for supporting the project from its very first suggestion; and to Project Editor Bob Marriott, for his work in preparing the text for publication, and for spending many hours scanning and processing the illustrations.

A vote of thanks is also extended to the ground-crew of hundreds of contract workers, support teams, trainers, launch crews, controllers, scientists, students and recovery teams that not only made Skylab possible, but also made it an outstanding success.

My warmest thanks are offered to the families of all the astronauts of Skylab, especially those of Pete Conrad and Karl Henize. Each member of the astronauts' families provided the all-important 'domestic support crew' while dad was having fun out of town. Skylab was and still continues to be very much part of their lives too.

Finally, and on a personal note, I wish to acknowledge the continuing support and encouragement of my parents, who for more than forty years have tolerated my personal involvement with the space programme – from drawing rockets at the age of five, to seeing the results of my writing endeavours. I must also include my aunt, Gwen Waldon, who still finds it difficult to believe that anything can fly in space – especially something as large as a house! Therefore, to my personal back-up crew, mum and dad – this one is for you.

List of illustrations, plates and tables

The human element

Flight operations

Colour plates (between pp. 280 and 281)

Tables

Front cover

Skylab sails serenely in Earth orbit, as seen from one of the Apollo spacecraft carrying the astronauts who lived onboard the station. Skylab supported nine astronauts in three crews of three between May 1973 and February 1974. They spent 28, 59 and 84 days in space, and set American space endurance records that were unsurpassed for 21 years.

Back cover

Each of the three-man crews' colourful mission emblems are displayed with the Skylab programme emblem. Following a tradition which began during the Gemini programme, and which continues to the present day, American crews designed their own individual mission patches that identified their flight and objectives, although no individual call-signs were adopted.

Acronyms and abbreviations

A7LB	Pressure suit
AAP	Apollo Applications Programme
ACS	Attitude Control System
AES	Apollo Extension System
AFB	Air Force Base
AL	Airlock
ALSA	Astronaut Life Support Assembly
AM	Airlock Module
AMU	Astronaut Manoeuvring Unit
AOS	Acquisition Of Signal
APCS	Attitude Pointing and Control System
ARIA	Apollo Range Instrumented Aircraft
ASTP	Apollo–Soyuz Test Project
ATM	Apollo Telescope Mount
C&D	Control and Display
CapCom	Capsule Communicator
CCS	Command Communications System
CDDT	Countdown Demonstration Test
CDR	Commander
CM	Command Module
CMG	Control Movement Gyro
CSM	Command and Service Module
CWS	Central Work Station
DoD	Department of Defence
ECG	Electrocardiogram
ECS	Environmental Control System
EDS	Emergency Detection System
EDT	Eastern Daylight Time
EEG	Electroencephalograph
EMU	Extravehicular Mobility Unit
EPCS	Experiment Pointing Control Subsystem

EPS	Electrical Power System
EREP	Earth Resources Experiment Package
EST	Eastern Standard Time
ETC	Earth Terrain Camera
EVA	Extravehicular Activity
FAS	Fixed Airlock Shroud
FCC	Flight Control Computer
FCMU	Foot Controlled Manoeuvring Unit
FD	Flight Director
FDF	Flight Data File
FOMR	Flight Operations Management Room
GET	Ground Elapsed Time
GSE	Ground Support Equipment
HHMU	Hand Held Manoeuvring Unit
HOSC	Huntsville Operations Support Center (MSFC, Alabama)
IMU	Inertial Measuring Unit
IU	Instrument Unit
IVA	Intravehicular Activity
JSC	Johnson Space Center (Houston, Texas)
KSC	Kennedy Space Center (Florida)
LBNP	Lower Body Negative Pressure device
LC	Launch Complex
LCC	Launch Control Center (KSC, Florida)
LCG	Liquid Cooled Garment
LES	Launch Escape System
LM	Lunar Module
LOS	Loss Of Signal
LSS	Life Support System
MCC	Mission Control Center (Houston, Texas)
MD	Mission Day
MDA	Multiple Docking Adapter
ML	Mobile Launcher
MLP	Mobile Launcher Platform
MOCR	Mission Operations Control Room
MOL	Manned Orbiting Laboratory
MSC	Manned Spacecraft Center (Houston, Texas)
MSFC	Marshall Space Flight Center (Huntsville, Alabama)
MSG	Mission Support Group
NASA	National Aeronautics and Space Administration
NASCOM	NASA Communications Network
NBS	Neutral Buoyancy Simulator
O&C	Operations and Checkout
OA	Orbital Assembly
OWS	Orbital Workshop
PAD	Pre-Advisory Data *or* Payload Application Document

PAO	Public Affairs Officer
PGA	Pressure Garment Assembly
PI	Principle Investigator
PLT	Pilot
PS	Payload Shroud
R&D	Research and Development
RCS	Reaction Control System
SA	Solar Array
SAL	Scientific Airlock
SAS	Solar Array System
SEVA	Skylab EVA Visor Assembly
S-IB	Saturn 1B, first stage
S-IC	Saturn V, first stage
S-II	Saturn V, second stage
S-IVB	Saturn V, third stage
SL	Skylab
SLA	Spacecraft LM Adapter
SLS	Skylab Simulator
SM	Service Module
SMEAT	Skylab Medical Experiment Altitude Test
SML	Skylab Mobile Laboratory
SPS	Service Propulsion System
SPT	Science Pilot
STDN	Spacecraft Tracking and Data Network
SUEVA	Stand-Up EVA
SWS	Saturn Workshop
TACS	Thruster Attitude Control Subsystem
TCS	Thermal Control System
UCTS	Urine Collection and Transfer Assembly
UV	Ultraviolet
VAB	Vehicle Assembly Building
VOX	Voice Operated Relay
VTR	Video Tape Recorder
WMC	Waste Management Compartment
WR	Wardroom

Prologue

When the Skylab space station was launched in 1973, it was by far the largest spacecraft intended for manned occupation that had ever been placed into long-term orbit. For more than fifteen years the American space station would hold the space station mass record of approximately 90 tons, until the modular 130-ton Mir space station was assembled in the 1990s – and Mir has been surpassed by the 470-ton International Space Station. The popular press of the day described Skylab as 'a house in space', and essentially, that is what it was.

American astronauts had previously spent only a few days in space, onboard rather cramped spacecraft, with defined mission objectives of sustaining the astronauts, testing new procedures and equipment, rendezvous, docking, or landing on the Moon. Although many experiments and observations had been undertaken during these missions, engineering and technical objectives had always taken precedence over pure science.

Skylab was to change all that. As the Apollo programme reached its end, so Skylab became the focus of manned spaceflight in America, as 'the next logical step in space', towards the larger goal of a permanent presence in space. The four missions of Skylab were unprecedented in the American space programme. The mighty Saturn V, in its three-stage version, had allowed men to fly to the Moon; but now it was a two-stage version that would lift the unmanned space station into Earth orbit. The three crews would ride on the smaller Saturn 1Bs, performing rendezvous and docking with the station using the Apollo Command and Service Module. They would then spend from four to eight (and eventually twelve) weeks inside the modified Saturn S-1VB third stage workshop, to conduct a range of experiments, observations and research that had been impossible on previous American spacecraft.

The Skylab series was promoted as a major leap forward for science, medicine and technology that would reap a harvest of benefits for the people of Earth. The TV views of Skylab would be very different from those of Apollo astronauts in their confined spacecraft, demonstrating the wonder of weightlessness, and trying, with difficulty, to somersault in space. In an effort to provide Earth-bound TV viewers with appreciation of the sheer internal volume of the station, each of the Skylab crews provided a guided TV tour of their 'house in space', to demonstrate the

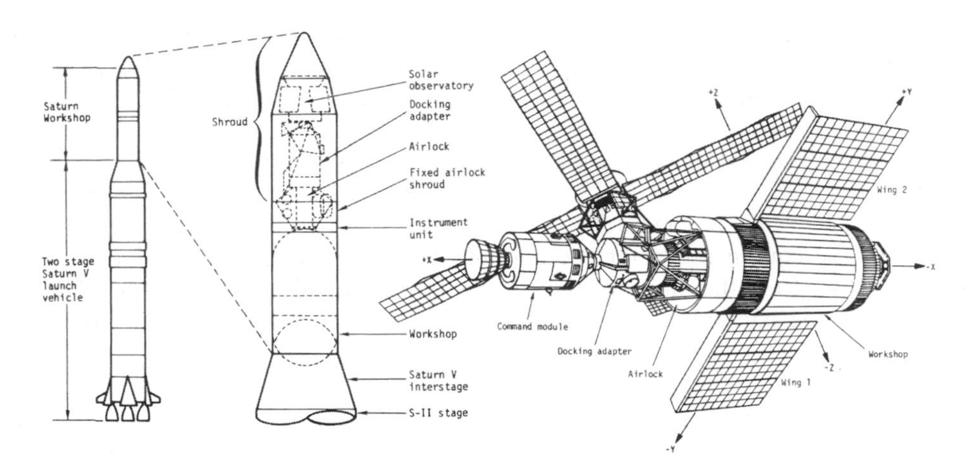

The major components of the Skylab cluster, showing (*left*) the launch configuration and (*right*) the orbital configuration. Orientation axes are also shown.

spacecraft's differences from those which had flown before. Using both these TV tours and later movie footage, the Skylab astronauts were able to convey a sense of the magical enjoyment of long-term life in microgravity.

To demonstrate the layout of the space station (officially called the Orbital Workshop (OWS) or, more commonly, the workshop) for TV viewers, the astronauts usually began in the docked Command and Service Module (CSM) at the forward end of the complex. The Service Module (SM) was attached to the rear of the Command Module (CM), and contained all the necessary equipment, power, consumables and engines to manoeuvre the spacecraft towards the station and initiate the return to Earth. The conical CM carried the three astronauts to their new home, and when docked to the Skylab the centre couch was normally folded away to provide more room. The CM was isolated from the rest of the workshop, and during the missions it was the location used for the astronauts' private calls and conversations with their families. At the end of the mission the CM brought home the crews through atmospheric re-entry, protected by the heat-shield at the base of the module. Three large ring-sail parachutes were used to decrease the descent speed for an ocean recovery.

To begin the tour, a gentle push from the middle of the couch area propelled the astronaut and his TV camera under the control panels and towards the apex of the CM, into the docking tunnel. On the Apollo missions this led to the Lunar Module crew compartment, but on Skylab it took the astronaut into the Multiple Docking Adapter (MDA), which was essentially a module for transferring from the Apollo spacecraft into the main part of the OWS. This cylindrical compartment was 17.3 feet long and 10 feet in diameter, with a mass of 13,800 lb and a habitable volume of 1,140 cubic feet. The main features of this element were the two docking ports, which allowed the Apollo spacecraft to dock with the complex. The primary docking port was located on the axial, forward end of the MDA, and was occupied by the resident crew CSM for the duration of their stay. The second port was located 90 from the first, on the –Z axis. This was an alternative port for docking with the CSM, and was available as a rescue port for a second CSM to use if the first spacecraft could not be detached from the primary docking location.

The MDA was also the location of control centres for major experiments: the solar observation control panel, the Earth resources cameras and controls, and the M-152 furnace/vacuum chamber used for the material processing experiments. The design of the equipment in the module allowed the Earth Resources Experiment Package (EREP) in the 'floor' of the –Z axis to be pointed towards Earth, while in the + Z axis the externally mounted solar experiments and cameras could be pointed at the Sun. Several experiments were externally mounted on this module, which also served as a major location for spare parts, tool kits and stowage. With such an array of equipment and disciplines, this was one of the most frequently visited areas during the three manned missions. From the MDA, the astronaut guide continued to float through its length from the primary docking hatch, deeper into the space station.

Floating through an inner hatch, he entered the Airlock Module (AM), which provided access from the MDA into the main part of the workshop via a flexible extension that connected the module to the top of the OWS forward dome. Measuring 17.6 feet long, and with a variable width of 10/5.5/22 feet and a mass of 49,000 lbs, it had a habitable volume of 624 cubic feet. The AM was a major structural element for the whole cluster, and was the location of electrical, environmental and communication control centres for the station. The module consisted of two concentric cylinders, the outer of which matched the OWS in diameter, carried the payload shroud during launch, and was the base for the telescope mount. The inner cylinder had hatches at both ends – which could be sealed for depressurisation – and it also featured a side hatch opening for EVA – a spare from the Gemini programme. The design of the hatch perfectly matched the curving shape of the AM, and as the pioneering American efforts of EVAs in Earth orbit began with Gemini 4 in 1965, it seems appropriate that nine years later they should end using a hatch of a similar design during the Skylab 4 mission. The AM – essentially the 'utility module' – also carried the automatic malfunction system, and manual controls for pressurisation and the purification of the atmosphere. Attached to the base of the AM and the top of the S-IVB was the unpressurised Instrument Unit (IU) – the electronic centre of the launch vehicle and workshop activation. The IU measured 3 feet in length and 22 feet in diameter, and had a mass of 4,550 lbs.

From the AM, the astronaut continued through the innermost hatch and into the main part of the station – the OWS. From the outside, it was the same size and shape as the third stage of the S-IVB stage that served as the second propulsion stage of the Saturn 1B and the third stage of the Saturn V. Located on the outside were to have been two wing-like solar arrays extending 30 feet on each side to produce a total wing-span of 90 feet. Both wings would have exposed 2,356.4 square feet of solar cell surface, generating 10,500 W of power at 131 F. One of these solar arrays had been ripped off during launch, leaving one jammed wing that was finally released by the first crew.

Protecting the workshop from meteoroid penetration was a thin metal shield wrapped around the main body of the station. During launch it was held in place against the workshop, and when deployed it would swing out six inches away from the skin of the station, to be held in place by tension bars. A particle which hit the shield would not penetrate the hull of the spacecraft, but would instead be reduced in energy and break up into thousands of smaller particles. However, part of this shield was ripped off when the solar array was separated from the spacecraft during ascent. An alternative solution was required to protect the interior of the spacecraft from the intense heat of the Sun and the cold of space. On the outside of the Sun-side of the hull, the first crew erected a parasol sunshade out of one of the airlocks, and the second crew constructed a twin-pole assembly that restored the internal temperature to optimum.

The workshop measured 22 feet in diameter and 48 feet in length, with 10,426 cubic feet of habitable volume and a mass of 78,000 lbs. The interior (which was originally formed by a large hydrogen tank and a smaller oxygen tank) was reconfigured to provide the hydrogen tank as the workshop living and working areas, and the oxygen tank as the waste storage container for all three crews. The two sections of the Workshop were separated by a perforated bulkhead, which became the 'floor' of the work area in the forward of the tank and the 'ceiling' of the crews' living quarters. A second 'floor' was added against the oxygen tank, providing the living quarters' 'floor'. By entering the top of the dome of the workshop hydrogen tank, the astronaut TV guide entered the 'roof' of the work area, and looked 'down' into the forward compartment floor approximately 20 feet away.

Entry into the cavernous forward compartment required a moment's adjustment, as the local vertical had changed and the astronaut was now on the 'ceiling' of the largest habitable volume in the station. A central handrail, resembling a fireman's pole, traversed the length of the compartment, and could be removed if required. Around the top of the compartment were storage containers and water storage containers arranged in a ring. These also served as a 'running track' for the athletic astronauts, using centrifugal force to keep them on the containers as they ran, somersaulted or leap-frogged around. The forward compartment was also the location of two scientific airlocks, although the Sun-side compartment was permanently obstructed by the support structure of the parasol deployed on the first mission. Experiments and large-volume equipment was stored here, bolted to the floor, as were the food lockers, freezers, film vaults, storage facilities and clothing stores. In the grid floor of the forward level was an open hatchway that led into the crew compartment at the aft end of the tank.

The aft section was described as the wardroom, and contained the food preparation and eating area, the three sleep stations, the waste management area and personal hygiene station, and a smaller experiment work area. Most of the biological and medical experiments were carried out in this section, which also carried the crew exercise equipment. Located in the centre was the trash disposal airlock that allowed the crew to transfer waste containers into the huge oxygen tank beneath the grid floor of the wardroom; and other dry waste storage areas were located between the grid floor and the oxygen tank structure. Positioned around the working areas were handrails, grips and foot restraints. The 'floors' of the upper and lower levels were constructed of a triangular grid that allowed shoes with matching triangular cleats to be twist-locked into the open grill. There was also a large circular window in the wardroom, providing a view either into space or towards Earth, depending on the orientation of the station. The TV tour usually ended at this window.

What the TV view could not fully show, however, were the external structures such as the refrigeration system radiator for the life support system, and the freezers that were located at the lower extremity of the station (in the area that a J-2 engine would be located on an operational S-IVB). Neither could it completely show the Apollo Telescope Mount (ATM). At launch, this had been folded and located inside the launch shroud, and upon entering orbit it was deployed at 90 from its axial position to a radial position, clearing the front hatch, after which its four solar arrays were extended. The central element was called the 'rack', and was octagonal in shape, measuring 11 feet in diameter and 12 feet in length, and weighing 24,453 lbs. It carried the four arrays, the electrical batteries, the electrical and mechanical gyros for the station's primary attitude control system and the ATM communication subsystem. With the arrays deployed, the wing-span was more than 101 feet. It supplied approximately half the electrical power to the station from a 120.5 square foot surface area. Each wing weighed 1,071 lbs, and consisted of four-and-a-half solar panels.

With the CSM docked, the cluster measured over 117 feet long, with a mass of 199,750 lbs and a habitable volume of 12,700 cubic feet.

Skylab was a large and impressive spacecraft, and represented a significant advance in habitation volume and scientific resources over any previous vehicle. It would become a milestone in space station development. The flown design of Skylab, however, was far removed from the original plan to extend the Apollo hardware beyond the goal of landing on the Moon.

Origins

Five months after Apollo 17 – the final Apollo flight to the Moon – splashed down in the Pacific Ocean in December 1972, NASA launched Skylab to begin a new phase of American manned spaceflight – space station operations. It was to be a short phase. Although there was the desire for a second Skylab station, and discussions on other missions, there were to be only three manned Skylab flights, followed by a joint docking mission with the Russians in 1975. No other American manned flights were planned until the advent of the Space Shuttle later in the decade. Skylab and the joint docking mission were to be the final American spacecraft of the one-flight design.

Adapting former Apollo hardware for new programmes seemed a logical step to take after the Moon flights ended in 1972. Development of the Skylab space station programme evolved from NASA plans, in the late 1950s, for an advanced spacecraft to follow the one-man Mercury, while the idea of a large, long-term platform in Earth orbit had been the subject of science fiction and study around the world for decades prior to the creation of NASA in 1958.

THEORIES AND DREAMS

A little over one hundred years before Russia placed Salyut – the world's first space station – in orbit in 1971, one of the earliest fictional proposals for a space station appeared in *Atlantic Monthly* during 1869–1870, in Edward E. Hale's story 'The Brick Moon'.

Hale put forward the theory that an artificial satellite 200 feet in diameter, placed in a polar orbit 4,000 miles above the Earth, and following the Greenwich meridian ground track, would serve as a visible navigational aid for sailors. This would enable them to accurately determine their exact longitude by measuring the angle of the Brick Moon above the horizon. The Brick Moon was so called because of its construction, as it 'must stand fire very well.'

During the early years of the twentieth century, Russian space theoretician Konstantin E. Tsiolkovsky proposed many ideas for exploring beyond the Earth's atmosphere. Some of his ideas included the construction of huge metal cities (space stations) that had gardens to produce food (closed ecological systems), research areas (scientific laboratories), and storage and living quarters (crew compartments). These 'cities in space' would be resupplied by smaller 'metal dirigibles' (logistic ferry craft) that would link to the main structure (rendezvous and docking) and keep the occupants (flight crew) sustained for long periods of time (extended duration spaceflight). Tsiolkovsky also proposed that work could be conducted both inside the structure and, wearing suitable protective clothing (spacesuits), also outside (spacewalking).

During the 1920s, several ideas for prolonged stays in space were put forward, along with the useful applications which such ventures could offer humanity. In 1923, German space pioneer Herman Oberth suggested that creating permanent manned research stations in Earth orbit could be part of a larger infrastructure of exploration. They could then be used for observing our own planet – especially for forecasting the weather or by serving as communications relay stations. He also noted that these objectives offered both scientific and military application. Oberth proposed that such stations could also serve as a refuelling port for other space exploration vehicles, and could be resupplied by smaller freighter vehicles. By making the station slowly spin, it would create an artificial gravity force in which the crew could work.

In 1928, Baron Guido von Pirquet suggested that a network of such stations in space, placed at different altitudes around the Earth, could be used for a wide range of applications. Those in lower orbits could be used for Earth observation, while those in higher orbits could be way stations for resupplying and refuelling interplanetary space vehicles. In the same year this idea was taken further by Hermann Noordung, of the Austrian Imperial Army. He put forward the case for placing stations in an orbit synchronised to the rotation of the Earth (each orbit taking 24 hours). This would allow the station to remain over the same point on the Earth for prolonged scientific study and communications. Noordung's idea for a space platform proposed a circular design, resembling a doughnut shape. By allowing this 'wheel' to rotate, artificial gravity would be created inside the station. On one side of the station would be a power generation plant, and opposite would be an astronomical observation facility.

The Golden Age of science fiction

After World War II, a variety of studies were put forward on the prospects for manned spaceflight using rocket propulsion on missiles or aircraft. Proposals for orbiting the first artificial scientific satellites were also suggested for the International Geophysical Year of 1957–1958. These plans evolved into the first spaceflight programmes of the USA and the Soviet Union, and would lead to the orbiting of the first satellites in the late 1950s, and to the challenge to place the first man in orbit a few years later.

The 1950s was a decade of significant change, and of adjusting to life after the hardships and devastation of the depression, a world recession and two world wars. There was a keen interest in looking to the future, and part of this future was in space. The decade reflected the mood of the times, with the publication of numerous articles, books and novels about exploring and exploiting space, and how benefits from space exploration and 'new technology' would greatly improve everyday life on Earth. Several feature films also portrayed life in the future, where exploration of space was commonplace, and where huge, wheel-shaped space stations orbited the Earth, as gateways to the Moon, planets and stars. At the dawn of the Space Age, the space station featured as a focal point in any future manned exploration of the solar system.

One of the leading figures in promoting space exploration at that time was Dr Werner von Braun, who had worked on the V2 missile programme in Germany during World War II, and who in 1945 had been relocated to the USA to work on the American missile programme. Adaptation of military missiles for space exploration had long been realised as being the most effective and quickest method. Leading figures of rocket and missile technology – such as von Braun – authored many books and papers, adding authenticity to some of the more fanciful ideas of spaceflight being shown in cinemas or in magazines of the day.

A 1946 study, by the Douglas Aircraft Company, into the creation of a manned satellite suggested that significant scientific results could be obtained in the fields of cosmic rays, gravitation, geophysics, terrestrial magnetism, astronomy and meteorology. Two years later, the *Journal of the British Interplanetary Society* published a paper by H.E. Ross on 'Orbital Bases'. Ross suggested that such platforms would be perfect locations for astronomical observations and zero-gravity research (microgravity), and would offer a vacuum research facility that housed telescopes, teams of specialists and support personnel, resupplied by smaller crews every three months.

In 1950 the associate editor of *Collier's*, Cornelius Ryan, had begun to assemble a team of space specialists to follow on from a US Army symposium on space problems. Ryan wanted to show the American public that life was not just casuality lists from the Korean conflict, and Cold War fears, but that by moving out into space, there was a brighter, better future to be gained. The team, consisting of Wernher von Braun (engineer), Dr Fred Whipple (astronomer), Dr Joseph Kaplan (physicist), Dr Heinz Haber (space medicine specialist), Oscar Schachter (space law expert), Willy Ley (science writer), and artists Chesley Bonestell, Fred Freeman and Rolf Klep, were to write and illustrate articles that promoted the promise of space exploration.

By 1952, *Collier's* was publishing a series of articles featuring papers presented at one of the first symposia on spaceflight, held at the Hayden Planetarium in New York on 12 October 1951. These covered a wide range of topics, including the manned exploration of space, the exploration of the Moon, and international space law and sovereignty. Many of these articles also featured the use of a space station as a key element in any extended and international exploration of space. The text was enriched by stunning artwork – notably by space artist Chesley Bonestell – which set the mood for the real space programme to begin.

Alongside these dreams and predictions came discussions about the problems in sustaining such ventures. A paper by H. Koelle, presented at the second IAF congress in London in September 1951, was a forecast of the future, and featured problems in creating a space station which have proven accurate fifty years later with the construction of the International Space Station (ISS). Koelle highlighted that a limitation in available payload weight would be a factor in any launch vehicle (still applicable in the construction of the ISS by the Shuttle), and the techniques of orbital rendezvous and docking would first need to be mastered (they were addressed during Project Gemini in 1965–66). Once the station had been established, the crew would encounter limitations in work and efficiency in the weightless environment. There would also be both national and economic factors to be considered with the growth of the station (an issue that would later affect both the American and the Russian programmes), which would restrict the operational budget (also a frequent spectre in the evolving development of any new high-technology programmes). Koelle also suggested that elements of a large space station could be launched by several smaller rockets and assembled in orbit, enabling it to become operational before it was completed. This also allowed for delays in budgets and the launch schedule, and adjustments to the flight operations – a factor encountered during early ISS construction. Koelle proposed that artificial gravity would offset any physiological effects of prolonged weightlessness on the crew, which could number as many as 50–65!

Acknowledging these hurdles in creating a space station led to alternative ideas. Three years later, during the fifth IAF congress in 1954, Krafft Ehricke delivered a paper which argued that an enormous station was neither necessary nor desirable from the point of view of construction, maintaining station operations, and overall cost. However, he suggested that a smaller crew of four could offer an equally productive scientific research programme of experiments, observations and orbital reconnaissance objectives. Moreover, efficiency would be maintained by a constant rotation of crews.

Early studies by NASA

During 1958 – the year after the launch of the Soviet Sputnik – the American National Aeronautics and Space Administration (NASA) was created from the National Advisory Committee for Aeronautics (NACA). That same year, Krafft Ehricke, then working for the Convair division of General Dynamics, proposed that an early space station could be created by converting the Atlas ICBM under development by Convair. The proposal, called ‘Outpost’, featured a four-man spacecraft that could pivot around to the side of a space laboratory to provide access through connecting side-hatches. The modified Atlas would have a Vega upper stage and a crew capsule for the ride into orbit and recovery at the end of the mission.

In June 1959, Wernher von Braun – by then with the Army Ballistic Missile Agency – stated in a report proposing the establishment of a military lunar outpost (Project Horizon), that a spent booster stage – that is, one that had exhausted its fuel in attaining orbit – could be used as a basic structure for a space station. He had first proposed this theory in the 1940s when working on the V2 development programme at Peenemünde. The idea of taking the final stage into orbit with the payload on an orbital mission was attractive, as it afforded an additional useful volume after it had

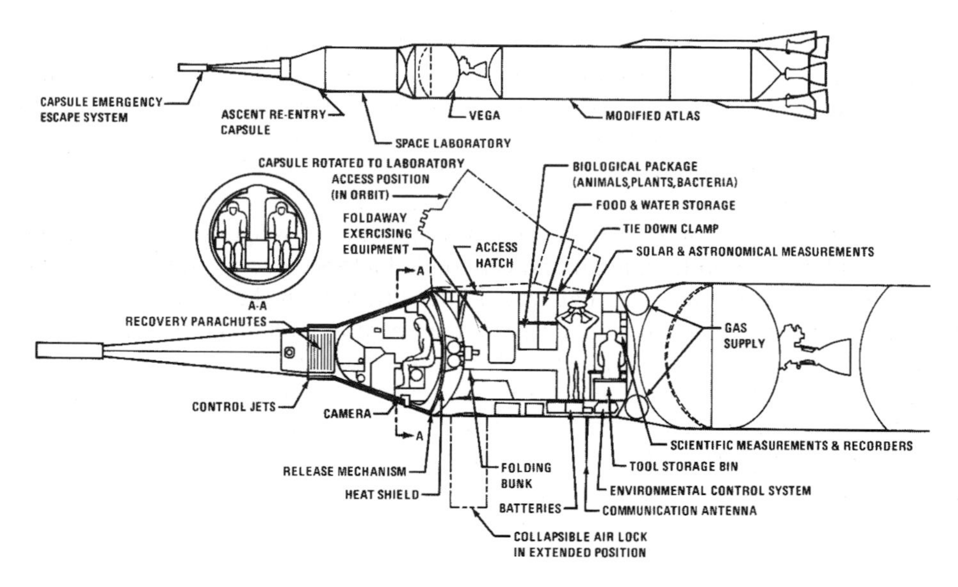

This 1958 design concept by Kurt Strass and Caldwell Johnson of the NASA Space Task Group at Langley Field, Virginia, features a two-man laboratory.

performed its initial task of placing the payload in orbit. The military application of orbital reconnaissance was already recognised, and by placing a rocket stage in orbit it would become an empty shell that could be employed as an observation platform once cleared of launch propellants.

When the designs of the huge launch vehicle family (Juno – later renamed Saturn) were begun in the late 1950s, long before the launchers were assigned to the Apollo programme, von Braun talked of using the spent upper stage from these rockets to create the first temporary space stations. This was but one of a range of objectives for the Saturn family, to be evolved and developed at the NASA Marshall Space Flight Center, in Huntsville, Alabama. This concept later became known as a 'wet stage' space station, and became the genesis of what would later evolve into Skylab.

After Mercury?

During 1959, NASA began internal discussions into the long-range goals of the agency beyond the one-man Mercury. This included a multi-crewed orbiting space station, a larger and more permanent space laboratory, manned lunar orbital and lunar landing flights and, ultimately, manned interplanetary missions.

From these proposals, NASA began to work towards a follow-on spacecraft to Mercury. They planned to upgrade and enlarge the existing one-man design to support two men for three days, and also fund a new two-man Mercury capsule (with a separate large cylinder structure) to support a two-week mission. In the 1960 budget request, the agency asked for $2 million to study methods of either constructing a new space laboratory or converting the Mercury design.

Enthusiasm for a manned lunar exploration programme was growing within

NASA, and during a steering committee meeting on manned spaceflight, caution was expressed over diverting funds and reserves away from achieving a lunar landing objective if the space station became an integral part of NASA's plans. The successive steps taken towards manned lunar landings would be severely limited by the number of missions that could realistically receive adequate funding. These comments, cautions and subsequent recommendations were to prove influential in shifting the focus of priority in NASA's long-term man-in-space programme away from the creation of a space station and towards achieving lunar landing missions.

Later that year, a conference at the Langley Research Center (LaRC) in Hampton, Virginia, aimed at concentrating research efforts to design, build, launch and support a space station. Proposed research fields were to study the reactions of man in prolonged spaceflight, the investigation of materials, structures, and systems for spaceflight of extended duration, and developing methods for observing Earth and stellar objects. Further discussion addressed the important issue of why man's unique abilities could enhance these techniques over unmanned space operations.

At this time there were two other events being conducted that would have a direct application in the later Skylab programme. The Douglas Aircraft Company participated in the 1960 *Daily Mail* Ideal Home Exhibition, under the theme of 'A Home in Space'. The American company had 'won' a competition to provide a full-size mock-up to be displayed in the Empire Hall, Earls Court, London. It was estimated that up to 200,000 people walked through the exhibit, which was based on the 'wet stage' concept, formed from a simulated hydrogen tank from a two-stage launch of an unspecified booster. The hydrogen tank was selected as it offered a larger internal volume than did the oxygen tank. The completed mock-up was 62 feet high and 17 feet across. The impression given was that of a manned space station as it would appear on orbit. The scenario suggested that once in orbit, a crew of four spacesuited astronauts would transfer from the nose cone of the spacecraft through a connecting tunnel, and would clean out the still attached second stage to begin the mission, setting up equipment stored in the tank before launch.

Secondly, in the United States, representatives from the NASA field centres were evaluating the concept of space rendezvous at a conference at LaRC held in May 1960. The conference recognised that space rendezvous and docking would be essential for future manned space efforts, and proposed that each centre should work on developing technologies for such manoeuvres. Although NASA had no funding for a rendezvous flight test programme at that time, this was an important decision for both the lunar and space station programme development.

Enter Apollo

By the summer of 1960, plans for an advanced spacecraft to follow Mercury had evolved into a three-man spacecraft, with capacity for circumlunar flights. On 5 July a House Committee on Science and Applications had declared that to place a man on the Moon within the decade of the 1960s should be a high-priority programme goal for NASA, and that the agency should draw up plans to meet this goal and submit them to Congress. To reduce costs, however, the lunar plan was to be completely integrated with other agency objectives. Just twenty days later, the name 'Apollo' had been officially adopted by NASA for this new programme.

On 29 July the Deputy Administrator for NASA, Hugh L. Dryden, announced future NASA programme plans to industrial management representatives. He stated that the name 'Apollo' would be used for a programme of manned Earth orbit and circumlunar missions prior to 1970, which would also include the creation and orbit of a temporary scientific research laboratory. Beyond 1970, the programme would involve the development of manned lunar landings and a space station, leading eventually to interplanetary flight and a landing on another planet – Mars. From the beginning of the Apollo programme, the creation of a space station was once more a key element in long-range programme planning, and was part of a much larger space infrastructure.

The 'current planning' for the space laboratory (which was then termed Apollo 'A') envisaged an adapter measuring 12.8 feet in diameter and 7.9 feet high, fitted between the base of the Apollo Equipment–Propulsion module (that would become the Service Module) and the top of the Saturn stage. This was to be a 'space

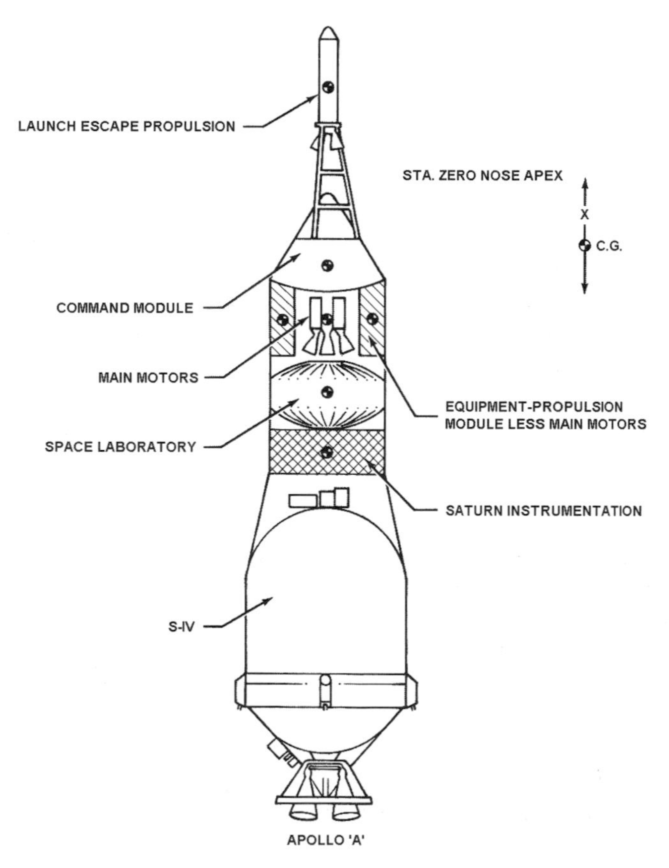

One of the earliest proposals using the Apollo spacecraft for a space station concept dates from autumn 1961.

laboratory' for scientific experiments that could be performed during a lunar flight, related to manned operations of spacecraft. These experiments included astronomical observations, monitoring the Sun, developing EVA techniques, and micrometeoroid impact studies.

On 25 May 1961, President Kennedy gave his historic 'before this decade is out' speech, committing America to a manned lunar landing. In the shadow of the Cold War, and in the recent wake of both the flight of Yuri Gagarin (the first man in space) and the Bay of Pigs invasion fiasco in Cuba, the young administration needed a new goal to unite and encourage the American people. That goal became the Moon, and Apollo became the programme to achieve it.

With a target to reach, efforts to develop the hardware gathered pace during 1961. The contract for constructing the Apollo Command and Service Modules was awarded to the Space and Information Systems Division of North American Aviation Inc. on 28 November 1961.

It was also becoming clear that the technique of orbital rendezvous and docking would play an important part not just in the lunar programme, but for any future manned space operations. Therefore, on 7 December NASA announced plans to develop a two-man Mercury spacecraft, the main objective of which was to test and develop orbital rendezvous and docking techniques with a previously launched unmanned Agena target vehicle. This new programme received the name 'Gemini' on 3 January 1962.

Although the choice of the spacecraft to carry the astronauts to the Moon had been made, the rockets to get them there were still under development at the NASA Marshall Space Flight Center (MSFC). The huge launch complex in Florida, and the astronauts' training facility – the Manned Spacecraft Center (MSC), in Houston, Texas – were also under construction. Exactly how to reach the Moon and return to Earth was still undecided.

NASA considered three main options. Direct Ascent would involve launching straight from the pad towards the Moon without first entering orbit, and would require a huge launch vehicle. Earth Orbital Rendezvous (EOR) would have components of the spacecraft assembled in Earth orbit before embarking on a direct flight to the Moon's surface. This was the favoured option at MSFC. The Lunar Orbital Rendezvous (LOR) technique would involve a spacecraft entering lunar orbit, and a separate smaller landing vehicle making the descent, before redocking with the parent craft for the flight home.

On 11 July, NASA announced the selection of LOR, due to the higher chance of success, the reduced costs (10–15%) over the other two methods, and the least amount of technical development required beyond existing commitments. Four months later, NASA awarded the contract for the Apollo lunar landing vehicle to the Grumman Aircraft Engineering Corporation. It was called the Lunar Excursion Module (LEM) – later shortened to Lunar Module, but still pronounced 'Lem'.

As the manned lunar programme accelerated, so studies into using hardware for other programmes began to emerge, based around the Apollo spacecraft. Despite being designed for lunar missions, its application to the space station studies was to soon be recognised.

Apollo X

Some of these early alternative mission studies being conducted were based on a concept proposed by Emanual Schnitzer of LaRC, which used a standard Apollo CSM with an inflatable spheroid and a transfer tunnel to create a space laboratory. This design study was designated Apollo X (Experimental – usually assigned to the rocket research aircraft of the day, such as X-1, X-2, X-15, X-20).

By the end of 1962 it was recognised that the future development of any space research station would clearly be restricted by the limitations of the overall NASA budget. Any authority to proceed would be reflected in this fact, and so the simplest design at minimal cost would probably receive the largest support for the first space stations to be flown. Another important requirement was to establish a clear definition of objectives and benefits that would establish the role of the station early in its development, and where this research would lead to in the overall programme.

The Apollo programme had Presidential approval to place men on the Moon by 1970, but no follow-up programme had been defined or authorised. As the Moon programme developed through the early 1960s, domestic welfare issues and an escalating war in south-east Asia began to dominate the evening TV news and morning newspaper headlines. These external issues mitigated against NASA seeking further funds for new projects.

NASA feasibility studies had established that a research space laboratory could be placed in orbit by 1967, but what was needed to achieve this was a justification for the programme for national goals, or in the fields of science and technology. It was also important to identify exactly what could be accomplished on a space station that could not be achieved by Mercury Gemini or Apollo, or by unmanned spacecraft. One of these studies proposed the modification of Apollo systems for a 100-day orbital duration capability, which could evaluate the feasibility of using Apollo hardware for a series of these missions. In a preliminary statement of work for a manned space station study programme, it was stated that keeping the design as simple as possible, not providing a permanent artificial gravity, and using existing hardware, was the logical approach for achieving a short-term and cost-effective station.

An early 1963 report on future manned spaceflight studies had indicated that a space station would be a requirement by 1970, to support the launch and repair of spacecraft, and to serve as a scientific laboratory. It was also proposed that this station could be launched in two sections using Saturn C-5 launchers, joined together in orbit, and could be supplied by modified six-person Apollo CSM ferry vehicles.

In a testimony before a House Committee on Science and Astronautics on 4 March 1963, NASA Deputy Administrator Hugh L. Dryden indicated that the most obvious candidate for a post-lunar landing programme (Apollo) was a manned Earth orbital laboratory. NASA and the DoD had already begun to discuss closer co-ordination between both agencies in space exploration, at a time when the military was evaluating the use of a manned orbital platform for its own reconnaissance objectives.

In another development on 1 June, NASA issued two space station study contracts to the Lockheed Aircraft Corporation and the Douglas Aircraft Company

An artist's concept – dating from the early 1960s – of a telescope mount on the CSM.

Missile and Space Systems Division. Requirements to be met by these studies included a capacity for crew rotation, resupply facilities, and an operational lifetime of around five years.

Parallel to overall programme discussion, talks were evaluating the ability of spacecraft to carry scientific instruments – notably telescopes – into orbit, and what would be required to achieve this aim. During a NASA Manned Spacecraft Center (MSC) meeting with the Bendix Eclipse-Pioneer Division in June 1963, the problems of including stabilisation techniques for incorporating high-resolution telescopes onboard manned space vehicles in different vehicles were explored. These included pointing accuracy, fields of view, and actual location onboard the vehicle.

As well as the paper study efforts, there were problems in developing hardware and systems for later programmes, which often reflected the difficulties in providing technology for prolonged use. In July, a planned 30-day engineering test of a life support system in the Boeing Company space chamber was abandoned after only five days, due to a faulty reactor tank. This test was to have evaluated America's first life-supporting equipment for a multi-person long-duration space mission, including environmental control, waste disposal, crew hygiene and food techniques, and habitability issues.

On 30 July 1963, North American started another important study in response to a request by MSC. An Apollo extended mission study report was to evaluate the potential for two- or three-man configuration to orbit at 100–300 miles altitude for 100 days. There would be no resupply during the 100 days, and the spacecraft, launched by a Saturn 1B, would either use a CSM on a solo flight, or a CSM with separate mission module as living quarters. It was an evaluation of whether an Apollo CSM could remain in space for 100 days, withstand prolonged exposure to the space environment, and then safely return the astronauts.

In November 1963, the resulting report indicated that the uncertainties over any prolonged exposure to long-duration flight for both men and machines would be answered on an Earth orbital laboratory on missions exceeding one year in duration. The report suggested that the modification of existing equipment, such as the Apollo CSM, would be more suitable than the development of totally new hardware. Based on the favourable results from the 100-day study, the proposed missions ranged from the adoption of basic Apollo hardware, to a separate laboratory with self-contained systems and life support. Both were technically sound, and were capable of achieving mission objectives within minimum cost and time frames.

USAF Manned Orbiting Laboratory

On 10 December 1963, the USAF announced the cancellation of the military X-20 Dyna Soar space-plane and the transfer of efforts in manned spaceflight to broader research with a Manned Orbiting Laboratory (MOL, pronounced 'Mole'). NASA would provide technical support, which, it was stated, could suit both USAF military and NASA space station requirements.

Each of the five planned crewed MOL missions was to consist of a 41-foot long, 10-foot diameter orbital pressurised laboratory that would be launched attached to a Titan III booster. On top of this, two military astronauts would ride a modified Gemini capsule (evolved from plans for a military 'Blue Gemini'). The combined length of the vehicle was 54 feet, with a mass of 25,000 lbs, including the 5,000–6,000-lb Gemini spacecraft. Once in orbit, the Gemini capsule would remain attached to the MOL, and the two men would transfer into the pressurised compartment via an internal hatch in the rear heat-shield. The station had the capacity to support the two-man crew, in a shirt-sleeved environment, for up to 30 days, with facilities to perform EVA. At the end of the mission, the two astronauts would re-enter the Gemini capsule, separate from the MOL, and return into the ocean. The MOL would be abandoned in orbit, with no capacity for resupply or reuse.

The primary objective of the programme was to determine a man's ability to perform useful military observations and experiments in space. Potentially, suitable research conducted from orbital height by humans could supplement or enhance the work of unmanned reconnaissance satellites to benefit national security. For crewing MOL, only serving military personnel would be selected, with no NASA astronauts transferred from the civilian programme. However, as those selected would be trained to operate the Gemini spacecraft and perform many of the activities for which NASA astronauts were trained, the selection criteria and training programme was similar to the first three NASA astronaut classes in the required number of jet flying hours, and the candidates' height, age, and academic qualifications. Although DoD experiments were to be carried, there was no indication that military scientists were to be selected or would fly on the MOL.

Extending Apollo

In December 1963, Wernher von Braun suggested a series of missions to extend the basic Apollo system and expand lunar exploration. These missions could bridge the gap between the Apollo landings and a more permanent lunar base, using short-term shelters under the Integrated Lunar Exploration System. Although the plan did not initially receive much favour, it did spur lateral studies for other possible applications of Apollo hardware, including that for a space station – but only to be attempted after the primary goal of landing on the Moon had been achieved.

By January 1964, the definition studies were followed by Phase II follow-on studies, with a plan for two contracts – one for the CSM-supported missions, and a second for all the other various concepts for laboratory modules. By February, Lockheed had proposed that a Saturn V could be used to launch a larger station, with an operational life of five years, and a crew of 24 launched by manned logistics vessels. The following month Edward Grey, the Advanced Manned Missions Director in the Office of Spaceflight, asked LaRC to prepare a Project Definition Plan for their study project, the Manned Orbital Research Laboratory (MORL). He also asked MSC to provide the same for their Apollo X, Orbital Research Laboratory, and Large Orbital Research Laboratory studies. It was an important step in defining the requirements for initiating a new programme (at an appropriate, but as yet undefined time), 'should a climate exist in which a new project can be started'.

The Lockheed study had indicated that a station could be launched by a two-stage Saturn V (a configuration later adapted for Skylab) between 1967 and 1970, with an optimum date of July 1968. Launched unmanned, it could be rotated to generate centrifugal forces, would support a crew of 24, and would be capable of autonomous operation for up to one month, with full operational status of between one and five years. The crew would be exchanged after three months to a year, with resupply missions occurring every 90 days. The logistics spacecraft would be based on a six-man modified Apollo, or twelve-man modified lifting body design. Maximum use of available or planned equipment and technology was proposed, although the increased pace of the Gemini and Apollo programmes was recognised as a limiting factor in the development of any hardware, experiments or research for the space station programme.

When NASA created the Apollo Logistics Support System Office in April 1964, formal investigation of extensions of Apollo hardware was promoted, including applications for the Lunar Module beyond the role of a primary manned lunar landing vehicle. Although these studies were initially orientated towards lunar applications, Grumman had already begun in-house feasibility studies into using the shell of the LM to outfit the vehicle for other tasks. Meanwhile, in July, Douglas submitted its study on a manned space research station that could be operated for one year by a crew of six, and could be placed in orbit within five years. This study envisaged the use of either modified Gemini or Apollo spacecraft, with rotation of the combination for artificial gravity.

The studies into the space station options had also begun to establish the kind of research for which the platform might be used. One of the most obvious areas was the use of an astronomical device above the contaminating layers of the Earth's atmosphere, for celestial and solar studies.

In September 1964, a background briefing by Nancy Roman, then Director of NASA Astronomy Activities, added that any such telescope would be designed to operate independently and autonomously, although adjustment of the focus, replacement of the film, and repairs could be carried out manually. With the NASA centres performing related engineering studies, the agency then offered the scientific community the chance to propose astronomical studies that could be the objective of such a facility. At the close of the year, Boeing announced a ten-month study into a manned telescope, how such a device might be operated, and what particular role a man would have in such a project.

A preliminary draft report of the Ad Hoc Astronomy Panel of the Orbiting Research Laboratory on the value of a manned astronomical observatory and defining objectives for such a mission was completed in August, and represented a major effort in proposing a manned scientific research facility with clearly defined objectives. Publicly supporting such a venture, the panel indicated that although sounding rocket and satellite programmes had merit, there was a case for a broader, more flexible, and ultimately more economical astronomy programme that required the presence of man in space.

In the opinion of the panel, this programme was to be started as soon as possible, and although they agreed that this should initially be based on manned orbiting platforms, the panel looked forward to facilities being placed on the lunar surface. Hurdles to overcome in achieving this objective would include the assembly of large, bulky, and very fragile equipment in space, on-orbit maintenance and repair, certain modification of equipment, direct monitoring of the scientific apparatus and support hardware, and immediate data feedback during critical and specialised operations.

It was during these discussions (which recognised the requirement for a flight-orientated astronaut) that a recommendation was also made to include a qualified astronomer in the crew to direct scientific operations onboard the laboratory.

On 14 August 1964, the MSC Spacecraft Integration Branch proposed that Apollo X should be used in Earth orbit for biomedical and scientific missions of extended duration. A first-phase mission, consisting of a two-man Earth-orbiting laboratory lasting 14–45 days and launched by Saturn 1B into a 230-mile orbit, would be the first of a series of missions that included:

- Configuration A: two astronauts, 14–45 days, no laboratory module.
- Configuration B: three astronauts, 45 days, single laboratory mission.
- Configuration C: three astronauts, 45 days, dependent systems, and a double laboratory module.
- Configuration D: three astronauts, 120 days, using an independent systems laboratory module.

NASA emphasised that Apollo X was not in direct competition with the USAF MOL, but was aimed at more technical and scientific objectives rather than military objectives. By the end of the year, the studies had initiated a move towards tentative spacecraft development and mission planning through Fiscal Year 1969, under a new, undefined programme.

Running what was effectively two separate manned space station development programmes – one of them military and the other scientific (civilian) – caused questions to be asked in Washington. On 7 December, Senator Clinton Anderson,

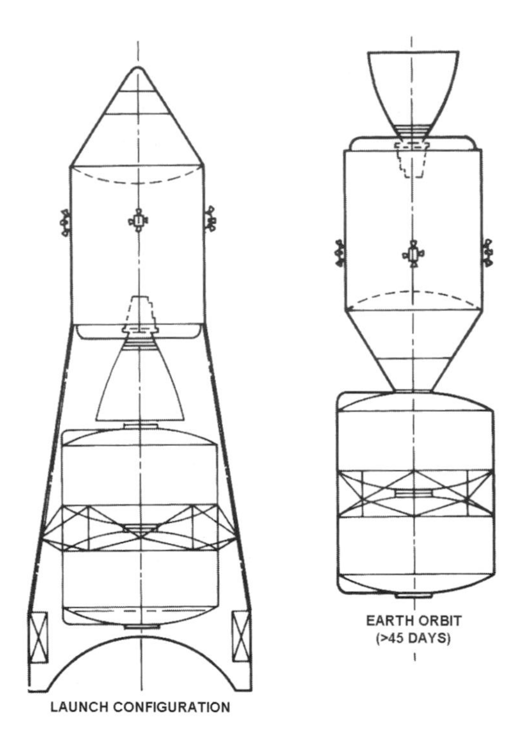

In August 1964 the Apollo X design was depicted in both launch and Earth-orbit configurations.

the Chairman of the Committee for Aeronautics and Space Sciences, wrote a letter to President Johnson, stating that duplicating MOL and Apollo X was inefficient use of 'national treasures'. He suggested that $1 billion (1964) could be saved over five years if the MOL was cancelled and its budget transferred to NASA's Apollo X programme. After a week in which the Senator was 'updated' on the differences between the two programmes, Anderson indicated that he had separately informed the USAF and NASA to take advantage of each other's technology in their own programmes (which would remain separate), and to aim all future studies towards a true national space laboratory.

Medical aspects of an orbiting research laboratory

NASA's increased interest in the creation of a space station resulted in the scientific community beginning studies into the types of research and scientific requirements that such a programme would demand. As the station would be manned – or, at least, man-tended – the biomedical aspects of prolonged spaceflight were of primary consideration. During 1963, the investigations within NASA were soon supplemented by contract industrial studies. Subsequent participation by representatives of American medical fields would provide guidance in directing future space efforts, and would also move space medicine studies forward as a whole.

NASA formed a group of twenty consultants, representing sixteen specialities in life sciences, and including some of the leading specialists in the nation. They would meet eight times between January and August 1964. This group was called the Space Medicine Advisory Group (SMAG), and it considered the current status of the programme, its applications, and various aspects of the biomedical programme of an orbiting research laboratory. Under three broad categories, the group was asked to consider its support for three fields of space medical research on a hypothetical space station. These fields were in life support, experiments, and design requirements.

In December 1963 (before the first meeting in January 1964) the USAF MOL programme had been officially authorised, and so several USAF medical officers joined the team for the meetings, in accordance with the policy of closer co-operation between NASA and the DoD. The group met on two consecutive days, normally in Washington DC. However, two three-day meetings were held in NASA field centres at Houston and in Florida, and included tours of the facilities. The resulting 144–page report was not issued until 1966. On release, it was recognised that a few concepts and findings may have dated in light of findings from the Gemini programme, but it still identified or underscored the most important baseline data, in controlled environment factors (life support), experiments that took into account those environmental factors, the effects of bodily functions in reduced gravity and how equipment could investigate and measure this data, and the most practical design and operational requirements for a facility to conduct such research.

The consensus was that medical questions would remain unresolved at the end of the 30-day MOL missions, and that a 90-day MOL, then under consideration, would also not fully answer them. For very long manned missions – such as a trip to Mars – flights of one year would be required if the ORL programme was to be effective. The group of life-science specialists was unanimous in proposing one vehicle of extend duration, rather than a series of different vehicles for different flights (as with the MOL approach). The suggestion was also offered to upgrade the interior as the flight progressed, reflecting mission results and latest technology. With no engineers as members of the group, however, the practicalities of achieving this were not discussed. The concept was based on the LaRC Manned Orbiting Research Laboratory study, with a crew of between six and twelve (optimally set at eight), with 400–500 cubic feet of space per man and an additional 1,000 cubic feet of volume for a laboratory area.

The report suggested that providing there were no adverse affects from the 30-day MOL flights, a 90-day mission was well within the capability of current technology; and if no insurmountable difficulties occurred from a three-month mission, 'we are justified in designing for a one-year or more ORL.' Other features proposed were the use of a high lift/drag vehicle as a recovery vehicle, rather than the ballistic capsule design, offering a gentler re-entry profile for the crew and increased land recovery range – a clear indication of cost effectiveness from the medical standpoint.

The orbit suggested was at 30° inclination and 200–300 miles altitude, but not a polar orbit in order to avoid ionisation radiation. It was also felt that the orbit should be as close to a 24-hour cycle as possible, to maintain circadian rhythm. What could not be agreed was whether induced gravity would be beneficial or harmful to the crew and the experiments. It was indicated that gravity might be helpful for housekeeping. The use of small laboratory animals (mice, rats and squirrel monkeys) for controlled experiments was also proposed; the smaller the animal the more advantageous for weight, volume and power requirements.

The flight profile suggested an eight-person crew with, two replaced at 90 days, two more at 240 days, and all eight at 360 days. Depending on rendezvous and resupply logistics, an alternative proposal was for two crew to be replaced at 60 days, two at 120 days, a further two at 240 days, and all eight removed at 360 days. This offered comparison sets of data for between 12 and 14 individual crew-members, and a similar number of animal test-subjects that would follow the same cycling.

A range of experiments was proposed to record ECG, temperature, blood pressure and EEG. The group also proposed a range of vision experiments, strain gauges, tape recorders for voice analysis, measurement of body sweat, mass measurement and body volume measurement devices, a bicycle ergometer, negative lower body pressure apparatus, measurement of expelled respiratory air, blood gases, blood samples, the study of the otholith sensory system, and a range of X-ray examinations. In addition, the group highlighted the need to provide for the recording and collection, storage and refrigeration, and transport and recovery of specimens and experiment results.

The report suggested direct contact with principle investigators and trained ground support personnel to offer communication support to the crew. It also suggested a procedure similar to that eventually provided for Skylab – the availability of a special 'ambulance vehicle' with an easier re-entry mode, perhaps with a specially trained medical attendant as a crew-member. Finally, it was suggested that a total ground simulation prior to the flight (again a procedure that was adopted for Skylab) should be completed by the flight crew itself, or 'astronaut-like subjects' (specialists, engineers, or back-up crew-members). This would be for 'the training and testing of ground equipment, systems and methodologies ... allowing all onboard equipment to be tested in advance of ORL flights.'

The Apollo Extension System (AES)

To win any additional support in Washington for a budget to create a space station, a strong ally was required. Those at the Marshall Space Flight Center recognised that after Apollo achieved its goal there would be a lack of future missions if those programmes were not started as Apollo gained pace. Von Braun's idea of using spent Saturn stages to create a space station seemed to be the way that Marshall could gain an advantage over the other NASA centres for a new post-Apollo programme.

To achieve this, Marshall sought out George Mueller, NASAs Associated Administrator for Manned Spaceflight, who had already demonstrated an interest in developing Apollo's potential, and had a strong desire to restrict the loss of personnel that would result from down-scaling Saturn launch vehicle development teams as Apollo neared its goal.

This idea caused outrage at MSC in Houston, where it was felt that Marshall was trying to grab Houston's speciality field in the development of manned spacecraft. Each centre at NASA had clear responsibilities, and though co-operation was essential, any programme was normally split between Washington (handling the budget and politicians), Houston (designing the spacecraft and training the astronauts), Marshall (building the rockets and engines) and Florida (sending the vehicles into space). According to MSC, it was a 'birthright' that Houston was in charge of manned spaceflight, and to let another field centre take that away was unthinkable. This competition and conflict between Houston and Marshall over management of manned spaceflight programmes was the beginning of a long dispute that continued into the Shuttle and space station programmes thirty years later.

To fund a new space station programme, the words 'new' and 'space station' had to be disguised under the shadow of Apollo. Explaining that it would use hardware developed from Apollo, after the first landing, and would extend the capabilities of Apollo systems without developing radically new hardware, allowed the space station idea to enter budget discussions in Washington. Several NASA officials explained that the idea of an 'orbital workshop' was not really a programme, as it used hardware already developed. It was a way to move the whole idea forward.

On 18 February 1965, Mueller testified before the House Committee on Science and Astronautics to outline NASAs post-Apollo objectives. During his presentation he stated: 'Apollo capabilities now under development will enable [NASA] to produce space hardware and fly it for future missions at a small fraction of the development cost. This is the basic concept in the Apollo Extension System (AES) now under consideration.'

The AES programme was being planned to include a number of follow-on missions after the initial Apollo lunar landings, using proven Apollo hardware and techniques. The series would feature both lunar and Earth orbital objectives that included missions lasting between four and six weeks in lunar orbit, up to two weeks on the lunar surface, and crews remaining in Earth orbit for up to three months.

This would require an early demonstration of AES capability of supporting later missions and systems development. With MSC having lead responsibility for study contracts, and both MSC and MSFC having responsibility for about half of the payload integration contracts (with co-operation from KSC, North American, and Grumman), the centre feuding gradually worsened.

On 12 May 1965, a presentation on the AES Experiment Program was prepared by the Space Station Study Office, Advanced Spacecraft Technologies Division, MSC, which provided background and condensed descriptions of 86 experiments or objectives for the fifteen station-related flights planned under the AES programme. Flights would be on either the Saturn 1B (200 series,) or the Saturn V (500 series), and several experiments would fly on more than one mission. These were grouped into sixteen categories: biomedical (twenty-one experiments), behavioural (three experiments), artificial gravity (two), living organisms (six), space environment (four), liquid/gas/solids behaviour (eight), astronomical observations (six), remote sensing of the atmosphere (five), remote sensing of the Earth (three), manoeuvrable sub-satellite (one), launch of an unmanned satellite (one), electromagnetic propagation (two), space structures (six), subsystems development and test (seven), EVA operations (seven), and manoeuvring and docking (four). The AES programme consisted of the following missions:

Flight	*Mission*	*Altitude (miles)*	*Inclination (degrees)*	*Duration (days)*
1	209	200	28.5	14
2	211	200	28.5	30
3	507	200	90.0	14
4	509	19,350	0.0	14
5	215	200	50.0	14
6	513	200	81.5	14
7	218	200	28.5	45
8	219	200	28.5	45
9	221	200	28.5	45
10	516	19,350	0.0	45
11	518	200	83.0	45
12	521	19,350	0.0	45
13	523	200	28.5	45
14	229	200	28.5	45
15	230	200	28.5	45

On 30 July 1965 a final report on modular multipurpose space stations was delivered. It contained a wide range of concepts, from a small, modified Apollo CSM, to semipermanent space stations. The range of proposals suggested six separate missions in four configurations, but emphasised the modular concept rather than an 'all up in just one launch' philosophy. This modular concept of building up a larger station from smaller elements was adopted on the Mir programme, and is currently being followed on the ISS. The 1965 missions were proposed as:

- 45-day mission in a 230-mile orbit at 28°.5 inclination, with three crew and a single compartment laboratory.
- One-year mission in a 230-mile orbit at 28°.5 inclination, with six crew and a double compartment laboratory.
- 90-day mission in a 230-mile orbit at 90° inclination, with three to six crew and a double compartment laboratory.
- 90-day mission in a 230-mile orbit at 30° inclination, with three to six crew and a double compartment laboratory.
- One- to five-year mission in a 230-mile orbit at 28°.5 inclination, with six to nine crew and six compartments – termed an interim station.
- Five- to ten-year mission in a 298-mile orbit at 29°.5 inclination, with 24–26 crew on a fully operational station.

That month, Grumman also issued its final study report of AES Earth-orbit missions for LEM utilisation. In a five-volume report, several lunar configurations were offered, along with 'LEM Labs' for extended stays in Earth or lunar orbit in conjunction with CSM, as part of AES.

THE APOLLO APPLICATIONS PROGRAM (AAP)

In August 1965, NASA established the Saturn/Apollo Applications (SAA) Office, within the Office of Manned Spaceflight. The SAA would assume responsibility for both the proposed Saturn 1B–Centaur launcher combination programme (liquid-

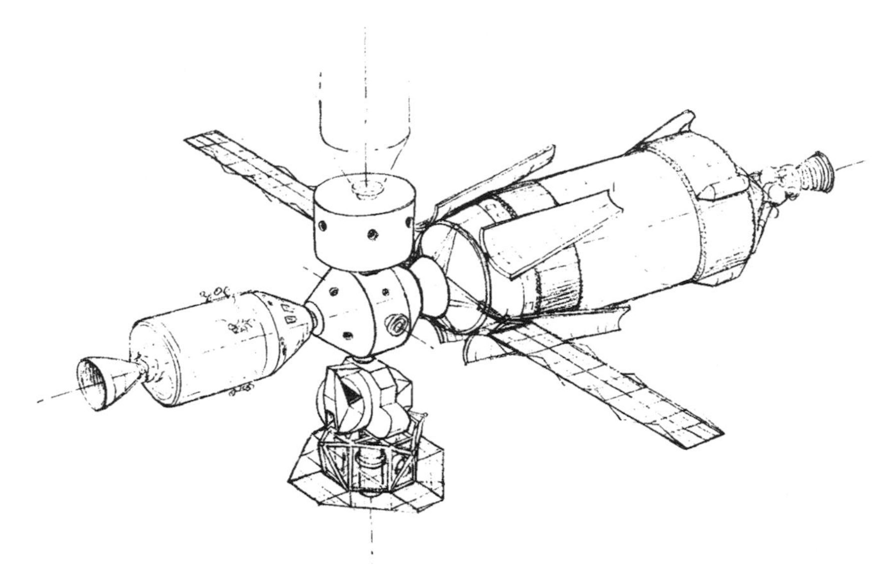

In late 1965, Willard Taub, of MSC, conceived a design which was later used as the basis of designs of the S-IVB stage as a workshop.

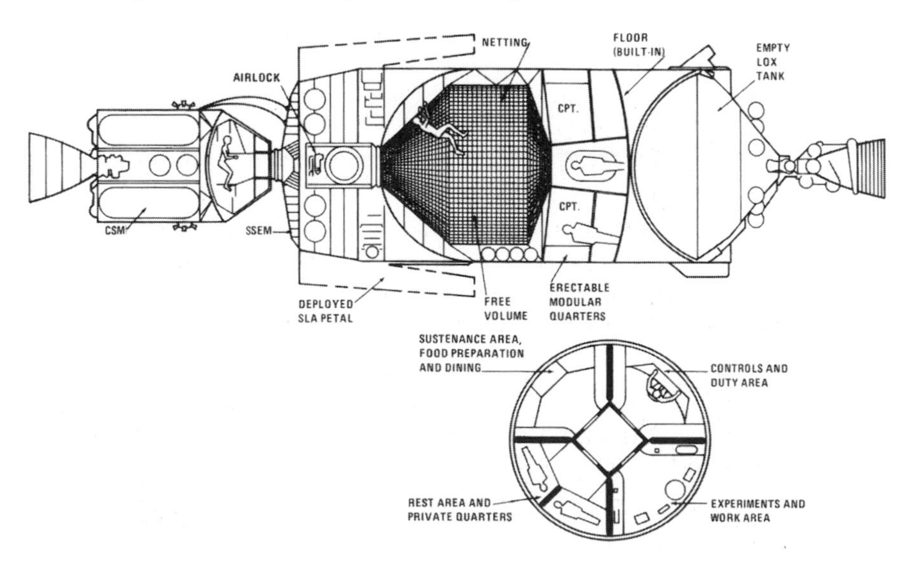

A 1967 concept for the AAP cluster S-IVB workshop wardroom area.

fuelled upper stage – subsequently cancelled) and the Apollo Extension System development.

At MSFC – where Mueller had allocated most of the planning for a possible space station design – serious consideration was being given to the concept of an S-IVB Orbital Workshop (OWS). This would involve the conversion of a 'spent' S-IVB stage to support an extended-duration mission in Earth orbit for the crew to perform useful experiments under the AES concept. Consequently, a four-month conceptual design study began at MSFC, with assistance from MSC and from the prime contractor for the Saturn stage, the Douglas Aircraft Corporation. The study evaluated previous spent-stage proposals, as well as attempting to keep the design as simple as possible to minimise costs and meet early launch dates.

One of the early requirements determined was the selection of suitable cabin atmosphere for expected longer-duration missions later in the flight programme. The nominal mission duration was to be about 45 days, but studies were conducted to increase this to 60, 90 and 135 days. Studies of single gas (oxygen) and two-gas (oxygen/nitrogen and oxygen/helium) atmospheres were conducted, both at MSC and with contractor AiResearch Corporation, and at the USAF School of Aviation Medicine.

A 100% oxygen environment at 5 psia was advantageous both scientifically and operationally, and was attractive because of the simplicity of the design of the system. But studies of vital body processes included problems in flights of 100% oxygen over 30 days. The full oxygen atmosphere was also highly inflammable. The two-gas atmospheres studied were 50% oxygen and 50% nitrogen, 70% oxygen and 30% nitrogen, and 70% oxygen and 30% helium.

An intermediate concept of the cluster configuration.

The following month saw the official division of responsibility between the three primary centres for manned spaceflight. The Manned Spacecraft Center (Houston, Texas) would be responsible for spacecraft development, the development of an airlock module to allow passage from the CSM to the station, flight crew activities, mission control and flight operations, and payload integration of the CSM. The Marshall Space Flight Center (Huntsville, Alabama) would assume responsibility for the development of the Saturn launch vehicles, payload integration of derivatives of AES modified vehicles (such as the spent-stage concept), and the LEM Labs. The Kennedy Space Center (Merritt Island, Florida) would handle all pre-launch assembly checks and the launch of all vehicles. The prime contractors for spacecraft (North American and Grumman) and Saturn (Boeing, North American, and Douglas) were also involved in definition studies. However, during a meeting later in the month, Jim Webb emphasised that despite his support for the project, all flight planning should remain extremely flexible and should in no way compromise the primary goal of the Apollo lunar landing.

The Manned Orbital Research Laboratory (MORL)

In September 1965, the final report into the Manned Orbital Research Laboratory (MORL) was issued. Completed by the Douglas Aircraft Company under NASA contract (to LaRC), it concentrated on the merits of a research laboratory focusing on the conservation of Earth's natural resources in providing surveys of the land, oceans and atmosphere. This was a definition study that projected a suitable follow-on programme for AAP operations and research potential. It would provide a six- to nine-crew laboratory, with a mass of less than 34,000 lbs, experiment capacity of 1,650 cubic feet, and 2 kW of power. Launch would be unmanned (by a Saturn 1B) into a low-inclination orbit, and it would be resupplied by Apollo ferry logistic craft consisting of a Saturn 1B, Apollo CSM and a 'cargo module'. In 1965, the projected launch date for such a facility was 1972 – four years after the planned first AAP missions.

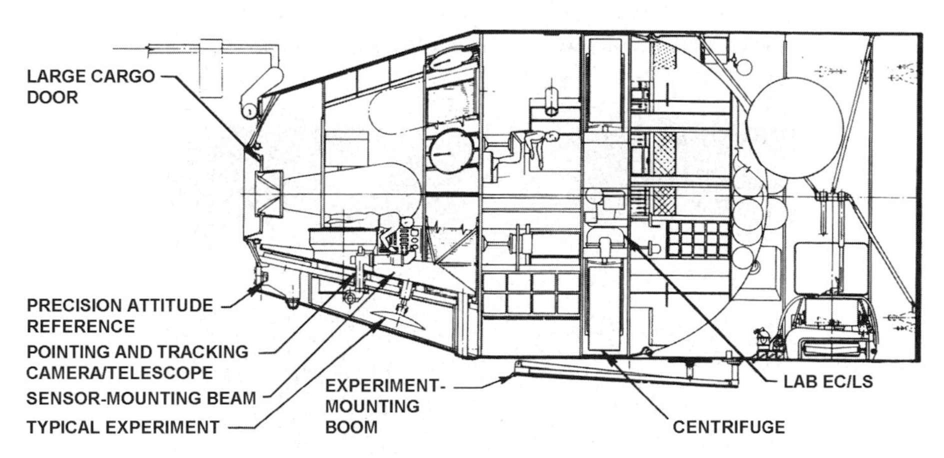

The manned orbiting laboratory concept. (Courtesy Douglas Aircraft Company.)

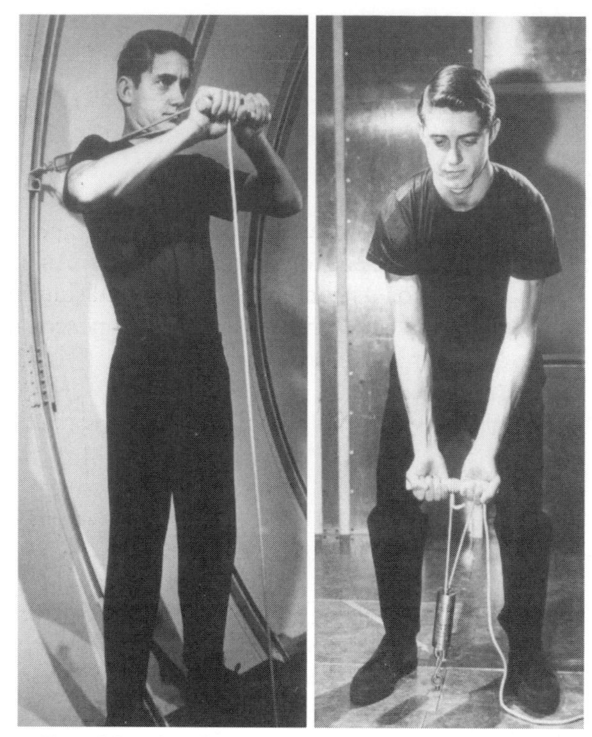

American athlete Dan Murphy demonstrates the Lockheed 'Exer-Genie' device for keeping fit in 'future space voyages lasting weeks or months'. This December 1965 photograph suggested that by using the device that pulls a nylon line through a metal cylinder that creates a forces between zero and 400 lbs, an astronaut could keep fit with 'as little as four six-minute periods per day'(!) (Courtesy Lockheed.)

Also during 1965, research objectives for the AAP itself were continually discussed at various meetings, symposia and venues. This clearly indicated that unlike Apollo – with its clear objective of landing on the Moon by 1970 – the AAP had much less defined objectives. The AAP was also not an end in itself, but a technological bridge to more extensive manned flight operations in the 1970s, 1980s and beyond. Programme redefinitions gradually evolved into a range of mission objectives in addition to the lunar distance missions. This included Earth-orbital missions of around 45 days at 0–90°, from altitudes of between 115 miles to synchronous orbits, and with resupply flights to extend missions to between three months and a year. The suggested research fields were also wide-ranging, and one area being examined was the creation of artificial gravity by rotating the S-IVB and CSM at each end of a cable. However, there would be a risk that the astronauts or the spacecraft could become entangled in the cable, and the idea was not well received at MSC, especially in the Astronaut Office.

By October, the lunar mission Block II spacecraft for Apollo CSM (over the earlier Block I Earth orbit engineering test flight CSM) and the first flight-qualified LM were being prepared. Programme definition contracts were expected in January

1966, followed by those for the CSM in October 1966 and the LM by the summer of 1967.

In December 1965, by way of deferring development costs for new hardware, Mueller requested that McDonnell should examine the feasibility of using some of the proven Gemini hatch and life support subsystems in an airlock, in conjunction with AAP EVA operations.

By the beginning of 1966, definition studies and proposed experiments were being evaluated prior to the budget statement for Fiscal Year 1967. As part of announced opportunities for study grants for astronomers (conceptualising instruments to be flown between 1969 and 1975), a description of the proposed Apollo Telescope Mount (ATM) was included for the first time.

Between 1 September 1965 and 28 February 1966, Ball Brothers Research Corporation (which was involved in the Orbiting Solar Observatory satellites) had completed a study that evaluated crew participation with an ATM. This design featured a three-axis facility located in Bay 1 of the Apollo CSM (as in the illustration on p. 10) and also a set of four 'typical' solar experiments: an ultraviolet spectrometer for acquiring high-resolution UV spectra from localised regions on the Sun; an X-ray telescope for obtaining images in selected wavelengths from quiet and active solar regions; a white light coronagraph for examining the intensity distribution and polarisation of the corona from 1.5 to 6 solar radii; and a UV spectrograph/spectroheliograph for profile spectroscopy and spectroheliograms of Fe15, Fe16 and He2.

Crew time would encompass the extension and stowage of the device from within the CM (primary, back-up and contingency jettison modes) and setting up the positions of the CSM and ATM experiments for thermal control and maximum data acquisition. They would also have to maintain the position of the CSM during the collection of data, and operate the data handling subsystem. During experimentation itself, the crew would select the active region for alignment, activate the experiments, and operate the ATM aperture covers in relation to RCS thruster firing and waste water dumps. The crew would also use EVA to assist in the final stowage, the recovery of film and tapes (similar to the deep space EVAs conducted on Apollos 15–17 with the Scientific Instrument Module's (SIM) Bay film retrieval on the way from the Moon), and exchange of cassettes during the mission. All recovered films and tapes would be stowed in empty food boxes – if the astronauts ate enough! Main observation experiments were to be monitored every sunlit orbit for 20–60 minutes.

The mission would begin observations one day into the flight. This would allow for the completion of outgassing from the spacecraft after entering orbit, and jettison of the cover panel. Extending the ATM would take approximately 90 minutes before the data could be collected, depending upon the experiment requirements. Deactivation at the end of the (unspecified) mission would include approximately 30 minutes of EVA to retrieve film canisters, and one hour to deactivate and stow instruments.

The report evaluated crew operations, digital data requirements, thermal control design concepts, contamination control, and the consideration of crew motion inside the CSM on ATM data collection (vibration and pitch/yaw pointing errors).

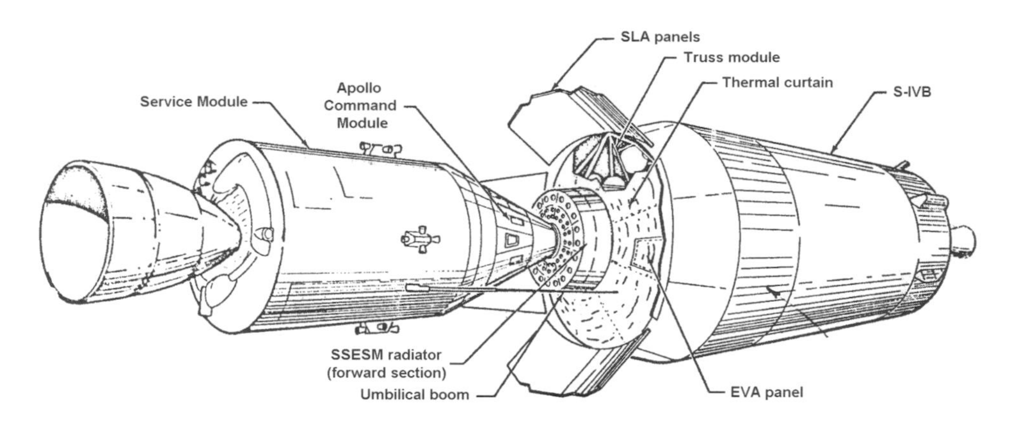

The SSESM concept.

1966: plans evolve

On 18 February 1966, NASA's Deputy Administrator Robert Seaman, testifying before the House Committee for Manned Space Flight, described the NASA AAP effort. It consisted of the extension of orbital flights of up to 45 days or more by minor modification to the basic Apollo system, the procurement of additional spacecraft and launch vehicles for follow-on missions beyond the initial Apollo lunar landing schedule, and the utilisation of Apollo vehicles during 1968–1970 as long as the primary objective of landing on the Moon was not compromised. He also stated: 'We cannot today look towards a permanent manned space station, or a lunar base, or projects for manned planetary exploration, until our operational, scientific and technological experience with major manned systems already in hand has further matured.' This indicated that any work towards goals beyond Apollo would first be evaluated within the scope of the Apollo programme in addition to the landing missions, and would be performed under the AAP programme.

In early March, a team of Douglas engineers presented a technical briefing and cost proposal to George Mueller on designs for AAP hardware and missions of up to 30 days. Mueller expressed strong interest in the spent-stage concept, which he thought could establish a solid base for much larger space station requirements. However, he cautioned that official approval of the concept had not been received from NASA administrators Jim Webb or Robert Seaman, and in order to make the programme official within NASA, he admitted that he still 'had some selling to do.'

The first AAP schedule

On 23 March 1966, NASA released its first AAP schedule. It projected a staggering 45 launches (26 Saturn 1B and 19 Saturn V) in both the lunar and Earth orbit phases of the programme by the mid-1970s, separate from the mainstream Apollo lunar landing effort. Among these were three Saturn S-IVB Spent Stage Experiment Support Modules ('wet workshops'), three Saturn V launched orbital laboratories, and four Apollo Telescope Mounts. The first AAP launch was expected in April

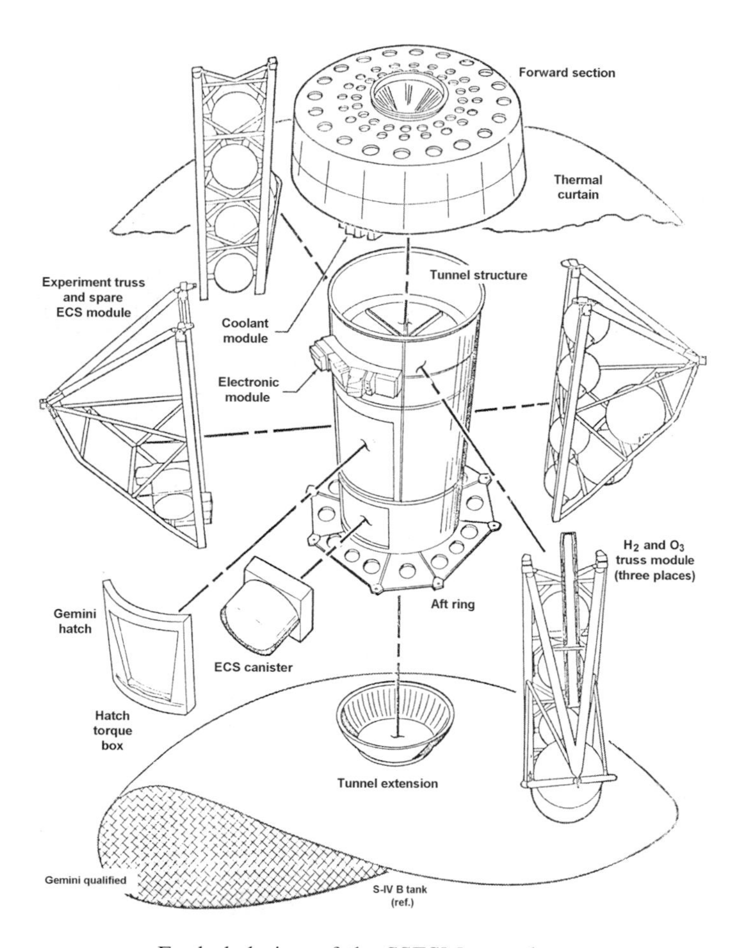

Exploded view of the SSESM tunnel structure.

1968, depending upon progress with the Apollo lunar landing, and assuming minimum modifications to the hardware and launch schedules.

The following day, MSC Director Robert Gilruth sent a long letter to Mueller, expressing certain misgivings on aspects of AAP planning. Gilruth questioned whether AAP represented the best approach to future manned spaceflight, and asked if it was wise to continue assigning hardware originally developed for lunar landings or to promote changes from the original use of the hardware without further study. He also pointed out a threat to the extensive use of Apollo lunar mission launch facilities, as the launch rate for AAP gradually exceeded the lunar schedule. Gilruth was also concerned that the constant change in the AAP programme as a result of the contracting budget could cause diversions that might seriously delay the Apollo lunar missions.

In the communication, Gilruth offered an alternative schedule, using Apollo hardware with nominal modification and a launch rate similar to the landing effort.

However, he pointed to a mismatch between projected AAP planning, the opportunities for manned spaceflight, and the actual resources available. He also suggested that by merely maintaining the production and flight rate of Apollo for AAP, 'without planning for a new major programme, and without significant research and development as part of AAP, we will not maintain the momentum we have achieved in the manned spaceflight programme.'

On 5 April, NASA had issued proposal requests to Douglas, Grumman, and McDonnell, to undertake 60-day studies on an S-IVB Spent Stage Experiment Support Module (SSESM). Four days later, definition studies (including costs and estimates for integrating the ATM with the LM structure) were issued to MSFC. The desired flight dates of the LM/ATM structure were given as April 1968, February 1969 and February 1970. MSFC would be responsible for development and modification of the Apollo lunar lander to carry AAP laboratory kits, including the ATM, and provision for orbital storage for three to six months between manned missions. With Marshall already handling the development of the S-IVB spent stage, giving them the LM/ATM only served to further enrage officials at Houston.

The following month, NASA Administrator Jim Webb stated that progress in the AAP was being hampered by lack of payload/experiment definition and funds. Unless Congress granted additional finds in the 1968 Fiscal Year budget, the programme would be in serious trouble. Any post-Apollo projects were already being hindered by the Vietnamese conflict, congressional discontent with escalating NASA administration costs, and an inability to completely combine the civilian NASA project with USAF MOL space station project.

Robert Gilruth summarised Houston's position on the need for a significant manned spaceflight goal using Apollo hardware after the lunar landing had been achieved. He argued that creating a space station was not enough, and suggested that a manned Mars fly-by or landing should become a NASA in-house focus for planning based on Apollo hardware. Although Gilruth pointed out the imbalance of AAP goals and resources, and the extent of engineering redesign and modification to hardware forced upon the project, he still expressed the desire to continue to be part of the AAP project at MSC, not wishing to lose all participation to Marshall. Gilruth's statement also warned: 'The future of manned spaceflight [at NASA] is in jeopardy because we do not have firm goals, and because the present approach appears to be technically unsound.'

Later that month, this point was underlined by the decision to defer funding of the ATM until the larger AAP funding became clearer. Meanwhile, paper studies for adapting the LM to a 14-day mission continued, along with plans to meet early objectives of ATM by mounting telescopes in an open bay of the Service Module (as proposed in the Ball study earlier in the year). NASA, the aerospace industry and the scientific community would continue to determine the final configuration of the telescopes when full funding became available. A final report from Ball indicated that 'significant space science could be achieved with the ATM system, and an astronaut crew could be very effectively utilised in the acquisition of scientific data. The mission could be performed without any major change in the spacecraft, and the use of the Apollo CSM does not preclude the successful accomplishment of the

A Lockheed Missile and Space Co. artist's concept of a possible (and crowded) AAP mission *c.*June 1966. It depicts an EVA astronaut dwarfed by huge solar arrays, while a second astronaut works on a large radar antenna to take radar measurements of the Earth's surface. The AAP LM was not planned to carry only an ATM, but also for early studies in manned radar topography mapping that thirty years later was conducted from the Space Shuttle. (Courtesy Lockheed Aircraft Corp.)

scientific objectives. Overall, the ATM concept [is] also applicable to other [unspecified] vehicles and targets, both celestial and terrestrial, using state-of-the-art technology with relatively little new development required.'

The report identified further problems, however, and the need for additional study in adapting the ATM to operate from an LM carrier. The most serious problem was contamination caused by outgassing from the CSM spacecraft. The report also suggested that further studies into the use of the ATM for stellar and geophysical experiments was still required, and that a clear ATM payload needed to be designed in the short term if a solar mission during peak solar activity years 1968–1970 was to be attempted.

A major technical planing session on the AAP was held at NASA HQ on 14 May 1966. It was decided that a dependent power supply would be produced from fuel cells in the Service Module of the parent spacecraft. Power for experiments in the SSESM should be fed by cabling from the CSM via an external umbilical, with the development of extended duration fuel cells sought by early 1968. When the Astronaut Office learned of this, they voiced operational and safety concerns that focused both on trying to connect electrical cables during EVA, the astronauts' exposure to exterior conditions, and on the AAP programme's apparent lack of experiment planning or hardware to define the operational requirements of the flight crew. There was also concern expressed over the suitability of a spent hydrogen stage for crew habitation, the pressurisation of oxygen, outgassing from the tanks, and the effects on the electrical cabling and other systems. From here on, the Astronaut Office became more involved with the design and operational requirements of the AAP programme. Following the establishment of an Apollo Applications Programme Office at both MSC and MSFC, the AAP Branch of the Astronaut Office was also established at MSC.

The meeting decided that the flight article should be a simple structure with 'no follow-on goodies' such as dual docking. The mission would begin with the unmanned SSESM launch, followed the next day by the launch of a 14-day orbital duration CSM. The crew would then rendezvous and dock to the SSESM on a minimum two-week mission, open-ended to a possible 28 days' capability.

An alternative design envisaged the Saturn 1B inserting the CSM, spacecraft adapter SSESM and S-IVB in orbit on just one launch. The crew would then separate the CSM, rotate it 180° (as in extraction of the LM on a lunar mission) and then redock for the experiment phase of the mission. Access to the S-IVB would be through an airlock in the SSESM, through a tunnel connection 65 inches in diameter and hatches 43 inches in diameter.

On 11 July, MSFC was selected as the lead field centre for ATM development, while the MSC became the lead centre for Earth resources. On 12 July, the strategy of post-Apollo space projects of the scientists within NASA focused on Earth-orbit missions, solar exploration, and astronomy missions, and how these could be applied to AAP. The following day the AAP Mission Planning Task Force was established to co-ordinate the exact definition of all proposed missions, the actual feasibility of these objectives, the extent of experiment compatibility between the spacecraft and launch vehicle, and the realistic capabilities of the crew to achieve these goals. On 21 July, the Kennedy Space Center created the AAP Branch within the Advanced Programme Office (as part of their existing Apollo Programme Office at the Cape) to handle the definition and requirements for the proposed, and often changing, AAP launch schedule.

On 25 July, Apollo mission 209 was approved to launch an unmanned OWS and docking module. A dual launch was planned with the Airlock Module on the manned 210, which would replace the LM and serve as a docking adapter at one end for the CSM. The other end would have a sealed hatch connecting to the S-IVB. This would provide a pressurised passageway into the stage. An oxygen supply would be provided for pressurising the AM, with the hydrogen tank to provide a shirtsleeve environment (after cleaning and outgassing the S-IVB) – a practical demonstration of the feasibility of the structure to support a space station spent-stage concept.

By the end of July the design of the ATM settled on an LM ascent stage, with the ATM forming half a rack in place of the descent stage, or a specially designed structure in place of the LM. To assist in the final choice of design, further work was requested to evaluate docking compatibility with the CSM in both formats, analysis of interfaces between the LM rack and ascent stage, and descriptions of subsystem installations for both the LM ATM and the rack ATM. Approval of ATM experiments would provide the US with a major new capability without interfering in the basic launch vehicle programme of the mainline Apollo programme. Robert Seaman emphasised the importance of the ATM: 'It would open up additional areas of knowledge we might need if the Russian programme accelerated, to the degree that we wish to add to our manned operations, [and] with the least lead time and maximum use of Apollo equipment.'

George Mueller had advised Seaman of the progress towards selecting the final design configuration, and recommended commencing development work immediately instead of waiting until the end of the year, to maximise the chance of flying the structure during 1968–1970. Both of them agreed that a great deal of in-house development could be completed at MSFC, and although important industrial contracts needed to be secured for some components, development of the ATM within NASA would help complete the project on time and within budget.

The Orbital Workshop

On 9 August, new designations for some of the AAP hardware were adopted. The S-IVB spent-stage concept became known as the Orbital Workshop (OWS), while the Spent Stage Experiment Support Module became known as the Airlock Module (AM). As more defined plans for the programme became apparent, Chris Kraft, Director of Flight Crew Operations at MSC, began expressing grave concerns over potential difficulties in current AAP planning and hardware integration. From the plans issued in June 1966, MSC would be responsible for integrating only the CSM, while MSFC would integrate the S-IVB stage, the Saturn instrument unit, and the ATM/LEM. Headquarters would be responsible for the total payload. Kraft considered it illogical that the scheme had two independent and parallel efforts for spacecraft payload integration at MSC and MSFC, and argued that it was inconceivable that HQ in Washington could take on a complex and detailed operational role. He called for clearly defined roles, and suggested the traditional role for MSC of design and integration, with overall payload responsibly going to MSFC. The dispute between both main centres was beginning to become uncontrollable.

Lake Logan peace talks

These comments were addressed during 13–15 August at a meeting of the Manned Spaceflight Management Council at Lake Logan, North Carolina, where centre responsibilities were defined. It was confirmed that the Marshall Space Flight Center would have a role in developing manned spacecraft. Despite the existence of some judiciary arrangements – such as MSFC being in charge of propulsion, and MSC being in charge of the 'command post' – the idea of modularisation would be adopted. As long as it was not too complex, complete parts of spacecraft could be assigned to separate field centres for development. This concept developed into the lead/support centre arrangement, where the lead centre held overall responsibility for management issues, and set hardware requirements for the support centre, which oversaw direct module development.

For the AAP, MSFC would become the Lead Center, overseeing the living quarters and laboratory component of the space station. MSC became Lead Center for mission operations, while KSC remained Lead Center for payload integration and launch operations. Experiments were divided between MSC and MSFC.

The idea of having two Lead Centers was intended to reflect the two phases of development and operations, but in reality they rarely became distinct from each other as the programme progressed. What happened was that Marshall became a contractor to MSC in supplying the OWS hardware, and developed and customised it according to MSC's recommendations for mission operations. The result was that the preliminary planning into the spent-stage concept at Marshall evolved from strong debates within NASA, to emerge as the best design to receive political acceptance and, in turn, funding. Although the meeting resolved centre roles on paper and in operation, it was still to be a long, difficult road leading to launch.

Planning to fit a budget

On 19 August, McDonnell (manufacturer of the AM) proposed using existing Gemini technology to save time and development costs. Ten days later Seaman notified Mueller to approve development and procurement for one ATM flight on mission 211/212, as an alternative to the basic Apollo Earth-orbital mission previously assigned to those flights. As only one ATM flight had been approved to date, it was important to focus all efforts on it.

On 19 October, NASA HQ reduced Houston's AAP budget operating plan for Fiscal Year 1967 for both the experiments and the OWS ($8.6 million for each). The 1966 estimate for the workshop mission was $17 million ($14 million on hardware and mission support, with $3 million for currently assigned experiments). In a critical reply, Robert Thompson, MSC Assistant AAP Manager, wrote to the AAP Deputy

During an October 1966 simulation, a Marshall engineer wears a 'spacesuit' as he sets up individual living area plastic curtains at one end of a mock-up S-IVB stage, now void of hydrogen after boosting the crew into orbit for a 30-day mission

Director John Disher, pointing out that 'prompt and adequate funding is required if the current schedule is to be met.'

By early November, refined plans for the flight programme resulted in recommendations that the LM ascent stage/half rack ATM should become the baseline configuration for development, as this configuration had desirable characteristics over the CSM/ATM rack proposed earlier. These included maximum solar data collection through an easier mode of operation; the maximum use of the crew-members' capability to conduct astronomical observations; the ability to conduct manned operations docked to the OWS (or alternatively, a tethered flight from cluster); and a much lower development cost against a higher assurance of success.

On 8 November, discussions led to the approval of mission sequence 209 to 212:

- SAA-209 (formerly 210): manned Block II CSM flight of 28 days, with the CSM fuel cells providing primary electrical power and carrying a Mapping and Survey System (M&SS) module.
- SAA-210: launch of the unmanned airlock–OWS–multiple docking adapter combination, with solar cells as the chief source of power.
- SAA-211: resupply module and manned CSM flight of 56 days.
- SAA-212: unmanned Lunar Module Apollo Telescope Mount flight.

In a subsequent planning session on 16 November, a system of mission identifications for the first four missions (using Saturn 1B) was suggested to avoid confusion between the designation of lunar and workshop missions. The flight plan for the workshop suggested that mission 209/210 was essentially the manned workshop activation phase of 28 days, and that mission 211/212 was the solar astronomy and orbital assembly phase of up to 56 days. Of the two 'phases', the second was more difficult to achieve, and resulted in small study teams looking into the complexity of the two phases and how one phase could benefit from the other. The overall desire was 'to proceed in getting the job done together'.

An alternative mission for 209 saw the S-IVB spent stage launched on the same vehicle as the crew, rather than separately – a profile that had been suggested in earlier designs. This single-launch mission would early demonstrate the concept of the OWS, its function of providing a controlled atmosphere for crew operations in the CSM, airlock and H_2 tank, and would provide experience for a CSM mission over 14 days and a range of in-flight experiments for up to 28 days before progressing with a more complex series of longer missions.

The CSM 105 (Block II lunar distance design) would be used on SA209, as it was not required for the lunar programme. After entry into orbit, a transpiration and docking manoeuvre would be accomplished with the airlock carried in the spacecraft adapter. On the first day, the crew would perform internal and external EVAs to reconfigure the vehicle from a spent stage into the OWS. A total of twenty-one experiments were to be assigned to the mission (one technology; six DoD; eleven engineering; three medical).

The adoption of the AAP cluster concept

On 18 November it was finally decided to designate AAP missions in numerical sequence, beginning with AAP-1 (SAA-209), although the programme planning document would still identify tentative hardware assignments pending firm vehicle allocation.

The 'cluster concept' was finally introduced into AAP design designations on 5 December. The flight plan projected an unmanned OWS launch the day after a manned CSM launch for the 28-day mission, and the OWS would then be placed in unmanned orbital storage. Six months later an LM/ATM launch would follow a second manned CSM launch. The CSM would then rendezvous and dock with the LM/ATM and take it to the waiting OWS. According to this plan, the first launch would occur in June 1968.

This was a radically new approach – unlike the SSESM design in which there would be limited rendezvous, no docking and no habitation equipment. Although the plan still used the S-IVB to put the payload in orbit, followed by the cleansing and pressurisation of the hydrogen tank, the new concept also included habitation of the module. The OWS would contain crew quarters, two floors, and walls, installed on the ground prior to launch. The AM would also be an integral part of the OWS, and would not be launched separately. The Multiple Docking Adapter (MDA) would feature five docking ports, allowing a CSM and up to four science modules to dock at any one time, and would also house most of the astronaut habitation equipment and experiments. This AAP schedule called for twenty-two Saturn 1B launches and fifteen Saturn V launches, including flights of two Saturn V workshops and four LM/ATMs.

This resulted in even more 'further action items', including cost and scheduling, twenty-seven priority items of configuration changes, workshop design, a decision on solar panels over fuel cells, the selection of a two-gas atmosphere over 100% oxygen, emergency procedures, EVA requirements, experiment definition, Apollo vehicle modifications, and a definite plan for follow-on hardware procurement. In addition, a plan for test requirements, reliability, quality assurance, organisation, and manpower requirements.

On 22 December, Mueller wrote to Gilruth and von Braun, advising MSFC of a joint MSC/HQ medical position regarding gaseous atmosphere selection. The medical position was based on retaining the 100% oxygen environment in the CSM and a shirtsleeve environment in the workshop of 69% oxygen and 31% nitrogen at 5 psia. The 100% oxygen atmosphere would still be required for suited emergency operations and EVA.

As the year closed, the question arose as to whether AAP should be portrayed as an open-ended programme, or as having a defined goal for actively marking its completion. There were discussions as to whether the programme should include the capability of space rescue, the definition of its Earth-orbital tasks, and emphasis on the reuse of hardware from previous missions. There were also continuing discussions concerning the inclination at which the vehicle should be flown – a 28°.5 orbit versus conducting AAP-A meteorology experiments inclined at 50° at least. By the year's end, mission objectives for AAP 1 and 2 were promoted as a low-altitude, low-inclination orbit for Earth resources studies and biomedical research, with a three-man crew for a maximum of 28 days, using a spent S-IVB. This could also provide a base for a second crew to reactivate the stage within one year to continue the medical studies, as well as solar and celestial research, and to test Earth-observation passes of a proposed lunar mapping and survey system.

Tragedy at the Cape

On the eve of a ground test for the first manned Apollo flight, planned for February 1967, George Mueller conducted a briefing, explaining that NASA planned to form an 'embryonic' space station in 1968–1969 by clustering four AAP payloads launched at different times. Firstly, a manned spacecraft would be launched, followed several days later by an S-IVB that had been converted into the OWS. After docking, the crew would enter the OWS through an airlock. At the end of their 28-day mission, the crew would power down the OWS and return to Earth. Three to six months later, a second manned spacecraft would be launched on a 56-day mission to deliver a resupply module to the OWS and also to rendezvous with an unmanned ATM, the fourth and last launch of the series. Emphasising the importance of a manned ATM, Mueller stated: 'If there is one thing that the scientific community is agreed upon, it's that when you want to have a major telescope instrument in space, it needs to be manned.'

A day later, on 27 January 1967, during a simulated countdown, a flash fire occurred inside Apollo spacecraft AS-204 (Apollo 1), claiming the lives of the prime

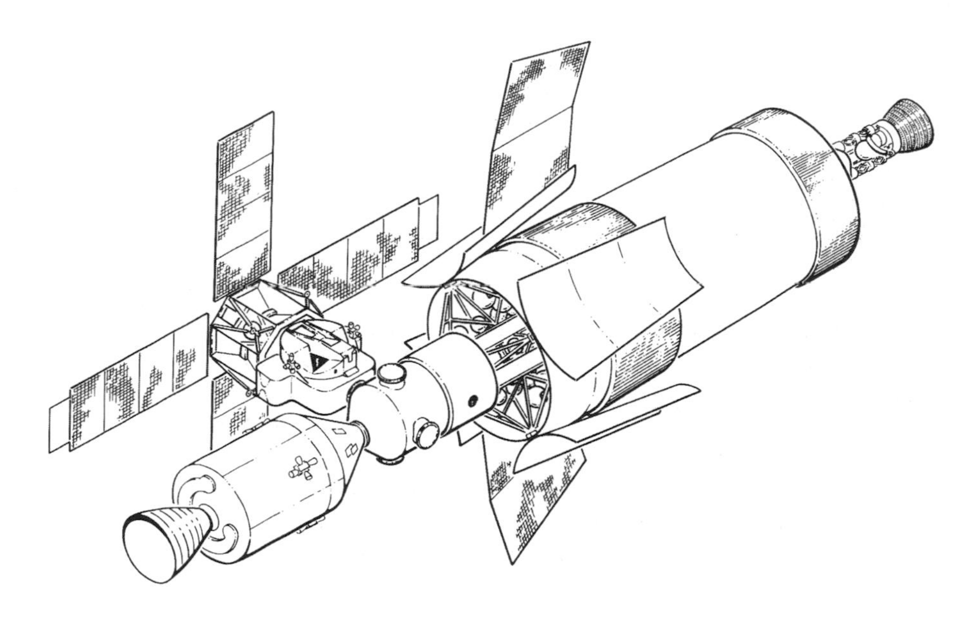

A 1967 concept for the AAP cluster.

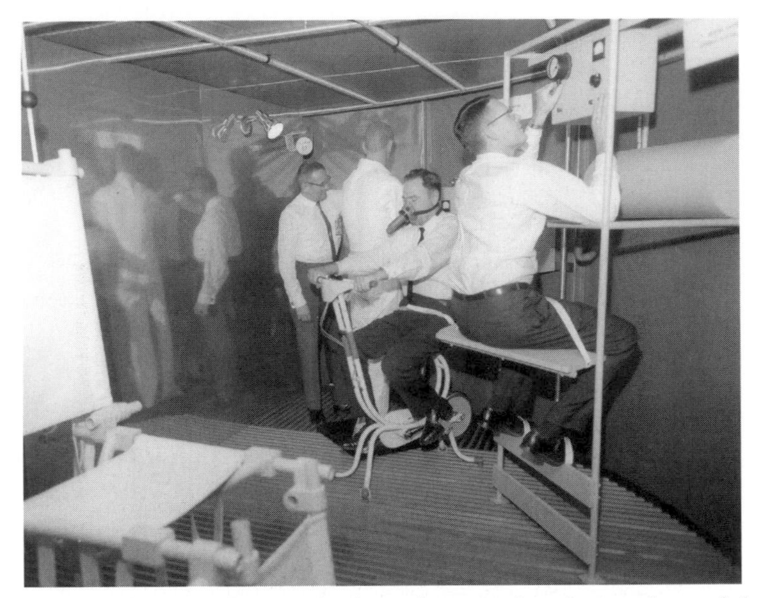

Marshall staff performs equipment check-out in an engineering mock-up of the OWS during February 1967. The living quarters feature a metal floor, installed before launch, allowing propellant to flow through, and crew installed fabric compartment dividers on to pre-installed metal poles. (*Left–right*) Robert Schwinghamer, Doyle Estep – both from Manufacturing Engineering Laboratory – and William Ferguson, OWS project manager in Industrial Operations.

crew, Virgil Grissom, Ed White and Roger Chaffee. As a result, the effort to reach the Moon by 1970 was delayed by the inquiry and the implementation of the recommendations. Consequently, the AAP slipped further into the background as a stunned nation recovered from the fire and the price that the three astronauts had paid to achieve the President's goal. But Grissom's earlier comments on the risks of spaceflight reflected not just the spirit of reaching for the Moon with Apollo, but also all other aspects of space exploration: 'If we die, people should accept it and get on with the programme.'

The aftermath

While investigations continued into the loss of the three astronauts and the effect on the space programme, background work continued in other areas of Apollo and the AAP. Four days after the loss of the three Apollo 1 astronauts, AAP Programme Director Charles Matthews recommended that although no specific crew assignments had been included in the programme, the scientist-astronauts who had been participating in the ATM programme should be given the opportunity to visit a number of leading astronomical observatories in the country. Potential crew-members could then derive a better understanding of the equipment being employed, the operational techniques, and the nature and type of observations that would be undertaken.

Standing in front of this full-size LM/ATM mock-up in February 1967 is Rein Ise, ATM manager at Marshall. This mock-up was used to determine instrument placement for five major experiments containing three separate instruments in the lower experiment rack (where the LM descent stage was located on lunar missions).

About six weeks later, Deke Slayton, Director of Flight Crew Operations at MSC, expressed concern about the proposed workload on the AAP crews when they finally reached orbit. According to Slayton, there remained an excessive number of experiments assigned to the first AAP manned mission. The planning required a total of 672 experiment man-hours; but only 429 hours were available, leaving a deficit of 243 in-flight man-hours. The same problem existed in the training programme, in which experiments required 485 hours per man, with only 200 hours per man available – a deficit of 285 hours per man.

Slayton was not only concerned with the overall hours required for experiments, but also with the investigations themselves. On 6 April he requested that the proposed 'jet shoe' astronaut manoeuvering experiment (worn at the feet) be deleted from all AAP flights. During January 1967 the engineering development model of the experiment had been tested by several astronauts at the air bearing facility at MSC, with the principle investigators (PIs) in attendance. The astronauts tested the system while wearing shirtsleeves, and they were fine; but when an LaRC test subject completed similar runs in an inflated pressure suit, the results were unsatisfactory. Wearing the full suit and life support system offered the wearer extremely poor visibility, and created a safety risk.

Rein Ise examines the ATM on a model of the OWS in February 1967, indicating the location of the LM ATM on the model.

REFINING THE FLIGHT SCHEDULE

In March 1967, several days of discussions concerning increased electrical power requirements to make the OWS habitable and to carry an increased experiment payload, resulted in the decision to add solar arrays to each side of the S-IVB, instead of relying on additional fuel cells in the Service Module. The ATM would continue to carry its own solar array panels and power system, supporting the experiment's own power requirements.

In the flight mission directive for AAP-3/AAP-4, dated 27 March 1967, definition of the objectives for those missions were highlighted once AAP-1/AAP-2 had performed the initial activation. The primary objectives of the second pair of missions were to deploy and activate the Apollo Telescope Mount to begin the collection of solar data, and to demonstrate the ability to reactivate the OWS, left unattended in orbit for several months. AAP-3 (CSM-108) would be launched on Saturn 1B 209 from LC34 to rendezvous with the OWS. AAP-4 would feature an SA-210 launch, carrying the ascent stage of the LM6 and the ATM. This launch would be from LC37B, and after CSM rendezvous the next day, the combination of CSM 108/LM6/ATM would transfer to the OWS. Two of the crew would transfer to the LM and separate it, to be docked radially to the OWS, opposite to the previously docked LM&SS hardware. The solo pilot in the CSM would then attach that spacecraft to the axis docking port on the station.

Ed Gibson simulates one of the tasks (fill and drain line plug installation) planned to be performed on the Saturn S-IVB orbital workshop. This photograph was taken in May 1967, during a walk-through of the mock-up station at Marshall. Eleven other AAP astronauts joined him for the simulation.

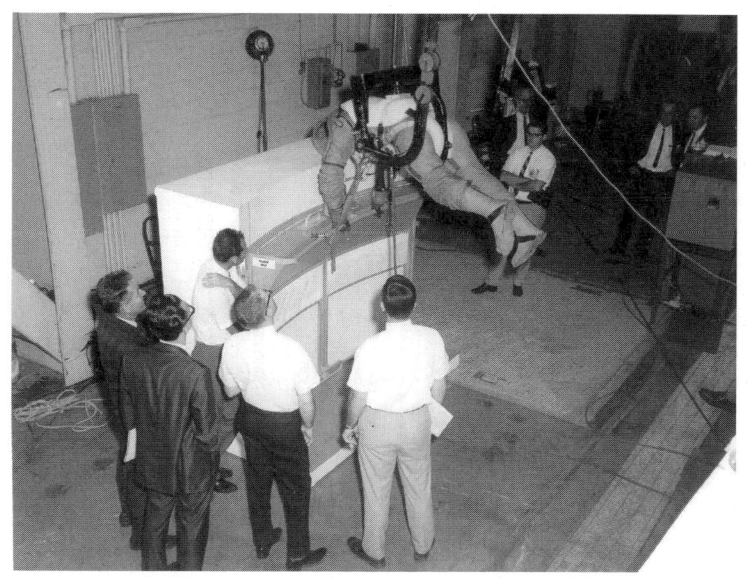

A second view of the May 1967 walk-through shows the suited astronaut evaluating fans that were to be installed inside the laboratory – one of several activities required to convert the stage from an LH tank into an early space station.

Following the Apollo 204 pad fire, several review boards re-examined the issue of flammability on the OWS, including conducting flammability tests of the S-IVB hydrogen tank, and insulating it with an aluminium foil flame-retardant liner.

By early May, flight schedules had changed again, indicating that the first missions would not be launched before the beginning of 1969. AAP-1/AAP-2 was now targeted for early 1969, to accomplish OWS set-up and 28-day mission objectives, but without carrying the LM&SS hardware. AAP-3/AAP-4 would fly in mid-1969 to accomplish the 56-day ATM objective and the reuse of OWS. The new plan also featured two additional 1969 flights (AAP-5/AAP-6) to revisit OWS and ATM, using refurbished CSMs that would previously have flown in Earth-orbit tests in 1968.

Other AAP missions were planned for low Earth orbit from 1970. These would see two dual launches (AAP-7/AAP-8 and AAP-9/AAP-10), each featuring a manned CSM and an unmanned experiment module to expand the facilities onboard the OWS. The series of missions to the first OWS would be completed by two long-duration missions (AAP-11/AAP-12), with CSM-only launches, to establish a near-continuous occupation.

This new plan effectively removed the Earth-orbit test of the Lunar Mapping and Survey System from the OWS phase of AAP. The hardware was remanifested on 8 May as AAP-1A for a solo CSM/Saturn 1B launch, unconnected to the workshop missions and planned for a 15 September 1968 launch.

AAP-1A: a mission added

AAP-1A would be launched by Saturn 1B into an 81 x 120 nautical mile orbit at 34°. Upon insertion, the CSM would perform the transposition and docking manoeuvre and withdraw the LM&SS. Upon extraction, the combination would be raised to a 141-nautical-mile circular orbit. The LM&SS test was planned for a minimum of five days and open-ended up to fourteen days. The trajectory maximised flight time over continental America, with highest Sun altitudes over Phoenix, Arizona (up to 50° latitude). In support of this additional mission, a 60-day study was conducted by Martin Marietta between 8 July and 5 September 1967. Additional studies were conducted between 6 September and 30 November 1967, to recommend mission and system requirements, the experiments to be carried, and the configuration of the experiment carrier.

Twenty-three experiments were to be integrated into the flight, to be carried in the recommended, but undefined, carrier configuration (see the diagrams on pp. 43 and 44). This was an aluminium truncated cone, 84 inches in diameter at the experiment mounting end, and 110 inches total length. The study suggested a launch at 10.00 EST on 1 April 1969, with a 14-day mission duration, each producing an average of six daytime passes over the United States and a daytime recovery in the Atlantic Ocean.

During the study period, five groupings of experiments were analysed, and after a mid-term review on 9 August 1967, NASA selected the final group. It included eleven early versions of AAP experiments and twelve scientific experiments transferred from the mainstream Apollo programme. There would be a maximum of eleven days of experiment operation (none planned for EVA), conducted from the LM&SS or CM.

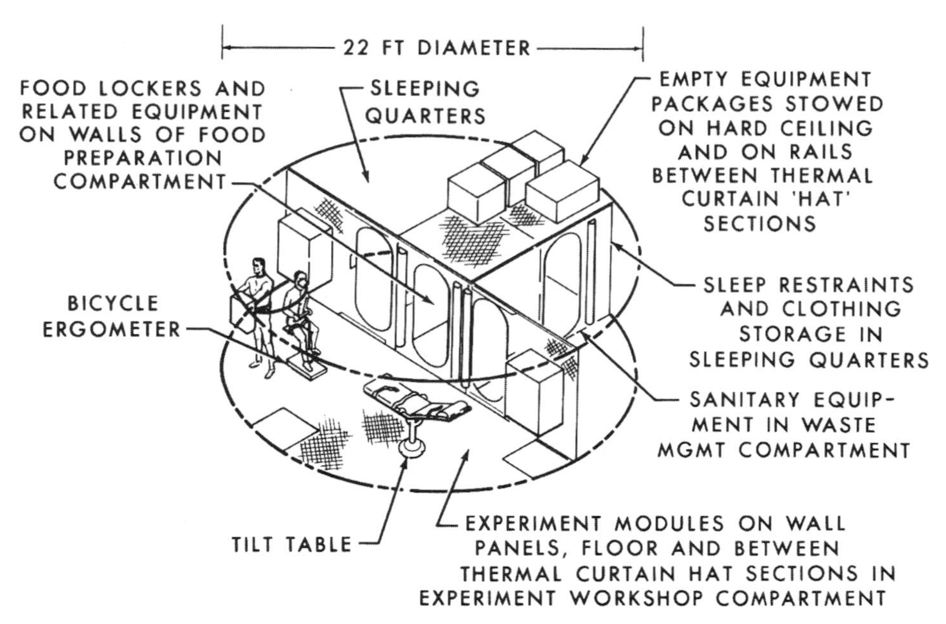

AAP cluster experiments for S-IVB Orbital Workshop.

The report summarised that the twenty-three experiments were an important contribution to the overall AAP. A selection of bioscience experiments and crew operations would assess man's usefulness in space, prior to the flight of the larger OWS. Major Earth resources and atmospheric science experiments would also provide baseline data, in order to develop advanced instruments and future sensors for OWS flights AAP-1–4. Astronomy payloads would study atmospheric effects on future stellar and solar observation and would provide early data for the later ATM programme. It was a bold plan – perhaps too bold.

Refining the design

On 12 May, J. Bollerud of HQ and Charles Barry of MSC released a staff paper, recommending a 69% oxygen and 31% nitrogen shirtsleeve atmosphere in the OWS. This generated further discussion between the centres on the impact that such a mixed-gas environment would have on engineering design, and the physiological response of the astronauts. The resulting consensus indicated that 5 psia would meet all the needs in the OWS.

Twelve days later, as a direct response to the delays resulting from the Apollo 204 accident and inquiry, NASA realigned its Apollo and AAP launch schedule. The new schedule called for an overall total of twenty-five Saturn 1B and fourteen Saturn V launches. This included two Workshops flown on the Saturn 1B, two Saturn V workshops, and three ATMs, with the first AAP launch planned for no earlier than January 1969.

AAP innovations were frequently highlighted to justify the need for and cost of the programme – particularly the re-use of the CM, the double use of the S-IVB as both propulsive stage and laboratory, repeated use of the same laboratory in orbit, and flights of increasing duration and complexity employing existing Apollo hardware and infrastructure. The whole programme was promoted as access to space at relatively low cost, with significant experiments that would benefit all mankind. It was also emphasised in programme planning documents that the AAP missions would draw upon previous flight experience, taking full advantage of the Apollo/Saturn system and continuing to offer significant contributions to a wider range of objectives.

The AAP Technical Summary of 1 June 1967 indicated that AAP missions were designed to, 'gain experience, test theory, perform experiments, and collect data.' The summary also stated that by modifying and expanding the basic Apollo systems, AAP would determine man's usefulness in space in areas of extended spaceflight and manned astronomical observations. The extended duration of the missions would enable the programme to gather much more data from each mission.

To help achieve these objectives while also saving costs, Slayton and Kraft at MSC suggested that it would be better to first launch the unmanned workshop and then the manned CSM. By doing so, if the OWS failed in launch or in orbit, the CSM need not be launched. The crew would then not be exposed to the hazards of boosted flight, retro-fire, or re-entry and landing, if their primary mission failed. It would also leave a usable CSM available for another mission. They also indicated the need to first flight-prove new equipment (essentially the OWS) before committing a proven system (the CSM) and finding that the new equipment was not yet ready to receive it. This was the method eventually chosen for Skylab.

In July 1967, Jim Webb testified before the Senate Committee on Appropriations

This 1967 photograph shows a model of the OWS with a 'serpentuator' (serpentine actuator) being employed to move bulky hardware from one part of the station to another as an early study for an EVA aid. Featuring hinged joints, and controllable from either end, it was a forerunner of the remote manipulator system flown on the Space Shuttle, and the advanced robotic arms planned for the ISS.

Subcommittee over the Fiscal Year 1968 NASA Authorisation Bill. He was asked to make further cuts in either the proposed Voyager unmanned landing missions on Mars, or in the AAP. Webb replied that both were vital to the US space effort. The AAP was a small expansion of investment after already spending $15 billion to reach the current stage of development, and he also supported the Voyager programme because of Russia's known interest in exploring Mars, which could be used for many purposes, to serve mankind or for military power. Criticised for his indecision, Webb said that he would neither support cuts nor endorse either programme beyond stating that both were needed.

This was the initiation of several budget cuts that affected NASA and its centres in the late 1960s and early 1970s. As the AAP struggled to survive, the spent-stage concept emerged as the strongest element and became a credit to the MSFC, which sustained the programme in the face of staff lay-offs and restricted budgets. By careful management of funds, re-using flight-proven hardware, and allocating significant amounts of the budget to extensive testing and back-up hardware, the programme managers were able to overcome shortfalls. Although the programme never obtained every dollar that was required, it made use of every cent it received, and ensured that it compensated for what it did not receive.

Changes to the changes

By July 1967, designers had decided to incorporate the two OWS floors into one common grated floor in the crew quarters, to save weight. The crew quarters would be located towards the oxygen tank at the far end, under one side of the grid. The other side (the larger volume) would be used for experiment operations.

In a letter to AAP Director Charles Matthews, dated 29 August, MSC AAP Assistant Manager Bob Thompson presented Houston's philosophy regarding major AAP reprogramming. There were two major factors in this alteration to mission planning: firstly, the very real likelihood of further funding cutbacks during 1968 and 1969; and secondly, the amount of surplus Apollo hardware that the AAP might inherit because of rescheduling as a result of the Apollo 204 pad fire. Thompson then recommended a new revision to AAP planning. This would be a manned Earth-orbital mission in 1969, and two manned flights (of 28 and 56 days) using the OWS during 1970. These would be followed by a manned AAP/ATM in 1971, and long-duration (two months to one year) manned flights in late 1971 and into 1972. He also stressed that until the goal of landing on the Moon been achieved, the AAP should be looked upon as an alternative, rather than an addition to, the main thrust of Apollo. Thompson also pointed out that it should be made clear throughout NASA that there would be no overlapping, and that the AAP must not compete with or detract from the main Apollo objective.

By October the budget squeeze was being felt, as NASA HQ reviewed the AAP schedule, reducing the number of AAP lunar missions to four to incorporate the budgetary cutbacks. There were now seventeen Saturn 1B launches and seven Saturn V launches in the AAP programme, with two workshops launched by Saturn 1B and only one on a Saturn V, and only three ATMs. The first launch of an AAP workshop was at this point manifested for no earlier than March 1970.

The beginning of the dry workshop

During meetings in Huntsville and at NASA HQ in Washington on 18–19 November, MSC proposed an alternative configuration for a dry workshop. This would be fitted out on the ground, be launched unfuelled, and not be used as a live stage during the ascent. It was suggested that this would overcome several problems that had been encountered during the last few months, and was based on the MSC idea of an experiment carrier that could be fitted out on the ground and launched on a series of Earth-orbital missions. MSC did not like the wet stage concept offered by Marshall, believing it was impractical to allow astronauts wearing suits to fit out the stage in orbit. In turn, the two-centre 'dogfight' continued, with Marshall labelling the MSC experiment carrier the 'Max Can' (after spacecraft designed Max Faget). After discussing this option, it was finally decided to proceed with the wet stage configuration for the time being.

On the other hand, there was considerable interest in the prospects for a manned astronomical observatory. On 1 December, after a NASA presentation, the Astronomy Mission Board expressed interest in crew participation in the ATM. They also recommended early crew assignment for the ATM so that adequate training in solar physics could be provided, and once again a recommendation was offered for scientist-astronauts to be assigned as members of any ATM flight crew.

The cancellation of AAP-1A

One of the casualities of the budget reductions was mission AAP-1A, which on 27 December was terminated as an unnecessary duplication of experiments that could be performed onboard the OWS. In the new year, 1968 budget restrictions required additional cuts in AAP operations, so that all that remained were three Saturn 1B launches and three Saturn V launches, with only one OWS on each launcher and just one ATM. The programme's first launch was then expected no earlier than April 1970.

By 29 January, NASA's descriptions of the Saturn V OWS in the Fiscal Year 1969 budget request featured a ground-outfitted OWS to be launched by Saturn V. It was redesignated the 'Saturn V Workshop' and was sometimes referred to as a 'dry workshop'. The Saturn 1B workshop (termed the 'wet workshop', or OWS-A) would be launched by Saturn 1B. By February there were two distinct Saturn V OWS concepts emerging. OWS-B was a relatively simple generic evolution of the Saturn 1B OWS-A, developed from the first AAP missions and retaining some of these early elements. The significant differences were in incorporating the ATM as an integral part of OWS, and not including the LM ascent cabin design. The controls for the ATM would be located inside the OWS. The second design, OWS-C, was a more advanced evolution towards extended operations in Earth orbit, and would providing living and working space for up to nine crew operating in orbit over two years or more.

Another revision to flight plans for AAP-1 and AAP-2 was issued in April 1968. Stringent funding restrictions faced AAP, and the most practical near-term programme that would be likely to succeed as a result of these restrictions was the Saturn 1B OWS. On 19 April the OWS unmanned activity was described at a flight

AAP-1A selected carrier configuration.

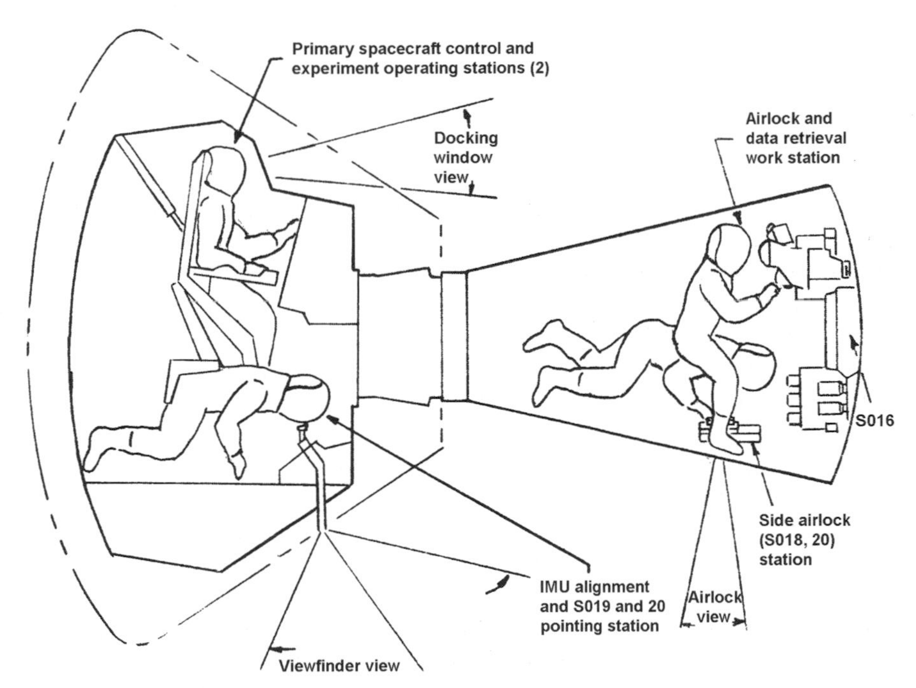

AAP-1A crew work-stations.

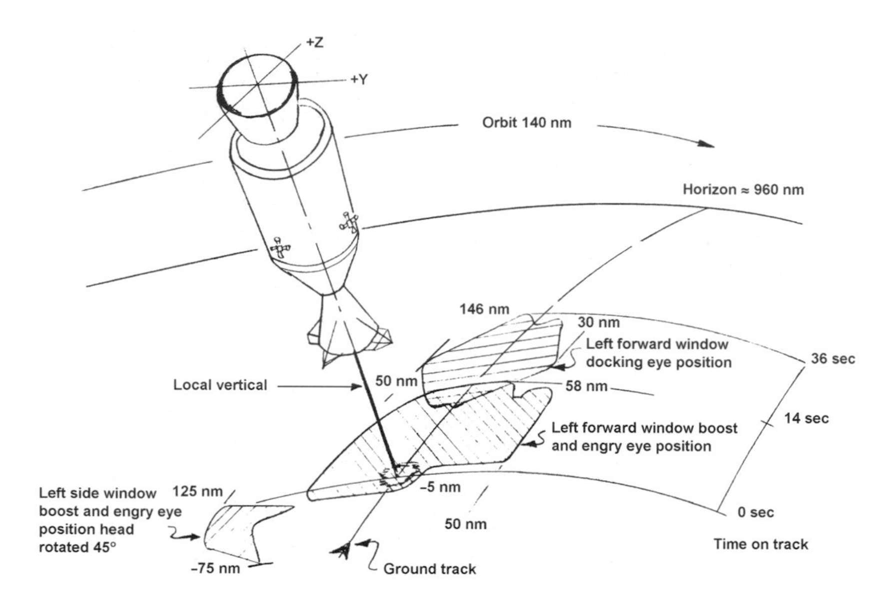

AAP-1A window visibility.

operations planning meeting. It was determined that the solar arrays would be deployed on the first Stateside pass after the liquid venting period was completed. Gaseous venting was expected to take 24 hours, and the micrometeoroid shield could not be deployed until it was completed, as it would interfere with the outgassing.

On 4 June 1968, yet another new AAP launch readiness delivery schedule was released. This time a total of eleven Saturn 1B flights and one Saturn V flight were allocated. Two flights of OWS – one launched by Saturn 1B and one by Saturn V – would be supported by a back-up OWS. There was still only one ATM included. The first launch would occur in November 1970, and the lunar missions were no longer considered as part of AAP planning, which solely featured the OWS operations.

In the summer of 1968, a movement began to rename to post-Apollo manned spaceflights (AAP), and recommended a new name for NASA's Earth orbital flight programme planned for the mid-1970s. It was suggested that a name to describe both the AAP and follow-on stations – then termed the Interim Orbital Workshop – would create a better identity for the programme, as had the names Mercury, Gemini and Apollo. It was urged that names such as AAP, Workshop, and Extension of Manned Spaceflight be dropped, as they did not accurately describe the major goal of manned spaceflight. The initial suggestion for a new name was the 'Space Base Programme'.

Whatever the new programme would be called, ensuring that it had hardware to fly was becoming more difficult. On 1 August 1968, NASA directed its contractors to limit the production of Saturn V engines for Apollo and the AAP, and effectively terminated the production of twenty-seven H-1, eight F-1, and three J-2 engines. Apollo had yet to fly with a crew on board, let alone complete a flight to the Moon, and already the stocks of hardware were being reduced.

An engineering mock-up of the workshop used for a five-day crew station design review held in February 1968. Note the triangular grid floor and the early design of the bicycle ergometer.

Onboard a KC-135 'vomit Comet', a suited engineer installs parts of the grid floor during a 30-second burst of weightless to turn a wet stage into a habitable station during October 1968. It was some of these time-consuming tasks that helped lead to the decision to change from a wet to a dry configuration the following year.

1969: a year of change

Early in 1969, during a meeting to discuss the feasibility of a large space station as a major post-Apollo effort, exactly two years after the loss of Apollo 1, George Mueller suggested that perhaps the proposed 'logistics shuttle system' should first be developed, before space station characteristics were decided upon. This was an influential early decision in trying to secure the budget for the Space Shuttle before funding of the interim space station it was designed to service.

Dry becomes wet

During May 1969, a review of the previous three years of delays, design changes, late decisions and cost cuts delayed the Marshall OWS wet stage design. But the early success in man-rating the Saturn V on the third (Apollo 8) launch meant that one of the Saturn Vs could be used for an AAP mission earlier than previously expected.

Looking forward revealed uncertainty in spaceflight programmes after the AAP, and it was difficult to visualise integrated OWS/ATM missions beyond 1972, to keep the programme sustained. Progress on much larger Space Station Phase B studies would rely on the success of the AAP to provide essential data for that design study. It almost became a case of flying the AAP to prove that it could be done, and then seeing what happened to the budget. The priority by then was to fly one AAP workshop before asking for funds to support any follow-on programmes.

During the Manned Space Flight Management Council meeting held at MSC on 21 May, Mueller sounded out each centre director and AAP offices with regard to the direction the programme should take. Discussions included wet over dry workshop configuration, as well as LM, ATM and CSM operations. MSC argued that it had originated the dry-stage option, while MSFC countered with the fact that the wet stage was the best that could be offered within the budget restrictions. It was also pointed out that since all of the Saturn Vs were committed to the lunar programme, only a launch by Saturn 1B was feasible, and this required a live second stage – the S-IVB.

However, presented with the greater lift capability of the Saturn V, even von Braun was beginning to warm to the dry stage concept, and he began to win over his reluctant colleagues at Marshall. In a letter to Mueller on 23 May, von Braun reiterated the use of the Saturn V and the benefits of launching a fully fitted out S-IVB by a two-stage variant of the larger booster. It also meant that the payload capacity of the CSMs could be reduced, as most of the consumables and equipment could be launched on the station using the Saturn V.

Three days later, Gilruth sent Mueller his recommendation for the change to the dry workshop, as it offered the best chance of an early completion of the AAP basic objectives of 56-day missions and solar research. The reduced cost of the programme would be crucial in obtaining approval from Washington to fund the proposed larger space station and Space Shuttle programmes for the late 1970s.

Losing the LM/ATM

All these changes meant that the ATM could be launched attached to the OWS on the Saturn V, removing the need for a Saturn 1B launch and the complicated

Dr Wernher von Braun, director of MSFC, explains the AAP programme to congressional subcommittee members during March 1969. This was still the wet concept first envisaged by the German rocket engineer more than 25 years earlier, during World War II. (*Left–right*) Von Braun, Congressmen Waggonner (Democrat, Louisiana), Cabell (Democrat, Texas), subcommittee chairman Teague (Democrat, Texas), Fulton (Republican, Pennsylvania) and Dr Ernst Stuhlinger, Associate MSFC Director for Science.

rendezvous and docking by the LM, and simplifying the design and the supporting rack. The solar arrays could also provide a back-up source of power, communications and control to the overall station, providing an added element of redundancy.

A few days later, representatives from NASA HQ, MSFC, MSC, McDonnell Douglas and primary contractors attended an OWS project meeting at Huntington Beach to discuss the proposal.

As AAP gained this new lease of life, the DoD cancelled the USAF MOL programme in June 1969. This was as a result of escalating costs in the Vietnam War, and the development of unmanned reconnaissance satellite programmes, with several MOL elements transferred to NASA, together with seven of the programme's former astronauts. Six years after efforts to combine the two space station programmes were abandoned, elements from both the MOL and AAP would merge into the new OWS programme.

One small step; one major change

On 16 July 1969, Apollo 11 left Earth for the flight to the Moon, returning on 24 July. On 18 July, as Apollo was two days out from Earth, the new NASA Administrator, Tom Paine, approved the change from wet to dry workshop, and followed it on 22 July by abandoning the idea of launching the OWS on a Saturn 1B, choosing instead to use a single Saturn V. At the height of Grumman's success in

MSFC employees evaluate equipment in the waste management (*left*) and food management (*right*) compartments in this early concept of the Saturn V dry workshop in October 1969.

placing Eagle on the Moon, they also received a letter of termination for the LM/ATM. This led to several other changes, including reducing the number of launches from five to four, all from LC39, rather than Pad 37 as previously allocated for the Saturn 1Bs. The launch date also slipped from November 1971 to July 1972. NASA's formal adoption of the dry configuration was announced on 22 July, two days before Apollo 11 came home.

The 13 August AAP delivery schedule reflected the altered programme, and also provided for back-up hardware. It called for two Saturn V launches, with the first flight of the workshop in July 1972. This would be supported by seven Saturn 1B vehicles, offering at least three manned visits to each workshop, as well as a back-up vehicle.

The AAP was gathering momentum towards becoming a fully fledged successor to Apollo, using the triumph of Apollo 11 to announce that the dry workshop would be the configuration that would give America the first of what was expected to be a series of space stations. There was still much to do before the workshop was launched, but as the 1960s ended, the programme finally received a new identity.

Apollo Applications Program becomes Skylab

On 17 February 1970, NASA announced that the AAP had been renamed, the new name being derived from 'a laboratory in the sky', first proposed by Donald L. Steelman, of the USAF, while working at NASA in 1968. At the time, NASA had decided to postpone naming the programme, due to budget restrictions, but now it formally approved the name. America's first planned space station would enter the history books as Skylab.

Preparing for flight

In a statement on 7 March 1970, President Richard M. Nixon listed his administration's proposed goals in space for the 1970s. The fourth item was an objective to extend America's manned capability for living and working in space. This would begin with a large orbiting workshop using hardware originally developed from the Apollo lunar flights. Nixon stated: 'We expect that there will be men working in space for months at a time during the coming decade.' Less dramatic than Kennedy's speech of 1961, it nevertheless appeared that there was Presidential approval for the Skylab experimental space station, as part of a larger programme to work in space 'for months at a time' through to 1979. However, once again in the space programme, although the interest and desire to achieve bold plans was at the forefront of the speeches and studies, the cash to achieve it certainly was not.

THE LOST APOLLOS

For most of 1970, NASA and the Skylab contractors were consolidating the design for the space station, and refining its launch plans to follow the final Apollo flights in three years time. In April, a rumour was circulated that after the cancellation of Apollo 20 in December 1969, NASA might also be forced to cancel two more manned lunar missions to divert hardware for the space station programme. The launch vehicle from one of these would be used to launch a planned second Skylab OWS. The Saturn V from the second cancelled mission might be held in storage to be used to launch a larger space base that could orbit for a decade and become the core module for a proposed 100-man space station.

There had been some concern about the safety of the astronauts in prolonging the use of the Apollo system, with President Kennedy's goal having been doubly achieved in 1969. There had been little indication that the Soviets were to attempt their own manned lunar programme, and public interest and media coverage of further lunar flights was beginning to wane.

Apollo 11 had proved it was possible to land on the Moon, and Apollo 12 had shown that the landing site could be restricted to a very small area, and so

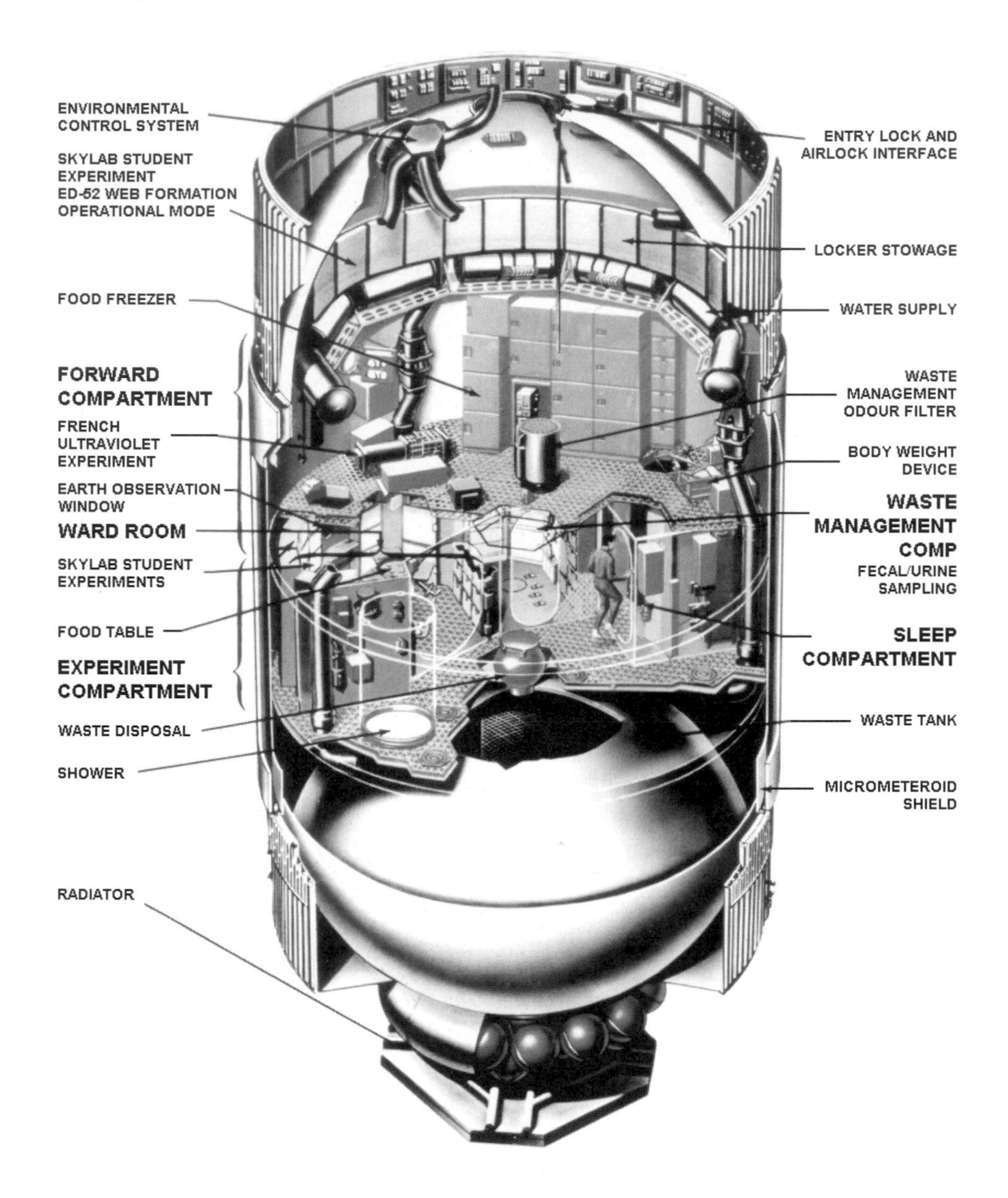

This cutaway of the Skylab OWS reveals the interior of the largest habitable volume on the station, compartments, equipment and experiment locations.

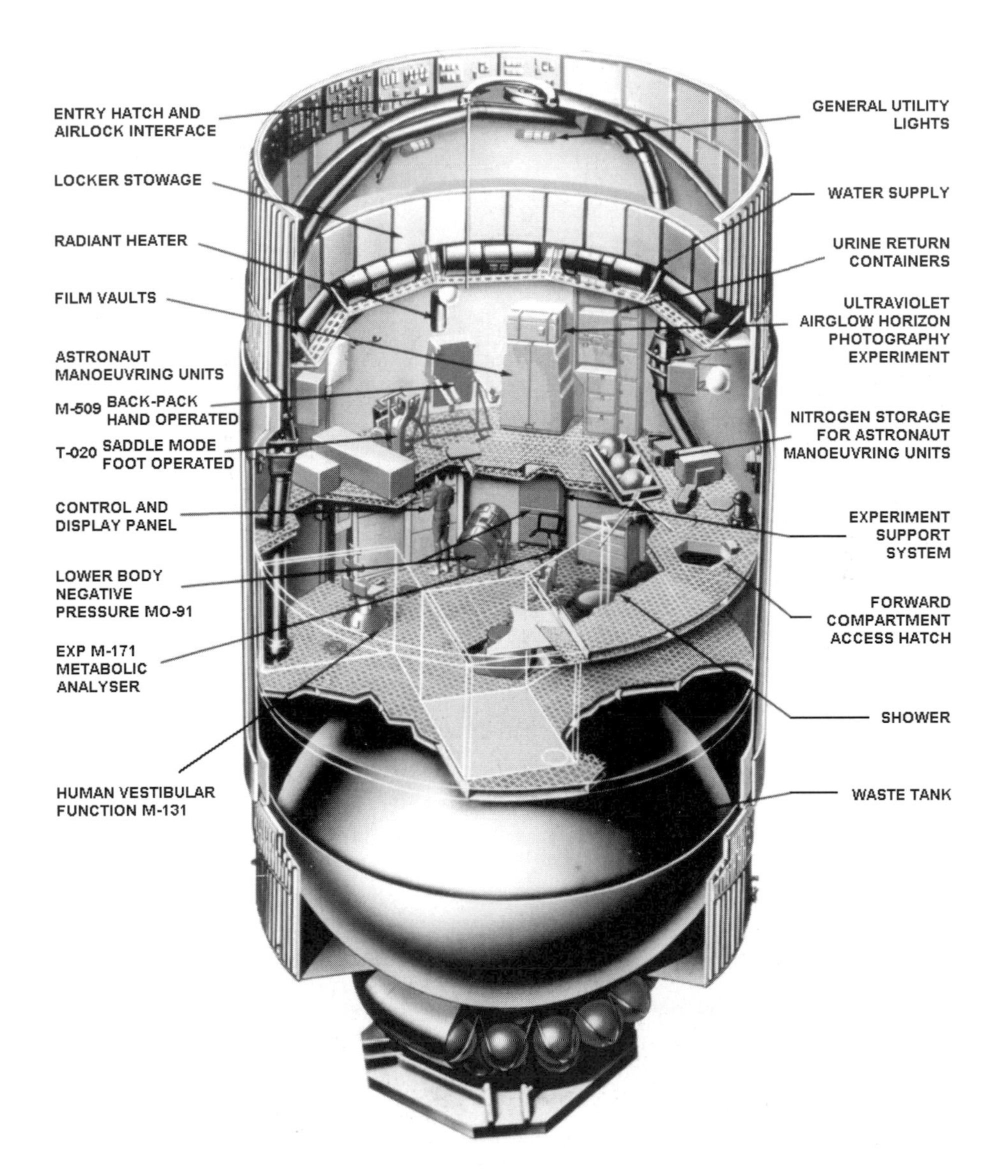

The opposite side of the cutaway reveals the remaining elements inside the OWS.

subsequent flights were aimed for much more difficult landing sites of greater scientific interest. However, in any Apollo lunar mission where the chances of rescue were slim, and at some stages non-existent, taking extra risks for 'a few more rocks' was no longer a compelling argument for either politicians or the public.

The aborted flight of Apollo 13 in April 1970 had revealed the vulnerability of the spacecraft and the astronauts. Had the explosion occurred after the LM had undocked from the CSM, or on the way home, the crew would not have had the resources and margin of safety that the lunar lander had provided during the accident. With further budget restrictions pending on NASA in order to meet any sort of longer-range plans, the agency announced in September 1970 that Apollos 16 and 19 were to be deleted from the programme, and that the remaining missions were to be redesignated Apollo 14 to Apollo 17. The Apollo lunar missions would end in 1972, and would be followed by Skylab at the end of that year.

Scheduling Skylab

The new schedule called for a launch of the unmanned OWS on 9 November 1972, to be operated for about eight months. The first mission would be launched the day after the OWS, and was termed the '28-day activation mission'. It would also study the physiological and psychological aspects of extended-duration spaceflight. The second mission of 56 days was to be launched on 19 January 1973, 70 days after the first mission, its primary objective being the operation of the ATM. The third crew, launched on 1 May 1973 (102 days after the launch of the second crew), would focus their efforts on Earth resources observations during their 56-day mission, ending the planned series.

As the design of the Skylab hardware was being finalised and the definitions of the scientific objectives and experiments were being discussed, the flight planning

An alternative location for the ATM is shown in this artwork. This OWS had three internal 'floors' in the hydrogen tank, access to the oxygen tank where the ATM console is located, and an airlock for EVA ATM film replacement. Was this the design for Skylab B?

was also reaching a definitive stage. By 1970 the astronauts were beginning to take a more active role in the programme, and although no crews were yet assigned, they participated in formulating how to prepare the crews for flying the missions. During this time (1970–1972) there emerged a possible fourth 'new' docking mission with Skylab – but not by an American Apollo spacecraft.

A Skylab–Soyuz mission

One other area of space exploration for which President Nixon had expressed his support was a cooperative programme with other nations – particularly Russia. Talks between NASA and the Soviet Academy of Sciences had been occurring infrequently for some time, discussing space technology and life sciences, space law, and the safe return of astronauts and cosmonauts who might be forced to land in foreign territory in an emergency. Several astronauts and cosmonauts had also met at different international venues over the years, but although there had been reports and statements about the theory of joint programme participation, there had been no serious discussions about actually flying in space together.

However, by the late summer of 1970 the idea of a co-operative mission with the Soviets was moving forward, due largely to the efforts of NASA's third Administrator, Thomas O. Paine, and the President of the Soviet Academy of Sciences, Mstislav V. Keldysh. After a series of letters and meetings between the two sides, it was established that a possible joint effort in space could be devised by using the current technology of the American Apollo and Russian Soyuz. Lunar-distance

Technicians simulate the storage of food containers in the forward area of the Skylab mock-up at the McDonnell Douglas facility in California during November 1970. The mock-up is already beginning to look like the actual flight version, 2½ years before launch.

missions were soon ruled out, as the configuration of Soyuz at that time was restricted to Earth-orbital flights. In fact, the Earth-orbital Soyuz was actually a derivative of the spacecraft that should have taken cosmonauts to the Moon. Another problem was that Soyuz and Apollo used different gas atmospheres to sustain the crews, and they also had incompatible docking systems. Whatever joint missions were proposed, there remained many basic techniques and operating procedures that would have to be resolved before anything left Earth.

With the Moon ruled out as an option, NASA explored Earth-orbital mission scenarios, and evolved two formats for what was proposed as a 'common docking mechanism for space stations', under planning for later in the 1970s. The Advanced Development staff at NASA HQ in Washington evaluated the possibility of either Apollo or Soyuz rescuing a disabled vehicle of the opposite type, or a separate mission to test the joint rendezvous and docking capabilities of the two spacecraft using jointly designed equipment that could be applied to later programmes.

It was soon realised that a direct rescue profile was impractical. Apollo was far more manoeuvrable than the Soyuz, and the cramped confines of the Soyuz meant that to recover all three astronauts from Apollo would be impossible. The only way that this could be done was by launching the Soviet spacecraft unmanned. On a real rescue mission, if the Apollo crew was disabled in some way, this would be pointless. Independent studies conducted at Rockwell indicated that Apollo could support five

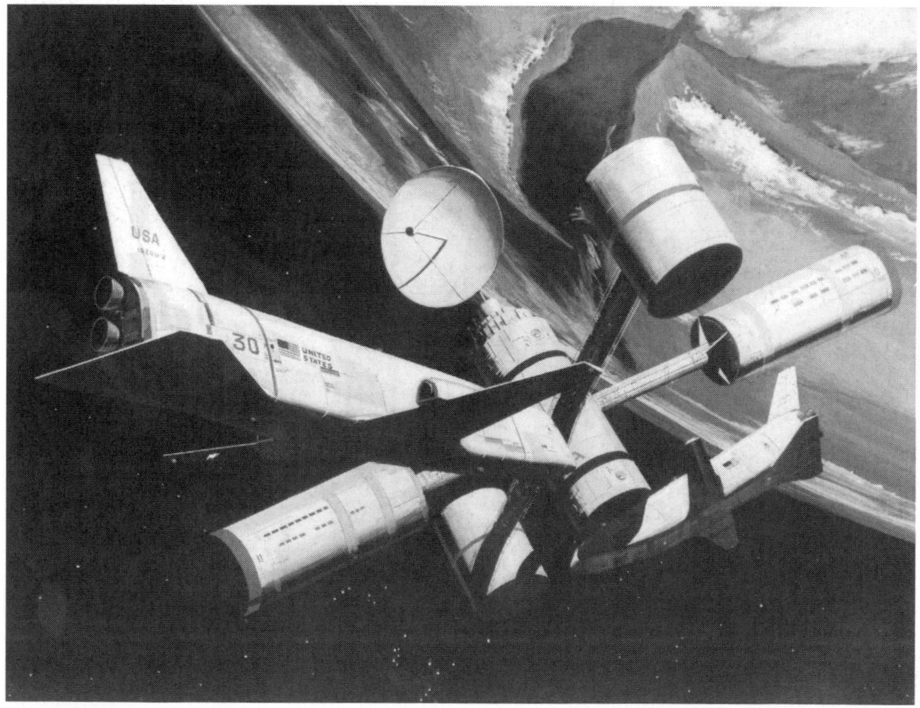

An artist's 1970s concept of Shuttles docked to an orbiting space base in the late 1980s.

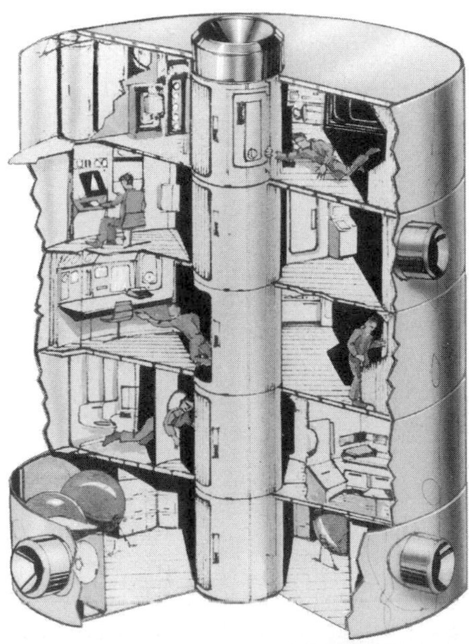

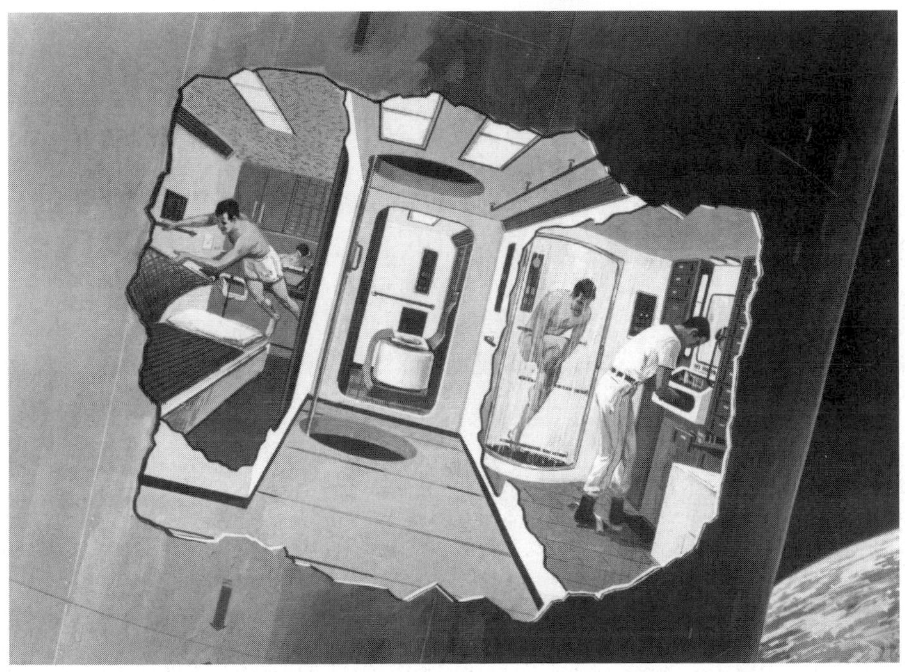

Twleve-man space station concept, *c*. 1970.

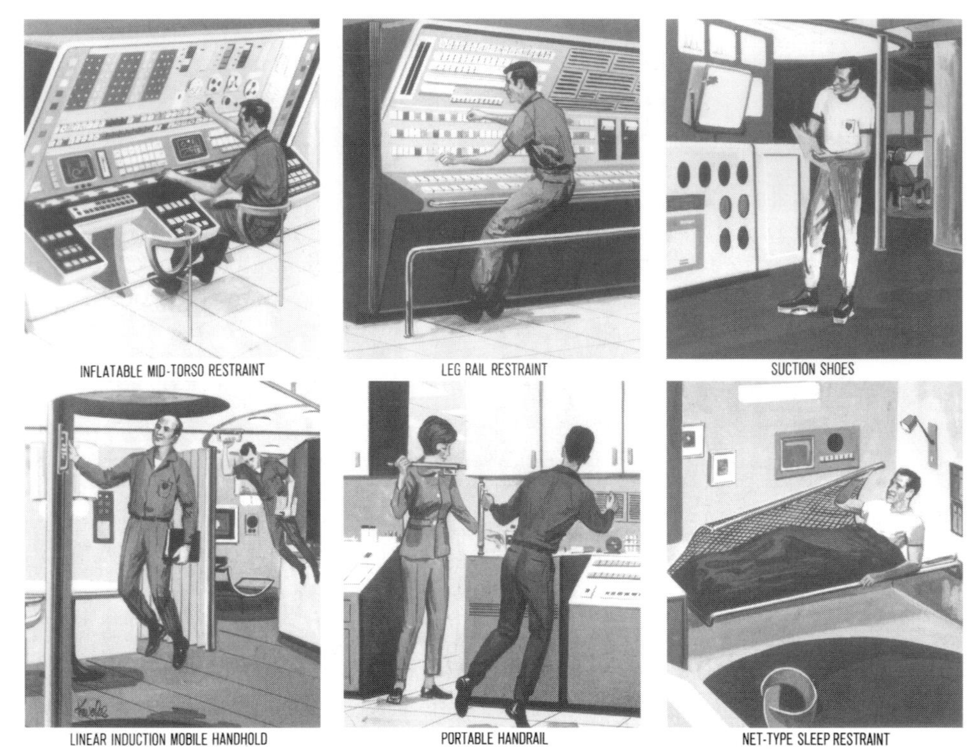

'Mobility and restraint device concepts for future manned space systems' – 1970s style.

people for a short period of time, and this meant that the American craft could rescue three cosmonauts from Soyuz by launching a two-man Apollo. Indeed, American spacecraft had the capability of being flown back from the Moon by one person if the two lunar explorers became stranded on the surface.

It quickly became apparent that providing an effective space rescue would be both costly and complex. It would become more effective when and if both nations adopted a reusable shuttle system rather than the one-mission spacecraft in service at that time. There were other options available, however, and one of these in the near term was a joint docking mission with the Skylab OWS.

Although the opportunity was there, it was also recognised that a number of hurdles would have to be overcome. The joint hardware had to be completed in time to meet the lifetime of Skylab (by 1973), the Soviets (and NASA) had to agree to it, and there had to be funds from current budgets to support it. But perhaps the biggest question lay in the uncertainty in whether the Americans would launch anything with a crew onboard after Skylab. The proposed Space Shuttle was still under study, and was far from becoming a definite programme; the development of larger space stations was not defined; and even a second Skylab was cause for doubt.

One factor in favour of the idea that was circulated at the Advanced Mission

Office in Washington was that the Soyuz could offer the capacity for rescuing a stranded Skylab crew. The near-disaster of Apollo 13 was clearly still fresh in the minds of those at NASA, and the idea that Soyuz might be capable of providing assistance to Skylab in the event of trouble in Earth orbit was attractive. This was also at a time when studies into the prolonged use of orbital storage of Apollo CSM were being conducted, as well as studies into the limitations of launching a 'rescue' Apollo quickly enough to render aid to a Skylab crew in immediate peril.

By late August 1970, after further discussion with Philip Culbertson, Chief of the Advanced Mission Planning Group in the Office of Manned Spaceflight in Washington, James Roberts, of the Advance Development Group, sent a memo addressing the three main areas of joint discussion: rendezvous, docking, and crew transfer. The memo was sent to the Kennedy, Marshall, and Manned Spacecraft offices, and to the Skylab, Space Station and Shuttle offices at Headquarters. In the memo, Roberts asked for their ideas for a joint mission and the technical feasibility of providing an American vehicle to support Soviet missions. KSC was also asked for an understanding of what would be involved in preparing a Saturn 1B or Saturn V for launch once information had been received that a Soviet manned flight was imminent, so as to render a rescue capability should NASA be called upon to do so. It was important to understand how long such an operation would take, how much it would cost in manpower, materials and cash, and, especially, how long a Saturn could remain launch ready. This had relevance both to the proposed joint programme with the Soviets, and also in determining the effectiveness of Skylab rescue capability from an Apollo.

In early September, one of the first responses came from Skylab Program Director William Schneider, who replied that although there did not seem to be anything to indicate that such a joint mission could not be achieved, there remained a 'potful of things' that would require much joint planning. These ranged from on-time launch, rendezvous and docking, co-operation between the launch and mission control centres, direct contact between the both spacecraft, tracking coverage, and a method of actually joining the two spacecraft. Although there was generally a cool response to Roberts' memo from the Skylab office, on 4 September 1970 NASA administrator Tom Paine wrote to the President of the Soviet Academy of Sciences, Mstislav Keldysh, proposing a Soyuz docking mission with Skylab.

While awaiting a reply from the Soviets, the Americans were evaluating three mission scenarios. In addition to a Soyuz docking with Skylab, it was suggested that Apollo could dock with a Soyuz, or that a future American spacecraft (possibly the Space Shuttle) could dock with a future Soviet spacecraft, whatever it might be.

The question was, would the Soviets even respond to the letter, let alone support such a joint mission? When the reply finally arrived on 23 September, the answer indicated that the Soviets were indeed interested, and suggested the first joint talks should begin as soon as October.

During a trip to Moscow in October 1970, NASA officials completed presentations to Soviet officials on the American docking systems that had been used on Gemini and Apollo. One of several systems presented, and that had not be used in flight, was a 1963 ring-and-cone concept, which was further developed for the Apollo Applications Programme, but was not adopted. This evolved into a 'double interrupted ring and cone', which featured twelve 'fingers' or guides on one cone, matching the ring of the second 'cone' to mesh together exactly. There were no docking probes or drogues to block transfer hatches, and as each half was exactly the same, it did not matter whether the spacecraft was the active or the passive vehicle, offering a fail-safe system. This was the basis of what became the Apollo–Soyuz Test Project androgynous docking system.

The meeting also reviewed plans for the proposed joint Skylab mission. One of the delegates on the Soviet side, cosmonaut Konstantin Feoktistov, wanted to turn to the future aspects of Skylab and his ideas of huge rotating space stations, rather than dwell on past programmes. Although no firm decision was made on a Soyuz–Skylab mission at this meeting, advances were made both on the possibly of some kind of joint mission, and towards the development of common docking hardware to achieve this mission. Both sides agreed to work together and to meet again.

An artist's concept of a Soyuz–Skylab docking mission. As the Russian Soyuz approaches Skylab to dock with the front port of the Multiple Docking Adaptor, the American Apollo monitors activity from a distance before docking with the second port. If this had occurred it would have created an international space station several years before the Soviets achieved the feat with Salyut 6. (Courtesy of the artist, David Hardy.)

In America, further studies were completed, offering three possible missions in the 'near-term', using the hardware that both nations were then currently flying:

- Soyuz docking to Skylab to demonstrate the feasibility of such an operation, and possibly including a programme of joint experiments with an American crew onboard the station.
- To occupy Skylab *after* the Apollo crew had departed.
- Soyuz docking to an Apollo directly to test each craft's rendezvous and docking capability in relation to the other vehicle.

At this time, of course, Skylab was the only *known* space station under development, but it was not due to launch until 1972. Although the Soviets had often indicated that they, too, had a space station under development, they had yet to launch one – although such an event was expected in the near future.

What was interesting about these proposals was exactly how were they to be achieved. In addition to the incompatibility of the spacecraft atmospheres and the design of suitable docking hardware, there remained the problem of where the Soyuz would dock on Skylab if the Apollo crew was on board. The only option would be in the radial second port using an Apollo-type docking probe. The Apollo crew could allow the Soyuz to dock first with the axis port (as shown in the illustration on p. 58) before docking the CSM to the radial port; but this still involved the fitting of an Apollo docking system to a Soyuz, and the provision of adequate adjustment of the atmosphere to allow the cosmonauts to enter Skylab.

Perhaps the most interesting scenario was to allow a *Soviet* cosmonaut team to dock to an *unoccupied* American space station – which was technically possible but operationally very demanding, politically uncomfortable, and unlikely ever to be realised!

Soyuz 9: an 18-day space marathon

On 19 June 1970, the Soviet Soyuz 9 mission ended with a 'dustdown' in a ploughed field, 47 miles west of the town of Karganda, Kazakhstan, Soviet Central Asia. On board were cosmonauts Andrian Nikolayev and Vitaly Sevastyanov, who had just completed a record-breaking spaceflight, in Earth orbit, of 17 days 16 hours 59 minutes, finally surpassing the 14-day American record set by Gemini 7 in December 1965. At the time, the Soviets indicated that the mission was a precursor of even longer missions in the future. At the post-flight press conference, Keldysh indicated that the future direction of Soviet cosmonautics was to be 'the establishment of prolonged orbital stations for scientific and economic purposes.'

Despite a longer than expected recovery from their prolonged flight in space (because they had not followed their exercise programme as advised), the two cosmonauts soon became ambassadors for the Soviet space programme, and were invited to the United States in October. On 21 October, during a ten-day visit to the American space centres, they were briefed on the Skylab programme at Marshall. At the time of their visit, Rusty Schweickart was to perform an underwater Skylab test in the large Neutral Buoyancy Tank, and permission was hastily secured to allow Sevastyanov (wearing an Apollo EVA suit) to join him in the tank. The cosmonauts' escort, Apollo 11 astronaut Buzz Aldrin, joined them in the tank, wearing scuba gear, while Nikolayev elected to observe from the outside. Afterwards, the press called the event the first joint American–Russian training session.

On 31 October, Deputy Administrator George Low wrote to President Nixon's science adviser, E.E. Davis Jr, in the wake of recent budget cuts, the loss of two lunar missions, and recent Soviet indications of creating a space station. Low expressed some concern that any further budget cuts could seriously threaten Skylab, stating: 'To forgo Skylab would leave the US without the database for any future manned missions, [and would] surrender to the USSR the option of having the first real space station in orbit.' He also indicated that such a move would leave the option of sharing manned spaceflight and the development of a common docking mechanism for future orbital spacecraft underdeveloped.

Salyut: the first space station

On 19 April 1971, the Soviets did indeed launch the world's first space station – Salyut – indicating that the Soyuz 9 record could soon be surpassed. The station had originally been called Zarya (Dawn), but that had been changed just prior to launch when it was learned that the Chinese had used this name for one of their satellites. It was then suggested that the name should be changed to Salyut, in a 'salute' to the tenth anniversary of Yuri Gagarin's flight on 12 April.

On 23 April 1971, Soyuz 10 was launched with a three-man crew on a planned 30-day mission. The crew managed to dock to the station, but not firmly, so after only 5 hours 30 minutes they were told to undock from the Salyut and then attempt a second docking. But they experienced difficulty in releasing from the Salyut, and it took some time before the cosmonauts managed to free their spacecraft.

Luckily they had not damaged the single docking port on the station, as this would have rendered Salyut unusable. With reduced consumables onboard, and concern about whether they could undock again, the order was given to not attempt a second docking, but to complete a fly-around inspection and then effect their recovery. The 24 April re-entry and landing went smoothly – apart from landing just 165 feet from a lake after completing a two-day flight.

The inquiry found that the docking equipment on the Soyuz had probably become damaged during the initial hard-docking approach. Once modifications had been implemented, plans were made to still fly two 30-day missions to the station, launching on 4 June and 18 July. However, by the end of the first mission, Salyut would have been in orbit nearly three months, and onboard supplies might not have been enough to support a second 30-day mission. The flight plan was therefore changed, with the first flight reduced to 25 days, to begin on 6 June.

Shortly before launch the prime crew was grounded due to the illness of one of the cosmonauts. Had this occurred prior to leaving for the cosmodrome, it would have been simple to replace him with his back-up crew-member; but they were days from launch and, following mission rules, the whole crew was grounded and replaced (after reported heated discussions with the original prime crew) by their back-ups. On 6 June 1971, Soyuz 11 launched with Commander Geogori Dobrovolsky, Flight Engineer Vladislav Volkov and Research Engineer Viktor Patsayev on board. They docked to Salyut on 7 June, and spent the next three weeks on board, becoming the world's first space station crew (call-sign Yantar – Amber).

Onboard the station the crew conducted about 140 scientific experiments in medical, biological, astronomical, Earth resource and technical fields. They experienced numerous difficulties not reported at the time, including the breakdown of equipment, disputes between themselves, and the strong odour of smoke. But they completed the mission that they did not expect to fly, although the difficulties added to their fatigue and their eagerness to return home.

While Salyut was in orbit, further joint discussions continued, and a Soviet delegation visited NASA MSC in Houston between 21 and 25 June. During these discussions, the Soviets commented that the idea of just docking an Apollo to a Soyuz with a simple probe and drogue device was not very productive to future operations, and would be seen by the world as nothing more than a 'space stunt'. The other options were a Soyuz docking to a Skylab, or, after the Soviet space station launch, an Apollo docking with a Salyut. The Soviets continued to indicate their keen interest in some sort of joint mission, incorporating a universal docking mechanism with application for future projects. As Salyut had been placed in orbit before Skylab, the Americans suggested that the proposals should focus initially on an Apollo docking with a Soyuz–Salyut configuration, and in a subsequent flight a Soyuz might attempt a docking with Skylab.

As the Soviet delegation returned to Moscow, the Soyuz 11 cosmonauts were ending their research programme to prepare for their return to Earth. After 23 days onboard the station, they closed down the Salyut, transferred experiment results and film cassettes to their Soyuz, and looked forward to a heroes return. After some difficulty in closing the hatch, they undocked and prepared for re-entry. The final communication from the spacecraft prior to re-entry indicated that all was fine. An automatic re-entry sequence – featuring spacecraft separation, parachute deployment and soft landing – was initiated, but no word was received from the crew inside the Descent Module. When recovery teams reached the capsule and opened the hatches, all three cosmonauts were found to be dead.

On 30 June, George Low expressed NASA's regrets over the deaths of the three cosmonauts, but did not anticipate any changes in the Skylab programme. However, it initiated concern from NASA that the Soyuz was the same type being used on the joint mission under discussion. Although the system that failed had no relationship to the integrity of the Apollo, there was some concern that perhaps the extended flight had had an adverse effect on the performance of the three cosmonauts in their preparations for the return. If this was the case, this could threaten the three long-duration manned missions to Skylab.

The subsequent investigation, however, indicated that one of two valves in the Soyuz respiratory system was in an open position when it should have been closed. As the cosmonauts were not wearing spacesuits, this led to their deaths as a result of rapid decompression during the descent. The extended visit to the Salyut had not caused their deaths, and although their physical and psychological condition at the end of a long stressful mission was probably not at its best, there was no reason why the Skylab missions should not proceed as planned.

Discussions for the proposed joint flight continued through the summer of 1971, and there were plans for a series of post-Apollo and Skylab missions using surplus hardware. Some of this would be utilised for the joint flight being planned for late 1974 or mid-1975, but by September, detailed costing for just one joint flight led only to thoughts of further post-Skylab Apollo Earth orbital missions being quietly dropped. Talks had continued with the Soviets on the idea of a CSM docking to a Salyut, with a Soyuz already docked to a proposed second docking port at the rear of the vehicle. NASA drawings reflected this option, which had not been possible on the first Salyut because of its solitary docking port.

Interest in an Apollo–Salyut docking continued into 1972 as talk of a Soyuz docking with Skylab quietly went away. At the meeting of officials in Moscow from 27 November to 6 December, in an effort to keep the joint flight simple, discussion progressed over a proposal to dock an Apollo directly to a Soyuz. A docking module would be used to achieve the physical union between the two spacecraft, and as a location to equalise the atmospheres between the two craft. The Americans were still looking at the possibility of flying an Apollo CSM to a Salyut at some time, but by then it was highly unlikely that a Soyuz would visit Skylab. Such a flight was being ruled out as being too complicated for the first Skylab OWS. If there were to be such a mission it would be with a second Skylab, or a different type of station that the Americans might place in orbit after 1975.

Discussions continued towards official acceptance of a joint mission by both sides, with the Americans hoping to dock with a Salyut. Then, in a surprise move in April 1972, the Soviets indicated that Salyut would not now be part of the joint mission, which would be a docking between an Apollo and a Soyuz. The reason given by the Soviets was that it was technically and economically not possible to convert the Salyut to have two docking ports. Unbeknown to the Americans, however, such a development was being evaluated at Soviet design bureaux, for a rear docking port instead of a forward one for the Almaz military space stations, commencing with Salyut 2 in 1973, and for the second generation space stations, beginning with Salyut 6 in 1977, to feature two docking parts.

Although Soyuz would not dock with Skylab (and now Apollo would not dock with Salyut), for a while there remained the possibility of the Apollo assigned to the joint flight visiting Skylab after completing the docking phase.

Skylab designations

Throughout the development of the joint US/USSR proposed missions in 1970–1972, work continued to prepare Skylab for its launch. On 16 July 1971, the official designations for the Skylab missions were issued in two forms. For public use outside NASA and for non-operational NASA documentation, the designations would be: workshop launch; first manned visit; second manned visit; third manned visit. For all operational use within NASA, the mission designations would be SL-1 for the unmanned workshop launch, and SL-2 for the first manned visit; and together these would be considered a joint mission. The second (SL-3) and third (SL-4) manned visits would be two separate missions. However, when the three crews designed mission emblems, the designs featured the numerals 1, 2 and 3. From then on,

A quarter section of an ATM canister is lowered into the sunspot vacuum chamber at MSFC in April 1968, to carry out thermal vacuum tests with two simulated telescopes during early tests of spacecraft systems and hardware under development.

confusion was frequent in deciding whether the first crew was Skylab 1 or SL-2, the second mission Skylab 2 or SL-3, and the third visit Skylab 3 or SL-4.

The origins of this confusion began around 1971 when the astronauts in the CB were first informed of the crew assignment to the missions. The official designations of SL-1 to SL-4 were adopted, and the crews began designing patches. Then, apparently, the flight crews began to receive internal documentation that was labelled Manned Mission 1, 2 and 3. This resulted in confusion in the CB mailroom as well as for the astronauts, as they received each other's mail!

As the three crew patches reflected the designations Skylab 2, 3 and 4, the astronauts approached Skylab Program Director William Schneider to clarify the situation. Were they Skylab 1, 2, 3, or 2, 3, 4? Schneider replied, Skylab 1, 2 and 3; and so the crew patches were reworked to the new designations.

The designs were submitted for approval at NASA Headquarters. This took several months, and the artwork finally arrived on the desk of Dale Myers (the Associate Administrator for Manned Spaceflight) in late October 1972, about six months before the first planned launch – whereupon he rejected them, stating that the 'official' designations for the missions were 2, 3, and 4! The crews now had to retrieve the original designs, and the process started all over again.

A full-scale mock-up of the Multiple docking Adapter (MDA), showing the location of the ATM control stations. The ATM console is shown to the left of the adapter used in the training of flight crews.

The flight model of the MDA during shipment from Marshall to Martin Marietta in Denver, Colorado, for final outfitting.

It was not long, however, before they were told not to bother, and the designs were left as 1, 2, and 3. The reason for this was that the flight clothing (already stowed on the OWS) and the hundreds of mission identification tags with the designs on them were completed several weeks earlier, and to replace them all again would be too costly and too disruptive to the pre-launch process. So although officially designated SL-2, 3 and 4, the emblems featured the alternative designation of 1, 2, 3 (which also easily identified the mission of the first, second or third crew).

Launch slips

On 13 April 1971, NASA issued an updated hardware delivery and launch readiness schedule that reflected the timeline of all the Skylab components. It indicated that the SL-1 OWS launch would be no earlier than 30 April 1973, followed by the SL 2 launch 24 hours later on 1 May.

The launch of the second crew was manifested for 30 July, and the third crew for 28 October. Despite some studies into delaying the first manned launch for up to two weeks to confirm that the OWS was in a stable orbit before committing a crew launch, a report issued in April 1972 indicated that it would be wise to leave the launch plan as previously scheduled, and to ensure the CSM was docked to the OWS as soon as possible to fully activate the station.

By 15 January 1973 it was clear that it would be difficult to maintain the original launch date of April, with delays in several items of hardware preparation, and so the launch date of SL-1 slipped to the middle of May. The exact date was to be defined after the evaluation of launch and landing constraints on the other two flights to determine whether the recovery date for last crew – 21 December 1973 –

McDonnell Douglas technicians fabricate one of two 22-foot 'floors' for the OWS. Note the triangular grid foot-restraint device integrated into the structure.

A full-scale non-flight structural test article of the Orbital Workshop at McDonnell Douglas Astronautics Company, in California, on its way to NASA Houston and Marshall field centres for verification of design. It was to be subjected to acoustic, vibration and static forces simulated for launch and orbit.

could still be maintained as planned, or if a slip past the festive season would be required.

On 14 February, Schneider reviewed launch schedule options and reduced the interval between the SL-2 and SL-3 launches by five days, and between SL-3 and SL-4 by ten days. There had been concern that moving the launch dates (as a result of the slippage from April 30) would affect the launch abort lighting conditions, would produce a less than favourable prospect of a night-time recovery, and would shift the circadian rhythm on SL-4 as a result of the reduction of the mission planning cycle. These changes were confirmed on 5 April 1973, with the optimum launch dates chosen to alleviate these concerns and eliminate any night-time recovery operations.

Mission	*Launch*	*Recovery*
Skylab workshop (SL-1)	14 May 1973	n/a
First manned mission (SL-2)	15 May 1973	12 June 1973
Second manned mission (SL-3)	8 August 1973	3 October 1973
Third manned mission (SL-4)	9 November 1973	4 January 1974

By the spring of 1973 all was ready to support the Skylab missions. The following is a summary of what was planned for the missions, prior to the actual launch of the first element in May 1973.

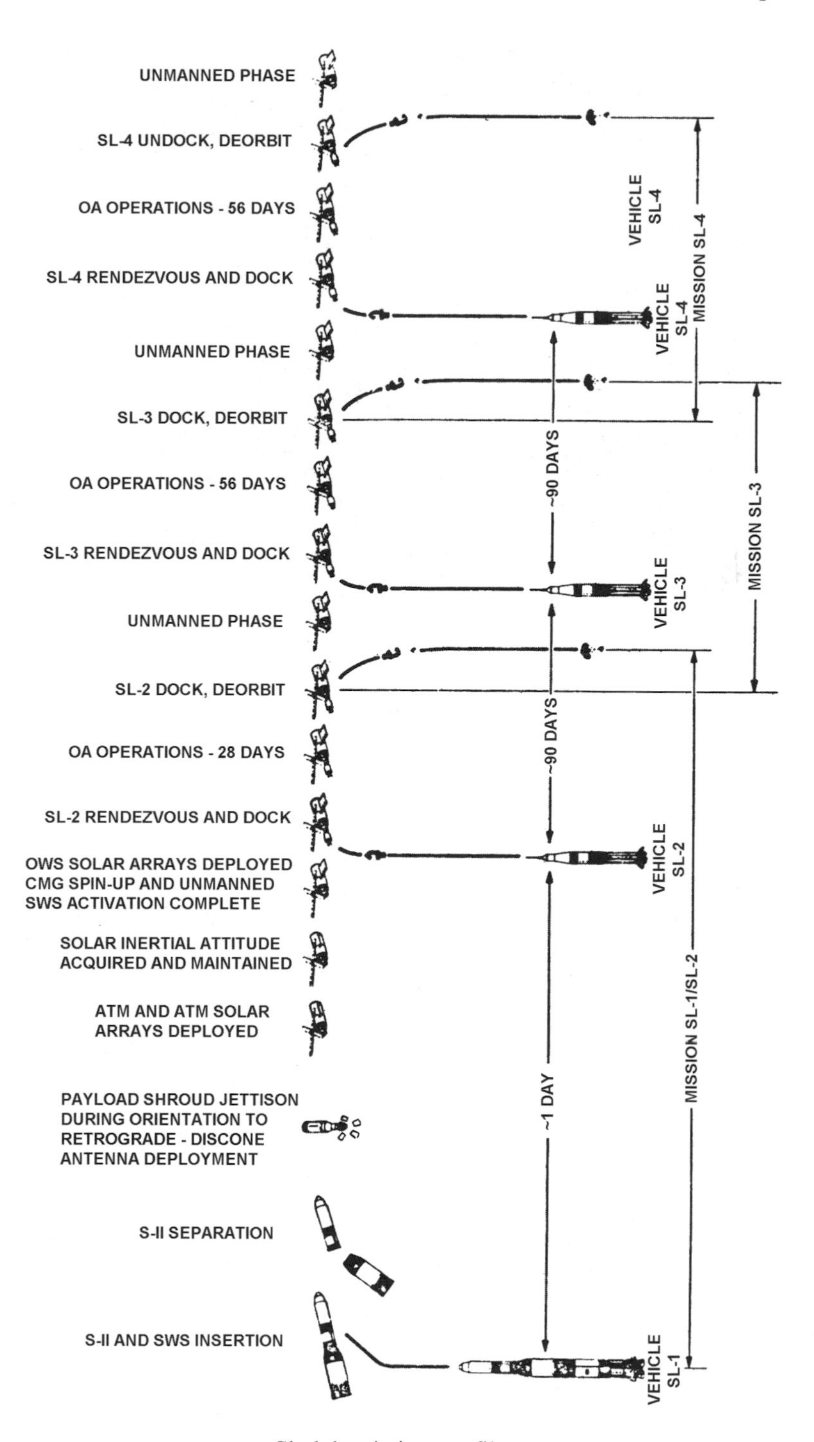

Skylab mission profile.

The almost completed MDA is shown during tests at Martin Marietta Corporation, Denver. It had 7,000 electrical connections and six miles of wiring, and the two CSM docking ports can be clearly seen.

THE SKYLAB-1 (OWS) MISSION

The mission profile for the Skylab series would begin with the 13.30 hours EDT launch of the orbital workshop (SL-1), on top of the last Saturn V from Pad 39A at the Kennedy Space Center, Florida. The launch of the unmanned Orbital Workshop would be the key to the success of the whole programme. The launch window in which to place the vehicle into the correct orbit was 1.5 hours, until 15.00 hours.

The two-stage Saturn V would consist of the S-IC first stage and the S-II second stage, carrying the payload of the Skylab OWS, ATM, AM, MDA and IU cluster combination. It would stand 333.7 feet tall on the pad, and would have a lift-off mass of 6,222,000 lbs. It would be launched on a flight direction (azimuth) of 40°.88 east of north, to follow a powered flight of 9 minutes 48 seconds. The second stage would insert the Skylab into a 235 nautical mile orbit, inclined at 50° to the equator.

Two minutes 20 seconds after launch, the first stage would burn out, and would be separated from the ascending vehicle. From an altitude of 43 nautical miles, the spent first stage would follow a ballistic trajectory to impact in the Atlantic Ocean 458 nautical miles downrange, 10 minutes 55 seconds after launch. The second stage would ignite shortly after separation, and would continue to boost the combination towards orbit. After attaining orbit, the S-II engines would be shut down, and the spent stage would be separated by retro-rockets. This was to be followed by the deliberate reduction of the velocity of the spent stage by vernier rockets, so that for safety it would gradually be left behind by the cluster.

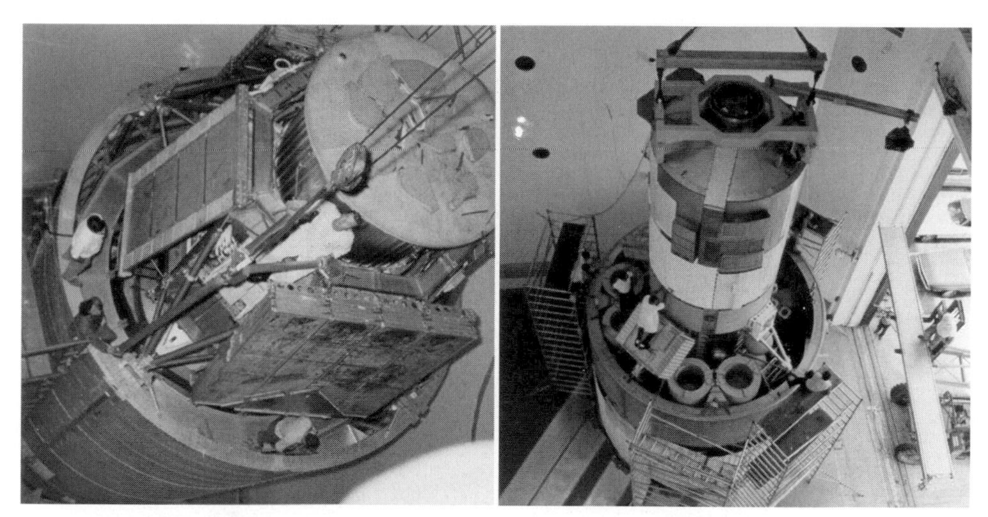

Technicians at MSC in Houston conduct vibration and acoustic testing of the ATM (*left*) and MDA (*right*) in July 1972, prior to shipment to Marshall onboard the ocean-going barge *Poseidon*.

With the second stage separated, the combination would continue in orbit as the deployment sequence was started. Soon after orbital insertion, the whole vehicle was to be pitched by the Thruster Attitude Control Subsystem (TACS) through to a gravity gradient attitude, to bring the nose of the complex pointing down towards Earth and to conserve onboard manoeuvring fuel supplies. As the complex passed 90° nose-down attitude, the Payload Shroud would jettison in four sections. The vehicle would then be orientated to a solar inertial attitude as the ATM and its four solar arrays were unfolded, which would also clear the forward port for subsequent CSM docking the next day. The deployment of the ATM arrays was to be followed by the deployment of the twin workshop arrays and the micrometeoroid shield, and the activation of the control moment gyro subsystem.

With the OWS safely in orbit, the ground controllers would perform a systems check and activation while preparations entered the final phase for the launch of the first crew the following day. The launch opportunity for the first manned mission was planned for 23.5 hours after SL-1 launch, at 1300 hours EDT. The ten-minute launch window would occur on the first two days of every five-day period (the first and second days, the sixth and seventh days, the eleventh and twelfth days, and so on) if the launch could not occur on the first attempt.

The launch of the first crew would be attempted only if certain criteria had been achieved, and would be determined by the events of the actual launch of SL-1. Before the first crew could be launched, the station had to be free of the second stage and separated far enough to pose no safety threat to the CSM's ascent and de-orbit trajectory. The payload shroud, the OWS solar arrays, and the ATM and its solar arrays, had to be deployed and operating. The pressure integrity had to be confirmed, the micrometeoroid shield deployed, and the attitude control and pointing system maintained at solar inertial attitudes. Communications and tracking

The ATM after arrival at KSC, Florida, for launch processing in the Manned Spacecraft Operations Building (MSOB) for check-out and systems testing in October 1972.

also had to be established, and radiation levels needed to be at a safe level for human habitation.

Supporting a human crew

Apart from the cluster elements described above, the Workshop included six major systems for the control and habitation of the assembly:

Habitability Support System This provided for the astronauts' everyday needs during the stay onboard Skylab. Provisions for food storage and preparation, personal hygiene, waste disposal and sleeping were stored in more than 400 locations around the station, totalling more than 1,800 items that were linked to databases. These were planned to be revised by real-time updates from the crew in their daily reports of food and equipment (such as film cassettes) usage. Every item stowed was tagged with a location identification so that it would be easy to find – in theory!

There was in excess of 800 gallons of potable water onboard, as well as 1,500 lbs of food in dehydrated, ready-to-eat, frozen or canned form. Each of the crew had changes of clothing – from socks to complete flight suits – stowed in the OWS before launch, totalling 450 individual pieces for three men on each of the three missions. There were also personal clothing provisions stowed on board the OWS for the late inclusion of a back-up crew-member on a flight crew. Each man also had his own personal hygiene and grooming kit, consisting of shaving equipment, wash cloths, hair comb, nail clippers, toothbrush and toothpaste.

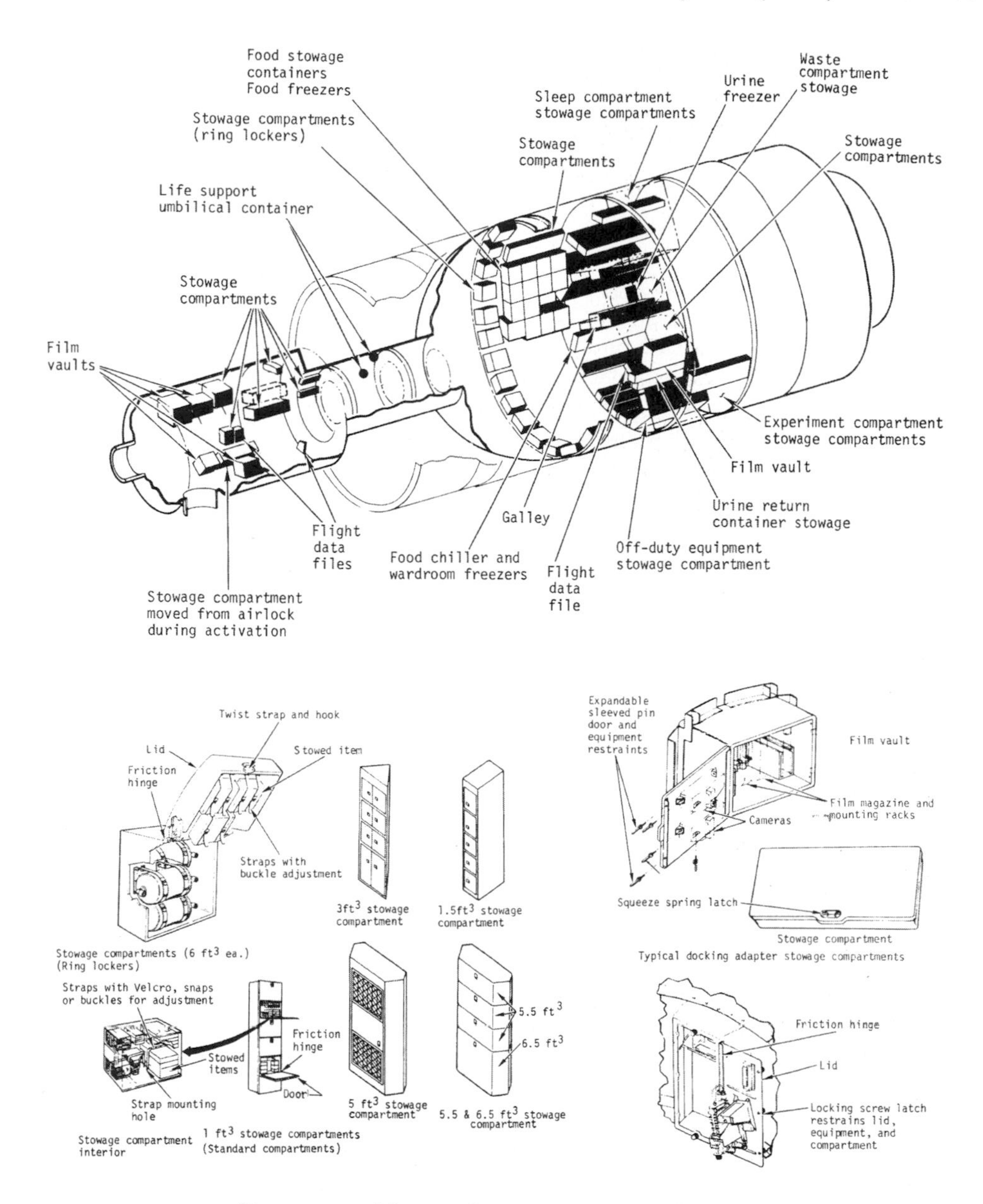

Stowage provisions and compartments on the OWS.

The personal hygiene station housed the wash facility, toilet system, and the separate shower facility. Due to the biomedical nature of the flights, all human waste activity required collection of both solid and liquid samples to analyse as part of certain experiments, or for the disposal of the remainder. All non-biological trash was transferred to the waste storage tank (the former S-IVB oxygen tank) below the wardroom floor.

Thermal and Environmental Control System Prior to the arrival of each crew, the OWS was pressurised to 5 psia with a two-gas mixture of approximately 74% oxygen and 26% nitrogen, to provide an environment similar to that on Earth.

Insulation and thermal coatings on the workshop, AM and MDA provided passive thermal control. Combined with the normal used of heaters, lighting, and internal equipment, a comfortable internal temperature could be maintained for operating experiments, supplemented by heat exchangers and duct heaters to control any extremes of temperature.

To keep the air inside the station clean and moist, the gases were passed through molecule sieve carbon dioxide removal equipment, water removal condensers, and coconut charcoal filters to remove odours, 'scrubbing' the air clean.

Electrical Power It was planned to have solar arrays on each side of the OWS, supplemented by the four array 'wings' on the ATM. After launch, one of the OWS wings was lost, but the system was designed with contingency and back-up procedures that the controllers and the astronauts used to compensate for the loss of the one array. The OWS and ATM solar arrays worked in parallel, and each had its own batteries, battery rechargers, and regulators. These provided for power requirements during each night-time pass, and were recharged during the daylight passes.

Attitude and Pointing Control System This system provided three-axis (pitch, yaw, roll) control to maintain stabilisation and manoeuvrability of Skylab throughout the duration of the mission. The station could be orientated using fine pointing control during Earth EREP and solar ATM data-gathering passes or astronomical observations. There were two subsystems for handling the movement of the station, each of them controlled via the ATM digital computer. The primary system was the three Control Moment Gyros, backed up by the network of nitrogen gas jet thrusters producing small amounts of thrust. This Thruster Attitude Control System (TACS) had its own separate redundancy system.

Instrumentation and Communication System There were thirteen internal speakers and intercommunication systems in the OWS. This included an EVA communication system, the network of measurement sensors to gather and process data around the active station systems and experiments, and a spacecraft-to-ground or ground-to-spacecraft communication system that routed audio communications through the CSM system. The TV system could route signals from any of five TV cameras on the ATM solar telescopes to the ATM control and display panel that was located in the MDA. Crew TV images from inside the station were also transmitted by the CSM system.

Caution and Warning System This was a monitoring system (voltage only), centred in the AM, which alerted the crew to out-of-limit conditions that had compromised (or were about to compromise) the safety of the crew or the primary mission objectives. If such conditions were not responded to in time, they could result in the loss of a major system or in a dangerous situation. Each subsystem was connected to an audible alarm and visual lights in the crew areas around the station. There were also fire extinguishers located around the station habitation areas.

The SL-2 flight plan

If all criteria were met, the launch of the first three-man team by Saturn 1B would occur the day after the OWS, at 13.00 hours EDT – the optimum time for a fifth orbit rendezvous with the workshop. The launch phase of the two-stage Saturn 1B would include separation of the S-IB stage 2.5 minutes after launch, and ignition of the S-IVB stage prior to ten minutes into the mission. The CSM and attached S-IVB stage would be inserted into an 81 x 120 nautical miles orbit, inclined at 50°. Six minutes later, the CSM would separate from the S-IVB and initiate a posigrade (fired with the direction of flight) manoeuvre, increasing velocity by 3 feet/sec using the RCS thrusters mounted in quads around the SM. This would change the orbit to 81 x 121 nautical miles.

A series of four manoeuvres would be completed by the CSM during its approach to the OWS. Then the terminal phase manoeuvre would be initiated by the Service Propulsion System 6 hours 47 minutes into the flight, with the two vehicles 22 nautical miles apart. At 1 nautical mile, the braking approach would begin, followed by station keeping and then docking with the forward (axial) docking port on the MDA.

The docking would use a probe and drogue system, similar to Apollo CM/LM docking. This system featured an extended probe at the apex of the CM that entered the drogue cone in the MDA. It then engaged through the capture latches (soft dock), and the probe retracted, pulling the CSM to the MDA and activating the twelve automatic latches (hard dock) on the MDA tunnel ring, forming a pressurised seal between the two vehicles.

After docking, the pressure in the tunnel would be equalised and the CM hatch opened, and a crew-member would visually and manually verify that all twelve latches had locked. Any that had not locked could be locked by a crew-member from inside the tunnel.

Days 1–2	Launch, rendezvous, docking, transfer and power-up of workshop
Days 2–25	Experiments
Day 26	EVA ATM retrieval
Days 27–28	Deactivation and preparation for two months' unmanned storage
Day 28	CSM undocks, deorbit, recovered

The crew would spend their first night in the CM and would not enter the MDA until the second day after pressure checks had been completed and the probe and drogue assembly had been removed. The Pilot would first enter the MDA to inspect the module, and he would be followed by the Science Pilot. They would then begin a series of system checks and housekeeping chores in the MDA and AM. The power and control umbilical in the CM docking ring and MDA tunnel would then be connected, and a transfer duct for the transfer of atmosphere into the CM would be installed to maintain the condition of the mothballed CM while docked. Meanwhile,

At 09.00 am EDT on Friday, 25 May 1973, Skylab 2 is launched, carrying astronauts Conrad, Kerwin and Weitz, to rendezvous with the unmanned Skylab OWS.

the Commander would place the CM in a power-down mode and activate the Caution and Warning monitoring system of 90 critical functions.

After checking the integrity of OWS, the hatches would be opened and the crew would enter the workshop to begin their resident crew stay. The primary mission objectives set for the first stay were:

- Establish the Skylab orbital assembly in Earth orbit.
- Obtain medical data on the crew for use in extending the duration of manned spaceflight.
- Perform in-flight experiments.

Skylab would repeat its ground track every 76 orbits (118 hours), which meant that the spacecraft would pass over the same part of the Earth every five days. For example, orbit 6 would be retraced on orbit 82 and again on 158, and on 234. This was perfect for the planned extensive coverage by the EREP science package on board. The space station would also pass through the South Atlantic Anomaly (a region of trapped radiation in the Earth's upper atmosphere) about nine times a day, for ten minutes each transition. Radiation exposure would play a part in the planning of the three crewed missions, and the scheduling of activities needed to be managed throughout the programme.

As the primary objective of the first manned period, the crew activation of the OWS would include hardware inspection, activation of parts of the environmental control system (ECS), and setting up of the food and water management systems. It was estimated that total activation of the OWS would occupy 2½ days upon entering the workshop. The experiments assigned to SL-2 (itemised in Appendix 3 and discussed in Chapter 5) included nineteen life sciences, ten space sciences, four astrophysics, four material science and manufacturing in space, six Earth applications, eight engineering and technology, and seven student investigations. Biomedical experiments would be emphasised on this 28-day mission, as would evaluation of the habitability of the vehicle. The overall objective was to establish Skylab as a habitable space structure, to evaluate the performance of the assembly, and to obtain data for evaluating crew mobility and work capacity both inside and outside the vehicle. By obtaining medical data on the crew in the first extended-duration American spaceflight since Gemini 7 in 1965, the first mission would determine whether the two subsequent Skylab missions of up to 56 days were feasible or advisable.

The crew flight plan allowed for eight hours sleep, eight hours rest, recreation, eating, personal hygiene, flight planning and housekeeping, and eight hours for experiment operations. Each day would include biomedical experiments to keep track of their health and provide a baseline of data for the other two missions. In addition, a total of ten man-hours per day was planned for monitoring the ATM and the fourteen EREP scheduled passes, and 17% of their experiment operation time would be taken up with space application experiments. One 2½-hour EVA – to retrieve and replace ATM film – was planned for the first mission on Day 26.

Skylab's 235 nautical miles and 93-minute orbit was determined as the most advantageous for scheduling ground-related planning (such as launch window planning), flight control, Payload Applications Document (PAD) experiment updating and messaging, and EREP support by aircraft flights. Once the repeating pattern had been established, atmospheric drag could be controlled to within a two nautical miles cross-range error, by additional TACS trim burns every twenty days.

Deactivation of SL OWS was to begin about three days before the end of the planned mission (Day 25) and would take about 36 hours. This included the transfer and stowage of samples, EREP data and film cassettes in the CM. The crew would secure food and water management areas, turn off crew quarters heating and circulating fans, configure the power and control distribution panel for autonomous flight, deactivate the environmental control system in the OWS and aft end of the AM, turn off the lights, and close the hatches.

Undocking would be a reversal of docking activities, and after Inertial Measurement Unit alignments, there would be a fly-around inspection to photo-document and visually describe the condition of the OWS. The crew would then perform separation, in a 5 feet/sec retrograde manoeuvre. The primary deorbit system was the 20,000 lbs thrust SPS with the sixteen (in quads of four) 100-lb RCS as the back-up system. The CM would be separated from the SM and orientated to point the heat shield against the direction of flight and begin the entry interface at 400,000 feet altitude. Drogue parachutes would be initiated at 23,000 feet to begin the parachute recovery sequence, heading toward a splashdown in the Pacific Ocean approximately 500 nautical miles south-west of San Diego, California.

Following splashdown the crew would remain in the CM, to be picked up by the prime recovery vessel. Once on board, time-critical experiments and results would be immediately transferred to a Mobile Laboratory (MOLAB) for preservation. The crew were to be examined as soon as possible after exiting the CM, completing the required post-flight medical experiments in the MOLAB. The CM would then be secured, and film and remaining experiments would also be unloaded into the MOLAB. Once the nearest land base was reached, the MOLAB, flight crew, CM, experiments and film and data material would be offloaded and airlifted to Houston for continued post-flight evaluation.

The SL-3 flight plan

The official mission of SL-3 would begin after the undocking of SL-2 from the OWS. During the unmanned part of the mission, status monitoring, system housekeeping functions and some experiment operation would be performed from the ground. Prior to the launch of the second mission, the workshop would be checked and confirmed by flight control, and the launch time set for optimum condition for rendezvous and recovery. Planned at 56 days, the ascent to and deorbit from the station would follow the profile of the first mission.

The mission objectives for SL-3 were:

- Perform unmanned operations of the workshop.
- Reactivate the Skylab orbital assembly in Earth orbit.
- Obtain medical data on the crew for use in extending the duration of manned spaceflights.
- Perform in-flight experiments.

Days 1–2	Launch, rendezvous, docking, reactivate Skylab systems
Days 3–4	Experiments
Day 5	EVA
Days 6–28	Experiments
Day 29	EVA
Days 30–53	Experiments
Day 54	EVA
Days 55–56	Deactivation, prepare one-month orbital storage
Day 56	Undocking, recovery

Launch would be based on the five-revolution optimum conditions. The mission could still launch up to 24 hours later if there was a hold due to technical or weather conditions on launch day, but further delays past Day 2 would force the mission to be recycled to Day 6 or Day 7 as necessary.

After docking, the crew were to transfer to the OWS and store the first-day medical samples in the OWS freezer. The full activation of the OWS would be completed after the first sleep period. Primary experiment operations on this visit would include eighteen life sciences, eleven solar physics, four astrophysics, six Earth

A splendid view of the S-IVB of the third mission, shortly after separation from the CSM.

applications (with 28 EREP passes), ten engineering and technology, and ten student experiments. A total of 230 hours of ATM observations were also manifested. Normal TV, systems housekeeping and other standard crew functions and systems operations would follow the eight hours work, eight hours sleep, eight hours rest pattern, and the crew were also to perform three EVAs to retrieve and replace ATM film.

At the end of the mission, deactivation and recovery would follow a similar profile to that of SL-2.

The SL-4 flight plan

As with the second manned occupation, the mission of SL-4 would officially begin with the undocking of the SL-3 from the OWS. There would be a similar unmanned operation phase, ending with the docking of the third crew at the MDA.

Originally planned for 56 days, the SL-4 mission would feature the same mission objectives as SL-3. The launch constraints and windows remained the same as for SL-3, as did OWS activation and the basic daily cycle followed by the first two crews.

Days 1–2	Launch, docking, transfer, activation
Day 3	EVA
Days 4–53	Experiments
Day 54	EVA
Days 55–56	Deactivation
Day 56	Recovery

Experiments assigned to this mission included fifteen life science investigations, six Earth applications, eleven solar physics, five astrophysics, nine student investigations, eleven engineering and technology, and twelve material science and manufacturing experiments. A total of 22 EREP passes were scheduled, as were two EVAs for ATM film retrieval and replacement.

At the end of the mission the crew were to deactivate the station for the final time, and follow the standard re-entry and recovery procedure, as with the first two missions. After two days of unmanned tests, the OWS would be placed in orbital storage pending a decision on its future use.

The crew schedule

Unlike previous American manned missions with durations of up to to 10–12 days, Skylab missions would last many weeks. Flight planning was therefore devised in which each crew was trained to operate a 24-hour day with an eight-hour sleep period, assigned work schedule and a rest and relaxation day each week. A typical duty day would run from 6 am to 10 pm. The MCC support of crew activities such as experiments was constrained by the Spaceflight Tracking and Data Network. Mission events were recorded as Greenwich Mean Time (GMT), mission day, and day of the year. Mission events for ascent up to docking were handled in Ground Elapsed Time (GET) and Mission Day/Day of the Year (MD/DOY).

Each mission day onboard Skylab was divided into 24 man-hours, split further into allocated percentages for different activities. Each working day was divided between operations, experiment performance, eating, personal hygiene, rest and relaxation, and sleep (planning time). The sleep and relaxation took most of the planning time at 33.6%, with mealtimes allocated at 12.3%. Each of the crew was allocated about 30 minutes daily for personal hygiene, and approximately 45 minutes for systems housekeeping.

A breakdown of the man-hour allocation of crew training and flight planning for the first 28-day mission produced 511 man-hours for performing experiments, 66 man-hours for housekeeping, and 177 man-hours for medical studies. There would also have been 166 man-hours allocated to work at the ATM console and 88 man-hours at the EREP station, with 80 man-hours for other experiments. Similar operations were planned for the second and third missions.

Using these data, a hypothetical crew working day on any of the Skylab missions would begin at around 6 am Houston time. The first task was to check the integrity of the OWS, followed by morning ablutions and breakfast. The first experiment to be performed each day was the M071 medical experiment that included measurements of residual food and the recording of water intake. The crew would also retrieve the daily flight update by teleprinter, and discuss this with the flight controllers.

Morning experiments would usually include the first scheduled EREP passes, involving two crew-members in the MDA, while the third crew-member would operate experiments from the SAL in the OWS. Later in the morning one of the crew would man the ATM console, while the other two would perform experiments in the OWS area. Lunch was normally scheduled for around mid-day in the wardroom, but if one of the crew was at the ATM then a snack could be eaten while monitoring the console in the MDA.

Afternoon experiments were programmed to begin at 13.30 hours, with one of the astronauts manning the ATM while the other two performed experiments in the OWS. Dinner was scheduled for about 18.00 hours, followed by a second recording of food and water intake for the M071 experiment. After dinner, the crew would report on the status of the day's work, commenting on accomplishments and on what they had not been able to achieve. These reports would usually take an hour, and would include an inventory of foodstuffs, used film, remaining film, unscheduled maintenance, any anomalies, and storage reports.

All garbage would be gathered into storage bags, and then passed through the airlock into the trash airlock in the oxygen tank below the lower floor area of the wardroom. Sleep periods were planned to start at around 10 pm each night, and included the M133 sleep-monitoring experiment, during which the Science Pilot would wear a skull cap to electrically record sleep patterns on to a tape for later analysis on the ground. Thirty minutes per day was also scheduled for physical exercise – usually in the evening prior to dinner.

The day off was scheduled to avoid conflict with EREP passes of primary interest. Little crew activity was planned for these days, with the exception of routine system housekeeping, real-time monitoring of solar flares at the ATM, the M072 mineral balance, re-entry simulations, debriefings, crew activities required for passive experiments, and any habitation observations onto the B channel tapes.

To assist in their recreation, Skylab crews personally selected reading material (off-the-shelf paperback books with flame-retardant covers) and taped music. Recreational games included darts with Velcro tips, four decks of playing cards made of flame-resistant paper, three handballs of different sizes, and a set of binoculars. The first crew had three ten-hour days off, the second crew seven, and the third crew raised to ten from the original seven.

Every day was a medical day, with each crew-member performing one series of M092/M093 or M092/M1712 every third day. If this conflicted with an EREP pass, the data collection was moved to 24 hours before or after the scheduled EREP pass.

EVAs were scheduled for each mission (one for SL-2, three for SL-3, and two for SL-4), and each was planned to last three hours from the start of egress to the completion of ingress. Two crewmen would perform each EVA with the third, wearing a soft suit (pressure suit without helmet and gloves), located in the forward compartment of the airlock. He would have access to the CSM, where he would perform system monitoring and read-out procedures to the other two. All EVAs were planned to take place over continental America, to provide maximum network coverage.

The routine maintenance chores that were planned to occupy the crews' time were scheduled at 100 man-hours for SL-2 and 200 man-hours for SL-3 and SL-4. These included domestic chores such as cleaning the dishes, putting out the trash, checking the air conditioning, dusting and cleaning, wiping out the bathroom, and even making the beds! Other chores would include systems checks, inventory of foods, film and supplies, changing teleprinter paper, vacuuming air-conditioning ducts, stowing loose items, management of food supplies, and setting up food trays and experiment hardware. A maintenance tool kit included 700 lbs of spare parts that were likely to be required – filters, fixings, hose clips, and so on – and 60 lbs of tools, including wrenches, screwdrivers, pliers and other work tools, a portable maintenance kit, and a waist belt tool caddie. Scheduled time for all expected maintenance chores was calculated to range between two minutes for verification of a certain system, to thirty minutes to clean and disinfect the waste management system.

All of the above were scheduled before Skylab flew, and were based on objectives to be accomplished, and minimum activities required to achieve these tasks and maintain the healthy status of the spacecraft and crew. They were based both on previous mission experience, ground training and simulations. What was found when Skylab was finally launched was that even NASA's best-laid plans can go awry in space!

Skylab rescue capability

The missions of Gemini 8 in 1966 and Apollo 13 in 1970 had clearly demonstrated the need to provide proper emergency procedures and training for any spaceflight crew. Faulty spacecraft thrusters on Gemini 8 initiated an immediate return to Earth when back-up control systems procedures were utilised to stabilise the spinning spacecraft after ten hours in space. On Apollo 13, the Moon landing had been lost when the Service Module oxygen tank exploded three days out from Earth. Having the LM still attached saved the crew, but it took a further four days for them to limp back to Earth, while MCC rewrote the emergency manuals as they went.

For Skylab, the design of the cluster allowed the CSM to remain as a 'lifeboat' in the event of major system failure on the station. In the event of a CSM failure while docked to the OWS, the crew had more than enough provisions to await the arrival of a 'rescue' CSM. The incorporation of a second docking port and the duration of the missions afforded the potential for rescuing astronauts for the first time.

Rescue capability studies

During 1970, three manned spacecraft centres (MSC, MSFC and KSC) conducted an assessment of the feasibility of crew rescue capability for Skylab. The Apollo CSM was designed to be activated by a large amount of ground support equipment, to attain operational status, and then fly (continuously powered) until the end of its 10–12-day mission. It would never intentionally be powered down or reactivated, and all onboard systems were designed as such.

With Skylab, the plan was to send the fully activated CSM to the station, and to then power down most of it until it was time to return home. This had been attempted only during the four-day emergency situation on Apollo 13, during which the CM was reactivated to recover the crew; but the SM had been disabled, and was of no further use. Rockwell indicated that the spacecraft was not designed to do this, and although they expected it to work after 28 or 56 days partially powered down, there was no guarantee. Therefore, a second CSM would be prepared for a rescue capability. At the same time, discussions with the Soviets had pointed to the possible use of a Soyuz as a rescue vehicle. The problem was that Soyuz could only carry up to three crew, and had both a different atmosphere mixture and a different docking system. Although the spacecraft could be prepared and launched quicker than an Apollo, its use as a rescue option for Skylab was impractical.

By December 1970, Rockwell's studies culminated in a NASA HQ decision to provide the crew onboard the Skylab with a *limited* rescue capability should the CSM return capability fail *while docked to the OWS*. To save time and cost overheads in trying to have a separate launch vehicle and spacecraft constantly on a launch standby, the rescue craft would actually be the next scheduled launch vehicle and spacecraft in the processing flow at the Cape.

The first crew's rescue vehicle would be the spacecraft and launcher prepared for the second crew, who in turn would have the third crew vehicle prepared for their rescue if required. The final crew would be supported by a back-up spacecraft and launcher. Upon receiving a rescue call, the in-flow CSM would be prepared for launch (after some minor modifications), to permit a two-man crew to go up and a five-man crew to come down.

The plan evolved over the following few months, based on the assumption that the stranded crew would be able to remain on the OWS with its ample supply of food, water, life support and power until the modified CSM arrived. If a failure

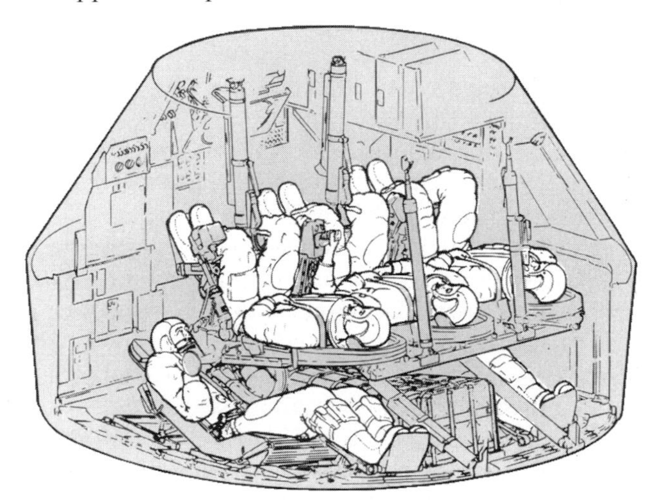

An artist's impression of how a three-man Apollo CM becomes a five-man Skylab Rescue vehicle. The two rescue astronauts (Brand and Lind) would have flown to Skylab to bring back the stranded astronauts.

occurred which stranded the CSM from the OWS, this rescue capability was not an option.

Skylab's mission rescue profile requirements were listed on 15 April 1971:

- Any trajectory planning for a rescue mission would be the same as for a nominal mission.
- Nominal rescue mission duration from launch to recovery would be limited to five days.
- The orbital assembly would manoeuvre to provide light acquisition support for the approaching rescue CSM.
- The rescue CSM would be capable of rendezvous without very high frequency (VHF) ranging.
- Landing and recovery would be planned for the primary landing area.
- The preferred transfer of the crew from the MDA to the CSM would be in shirtsleeves, although EVA was feasible.
- The KSC rescue launch response time would vary from 45½ days to ten days, depending upon the transpired time in the normal checkout flow.

By the end of that month, a Skylab Rescue Kit preliminary review was held at MSC. Here, it was determined that the 'rescue kit' should be installed in the CM within one shift (eight hours), that full pressure suits would be worn for entry, and that the centre couch would be ballasted for launch (the two-man rescue crew riding into orbit on the left and right couches). At the same time, studies were being conducted into the feasibility of jettisoning the unmanned disabled CSM from the axial port from within the MDA, freeing the main docking location where possible.

During 2–4 November 1971, an SL rescue vehicle preliminary design review was held at North American Rockwell. The anticipated re-entry mode for the rescue vehicle would be with the crewmen suited, with additional return stowage volume for programme-critical items. This was to be determined by North American (who would define the return volume and loading availability) and MSC (who would identify the returnable programme-critical items). The rescue CSM would be designed for both crew suited and unsuited re-entry, and for either axial or radial docking to provide maximum flexibility for real-time situations.

On 7 March 1972, the rescue capability became part of the programme flight planning, with NASA defining the hardware, procedures, documentation and training for a potential rescue mission that would need to be available immediately after the launch of the first crew. To accomplish this requirement, and to become a leading item on all future major meetings and panel discussions, the rescue mission was treated as a separate mission. Its status would be reviewed on a regular basis, along with the other missions as they each progressed towards launch.

By 25 April, KSC had updated the rescue launch response times from the launch date of each mission. On the day of the crew docking to the OWS, the rescue vehicle could not be launched for another 48½ days. At seven days into the mission, the rescue CSM would be 41 days from launch, and at fourteen days the SL rescue launch would be 36 days away. By three weeks into the orbital mission, the rescue craft would be 31 days from launch, and after a month, the rescue would still not be launched for another 26 days. For the first mission, of course, this implied a mission extension of up to 20 days if there was a problem early in the crew residency onboard the workshop.

For the second and third missions, at 35 days into their mission, the rescue craft would be 21 days from launch, at 42–49 days it would be 15 days from launch, and at 56 days it would be 13 days from launch. Any mission extension past these dates would be reviewed only if a rescue was required.

Rescue mission profile

The only failures considered for a rescue attempt were the loss of CSM return capability due to a systems failure, or the inability of the crew to enter the CSM while docked to the OWS. Failures elsewhere should not threaten the safe return of the crew in the docked CSM. The type of system failures considered in the CSM included the loss of entry or Earth landing capability; loss of all propulsion; loss of all electrical power; inability to undock from the MDA; loss of environmental control; or the loss of precise attitude stability.

The time to launch the rescue vehicle (45 days) from the same pad as the normal manned launch included the time required to refurbish the launch tower (22 days) and the installation of the specially developed kit in the CSM (eight hours). To convert the CSM to a rescue vehicle, storage lockers would be removed and two couches installed. The crew stranded on Skylab would continue to perform research until the rescue mission was launched. The life support capability of the workshop was sufficient for a total mission duration of about 150 days – well over five months of continuous manned operations.

Prior to rescue, the stranded crew would enter the MDA, seal it off from the rest of the cluster and then depressurise the OWS. They would then install a special spring-loaded device to separate the disabled CSM from the axial port of the MDA, at sufficient velocity to move it clear of the arriving rescue CSM. This was not absolutely necessary, as the arriving CSM could also dock at the side (radial) port of the MDA. This position, which was a contingency mode, was limited by the configuration, but was sufficient for full rescue operations.

Launch window opportunities could delay the rescue launch by five days at most, but the rendezvous and crew transfer times were only of secondary importance. For late emergencies, the time to prepare a rescue mission could be cut to ten days by eliminating certain procedures.

Modifications to the CSM included:

- Removal of storage lockers on the aft bulkhead.
- Installation of two couches in the resultant space.
- Life support and communications umbilical modified to accept two extra crewmen.
- Experiment return rack.

The rescue crew launched would fly with the centre couch empty. The returning CM would then carry five crew back – three on the normal couches and two in the extra couches under the main three (see the illustration on p. 81).

A modified drogue and mechanical triggering device for the mission CSM docking probe was included in the OWS portion of the rescue kit onboard the station. This would be used to trigger the undocking of the disabled CSM from inside the MDA while keeping the option open for future visits and re-use.

GROUND SUPPORT FACILITIES

As with any manned spaceflight operation, Skylab depended upon adequate ground support for the preparation and duration of the mission, and the immediate post-flight activities. Often related to an iceberg, the flight and astronauts are the 'tip you see above the surface'; but below this is the larger mass of the ground support infrastructure.

Launch facilities As with the majority of the hardware used on Skylab, NASA adapted former Apollo lunar launch facilities and operations to support Skylab missions. Launch Complex 39 at the Kennedy Space Center, Florida, was built specifically to support the Apollo Saturn V used to launch astronauts to the Moon. LC 39 had adopted the mobile launch concept, which was a departure from the fixed launch processing techniques used during the Mercury and Gemini programmes. Fixed launch site preparation tied up a pad for a considerable

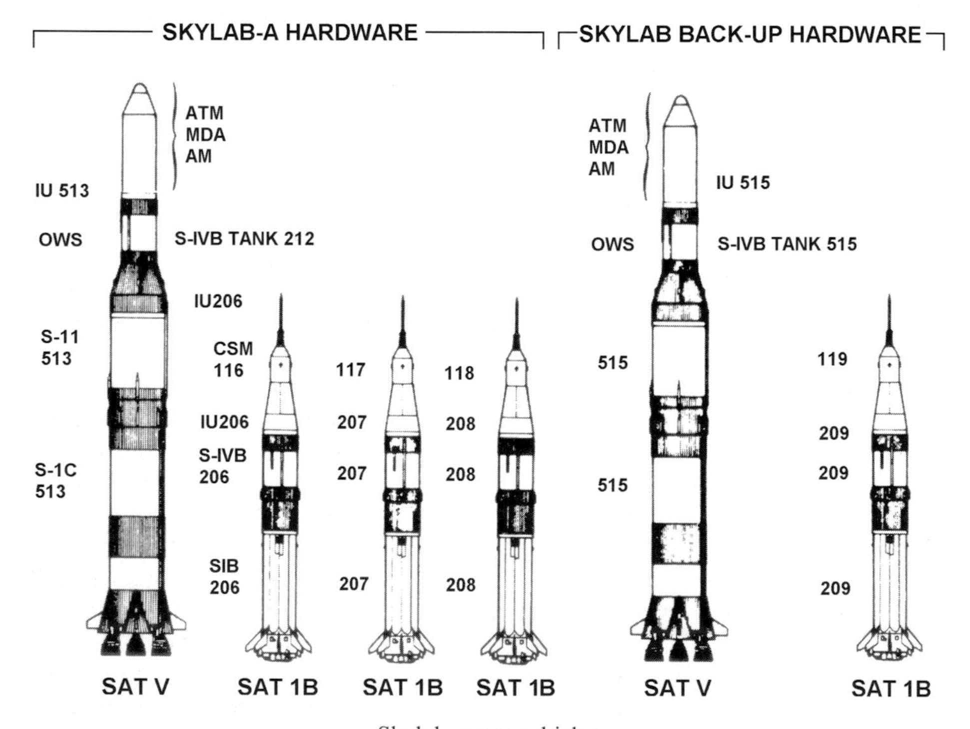

Skylab space vehicles.

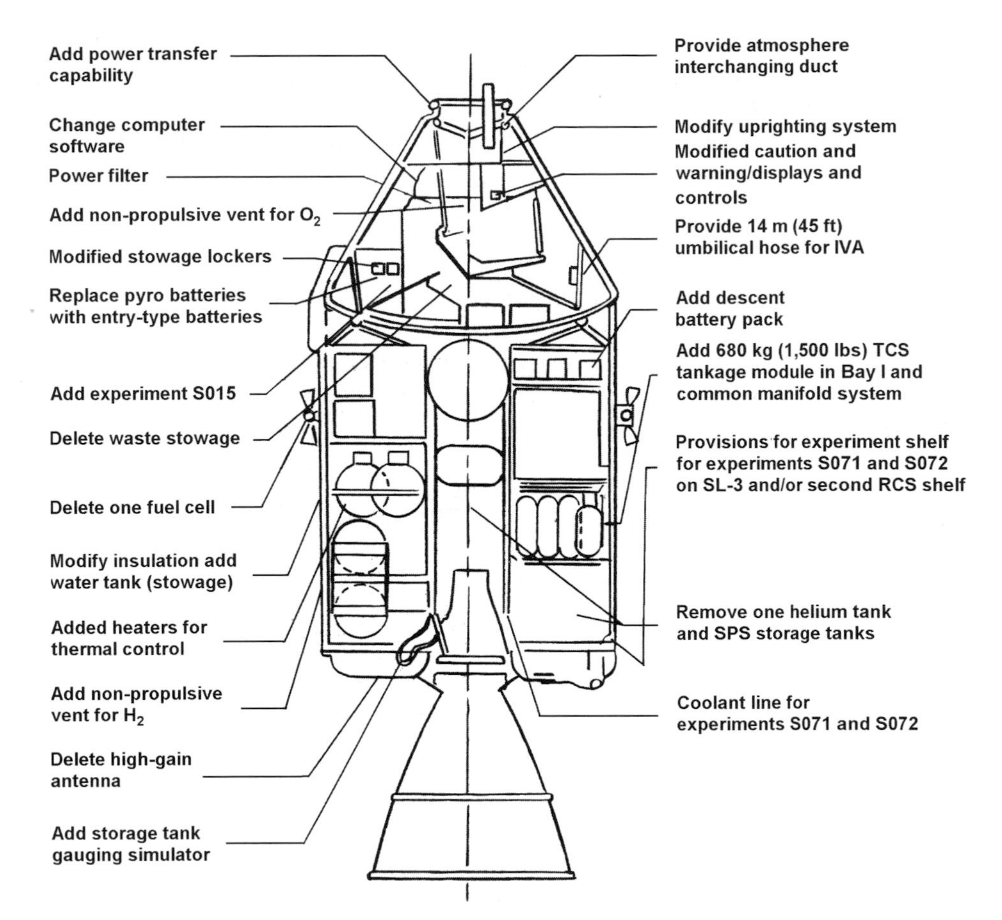

Block II CSM major Skylab modifications.

length of time, whereas mobile facilities allow preparation away from the launch area, offer greater protection from the elements, allow pre-launch checking, and minimise pad time.

The major element of LC 39 is the huge Vehicle Assembly Building (VAB), covering eight acres and a volume of 129,428,000 cubic feet. Measuring 525 feet tall, 716 feet long and 518 feet wide, it was for many years the largest man-made building on the planet, and it continues to dominate the relatively flat scrub area surrounding the launch site. Inside the VAB, the high bay (525 feet tall) and low bay (64 feet tall) are where the elements of the spacecraft and launch vehicle are checked and prepared for stacking. The low bay consists of eight stage preparation and checkout area cells. The actual stacking of the vehicle occurs in the high bay, where up to four Saturn Vs could be assembled at any one time. For Skylab, the SL-1/Saturn V and each of the Saturn 1Bs for the manned missions were assembled in the high bay areas. Several work platforms surrounded the vehicles and allowed personnel access during mating operations.

Adjacent to the VAB is the Launch Control Center (LCC), termed the 'electronic brain' of LC 39. The four-storey LCC was used for systems checkout and tests of all elements of Skylab during assembly, and then controlled the countdown and launch at the pad area. The LCC housed four firing rooms on its third floor, linked to one of the four high bays of the VAB. Three of the rooms were identical, allowing simultaneous launch of one vehicle and checkout of up to three others. Each of the firing rooms had 450 consoles manned during the countdown and launch of the Saturn vehicles, although not all of these were manned for the Saturn 1Bs.

To move the Saturn from the VAB to the pad area required the use of a Mobile Launch Platform (MLP) and the Crawler Transporter (CT). The MLP were three transportable launcher bases and umbilical towers weighing 12,600,000 lbs and measuring 446 feet tall when resting on the pad launch pedestals. The base section was constructed from steel in two storeys, and measured 25 feet high, 160 feet long and 135 feet wide. The Saturn was placed over a 45-foot square opening, providing the exhaust outlet into the flame trench and flame deflector out on the pad. On the umbilical tower were hydraulically operated service arms for umbilical connection and launch team access, as well as escape systems and lifts. They were retracted at various stages during the countdown. Skylab and the astronauts would orbit Earth at 17,500 mph, but the vehicles that took them there began the journey from the VAB to the pad along the crawlerway at a sedate 1 mph. The launch vehicle and MLP were lifted and taken to the pad on top of the crawler transports. These measured 131 feet long and 114 feet wide. The four double tracks on each crawler measured 10 feet high and 41 feet long, and each of the sixty shoes on the crawler track was 7 feet 6 inches wide, and weighed 2,000 lbs. The crawler's maximum speed unladen was only 2 mph, but loaded with a Saturn V it was 1 mph. A levelling system kept the vehicle vertical to within ± 10 arcminutes (which, at the top of the vehicle, was approximately equivalent to the diameter of a basketball). This allowed the vehicle and its payload to mount the 5° ramp up to the pad, where it deposited the MLP on the launch pedestals. The distance from the ground to the top deck on which the MLP was mated was 20 feet, and its area was 90 feet square. The 'road' that took the vehicle to the pad was the three-mile crawlerway, consisted of a 135-foot wide roadway – almost an eight-lane highway.

The two LC 39 launch pads (Pad A and Pad B) were constructed from heavily reinforced concrete to support the weight of both the ML and the launch vehicle, and also the Mobile Service Structure (MSS) and transports, and to withstand the forces of launch. Each pad covered a quarter of a square mile, and the pad top area stood 48 feet above the level of the nearby Atlantic Ocean. All propellants were stored in adjacent storage tanks, the propellant being transferred in stainless steel, vacuum-jacketed pipes up to the ML and across the access arms into the vehicle. On each pad was one liquid hydrogen tank, one light oxygen tank, and the RP-1 tanks. In the centre of the pad was the flame trench and deflector to spread the exhaust gases from the ignition of the first-stage engines, and a water deluge system to both cool the pad and suppress the sound waves, preventing them from rebounding onto the vehicle and damaging it.

The MSS was a 410-foot, 10.5-million-lb tower used to service the Skylab hardware at the pad, providing 360° of access from the 40-storey structure. It had five platforms – two self-propelled and three fixed but moveable. Two elevators carried personnel and equipment between the platforms.

LC 39 is enormous, and it produced volumes of 'gee-whiz' data that amazed reporters and spectators alike. From LC 39, twelve Saturn Vs left the pad – all but one (Apollo 10) from Pad A. For Skylab, the first dual launch from LC 39 was made possible by using Pad 39A for SL-1 and Pad 39B for each of the three Saturn 1Bs. Before Skylab could be launched, however, several modifications needed to be completed.

LC 39 modifications for Skylab operations

The major modification to LC 39 involved the construction of a 127-foot tall pedestal (termed the 'milk stool') on an MLP, for adapting the 223-foot tall Saturn 1B to accommodate a launch tower designed and built for the 363-foot Saturn V. There were also several modifications to the swing arms on both the ML, to adapt them to the particular needs of the two-stage Saturn V and the Saturn 1B on Pad 39. Modification of work platforms in the VAB was required for the smaller diameter of the Saturn IB, and 'clean rooms' were constructed for the ATM in both the Manned Spacecraft Operations Building (MSOB) and in High Bay 2 of the VAB.

Pad B and Firing Room 3 in the LCC required changes in propellant servicing control and monitoring areas to be compatible with a two-stage Saturn 1B. Pad A and Firing Room 2 required changes in propellant, environmental control, instrumentation, and control systems, to support a two-stage Saturn V and enlarged OWS payload.

Skylab 2 is rolled out of the VAB on top of the MLP and on its way to the pad. Behind is the huge VAB and the smaller Launch Control Center.

Launch configuration of the Skylab Saturn 1B atop the 'Milkstool' launch pedestal next to a launch tower designed for the much taller Saturn V.

The Saturn 1B launch pedestal was mounted on top of the ML, allowing the second stage (S-IVB) instrument unit and Apollo spacecraft to interface at the same stations as on the Saturn V. The 127-foot pedestal tapered from 48 feet square at its base to 21 feet 11 inches at the top. Mounted on the top of the structure was a 44-foot launcher table, with a 28-foot opening above the pad flame hole on the ML.

The pedestal was a free-standing tower with no horizontal sheer connection between the access bridge and the ML tower, simplifying load distribution on each leg. It weighed around 500,000 lbs, and had an additional 500,000 lbs of associated support equipment. It was designed to launch a Saturn 1B in maximum winds of 37 mph at the 30-foot level, and to withstand 125 mph hurricanes without a launch vehicle present.

Mission Control Center, Houston

Mission Control Center, Houston (MCC-H) was used for Skylab flight control operations. MCC-H is the former Manned Spacecraft Center (MSC) that was renamed after the death of former President Lyndon B. Johnson in early 1973, and is now known as JSC. The MCC for Skylab was located in the same building (#30) as that for Apollo, but called for significantly different manning arrangements than for the lunar programme. Apollo missions were flown for up to twelve days, and were

A group of flight directors and officials cluster around Flight Director Don Puddy's console in MOCR to consider the problems of the undeployed solar array and high temperatures inside the workshop early in the unmanned mission. JSC Director and former Mercury flight controller Chris Kraft (wearing a jacket) stands behind Puddy.

separated by several months. For Skylab, over eight months there would be three long missions of 28, 56 and 56 days running 24 hours a day, separated by low activity during two periods of unmanned flight.

To support this complex programme, facilities were needed within the MCC to provide real-time data for analysis and evaluation by engineers and scientists. MCC accomplished the in-flight analysis of trajectory, vehicle systems, experiment systems, scientific data and flight planning, as well as the control of the vehicle in-flight using voice, telemetry, tracking, and update facilities. This required monitoring, evaluating and updating in both real time and in long lead delayed time to achieve mandatory mission objectives, detailed test objectives, updated alternative and contingency flight planning (in the event of in-flight emergencies) and the safety and well-being of the crew.

MCC also participated in crew and flight control team training, simulations, pre-launch testing and countdown operations, SL countdown, launch, the initial activation of the workshop, crew launches, rendezvous and docking, station activations, manned orbital operations (including EVA), deactivation, undocking, entry, recovery, and unmanned operations.

The primary room for command and control was the Mission Operations Control Room (MOCR, pronounced 'Moker'), and a duplicate room was used for planning. Initially, four teams (shifts) of flight controllers working a 40-hour week were planned to support the unmanned launch and check-out of the orbital workshop and the first manned missions. For SL-3 and SL-4, five teams were trained, working a five days on/two days off cycle throughout the final two manned periods.

Houston took control of the missions from 'tower clear' to post-recovery with the

Controllers man consoles during Skylab activity. Despite the banks of computers, large volumes of paper were still required.

crew on the recovery vessel. The MOCR console layout for Skylab missions (shown in the illustration on p. 100) was as follows:

Front row (the 'trench')
GUIDO: Guidance Officer – responsible for monitoring the performance of the Instrument Unit on the Saturn launch vehicle, the CM computer, and ATM digital guidance and control functions, and for providing computer information about the onboard computers on the CM and ATM.
FLT DYN (FIDO): Flight Dynamics – responsible for computing and monitoring the required manoeuvre activities, maintenance of the orbital parameters and re-entry planning. During routine manned phases he also served as the Retrofire officer (RETRO).
LV/EVA/EREP: Launch Vehicle officer during launch operations, plus EVA procedures or EREP officer, depending upon the mission events at the time. A rotational position during the mission, according to flight requirements.

Second row
EGIL/EECOM: The Electrical, General Instrumentation and Life Support System engineer, responsible for monitoring and troubleshooting the orbital assembly's environmental, electrical and instrumentation systems, including the CSM systems when the spacecraft was docked to the complex. The EECOM position was the CSM Electrical, Environmental Control and Electrical System engineer.

GNS/GNC: The Guidance, Navigation and Control System engineer for the OWS and the CSM Guidance, Navigation and Control Systems engineer was another rotational console.
EXP: Experiments officer – responsible for monitoring the operation of onboard equipment for all corollary experiments, and for assessing the validity of the scientific data from those experiments.
MED OPS: Medical Operations officer – responsible for the analysis and evaluation of all medical activities, both operational and experimental, during the flight.

Third row
NETWORK: Network controller – responsible for the detailed operational control and failure analysis of the ground systems.
CapCom and FAO: Spacecraft (formerly Capsule) Communicator is an astronaut who is responsible for all voice communication with the crew in orbit. The CapCom astronauts in MCC are the 'extra' members of the flight crew in space. The Flight Activities officer was responsible for the development and co-ordination of the flight plan and all changes to the onboard flight data file (FDF). He was also the officer responsible for co-ordination of the experiment pre-advised data (PAD).
FD (Flight): Flight Director – the Boss – the chief console position in the control room. Responsible for all MOCR decisions and actions that concerned vehicle systems, dynamics, experiment operations, and MCC/STDN operations. All the other consoles in MOCR report to the Flight Director through an internal communications system (the Flight Director's Loop) that is not normally monitored by the general public. He is also responsible for the safety of the crew and the spacecraft, while implementing the mission objectives.
O&P: Operations and Procedures Officer – responsible for the implementation of mission control procedures, and for the monitoring and operational management of real-time data acquisition, retrieval and processing.
SKYCOM: Skylab Communications engineer – responsible for the management and troubleshooting of all communications up to the station.

Fourth row
DoD: Department of Defence representative – responsible for the co-ordination and direction of all DoD mission support.
HQTRS: Headquarters management representatives.
FOD: Flight Operations Director – responsible to the Director of JSC for all operational aspects of Skylab, providing a management interface between the Flight Director and the JSC, KSC and MSFC Program Managers.
PAO: Public Affairs Officer – responsible for keeping the public aware of the progress of the mission. This console was manned by a member of JSC's PAO staff. Known as the 'voice of mission control', it is this voice that is heard by the public during the air-to-ground communication between the CapCom and the crew during radio and TV coverage. The PAO also explains to the public what is happening on the mission, and provides other background details.

Many of these console positions were directly supported by several Staff Support Rooms (SSRs) adjacent to MOCR, which included experiment and system experts in areas of mission operations; the Saturn (booster) launch vehicles; ATM experiments; OWS/CSM vehicle systems; aeromedical/medical experiments; EREP experiments; corollary experiments; and the recovery operations control room.

Huntsville Operations Support Center

The HOSC was located at Marshall, and was manned by staff from the centre with experience in the Saturn launch vehicles, the Orbital Workshop systems, and the experiments for which the centre was responsible. This support was available during launch preparations and operations at the Cape in Florida, and during mission operations controlled from Houston. High-speed computer connections linked the HOSC facilities with those at KSC and at JSC.

Mission management

A Skylab programme management responsibilities structure existed at Headquarters in Washington, and at field centres in MSFC, JSC, and KSC. One management group and two management teams provided a means for the Skylab organisation to rapidly exchange information and decisions affecting real-time mission operations.

The Skylab Advisory Group for Experiments (SAGE) was responsible for experiment objectives, policy, priority and changes and consisted of senior HQ personnel from experiment sponsoring offices (see Appendix).

A symbolic smoking of cigars after a successful mission and recovery of the SL-2 crew. (*From left*) George Low, NASA Deputy Administrator, Jim Fletcher, NASA Administrator, Ken Kleinknecht, Skylab Program Director, Dale Myers Associate Administrator for Manned Flight (partially hidden), and Charles Berry, Director for Life Sciences at JSC.

The Flight Management Team (FMT) provided overall management, and consisted of the Programme Director, the JSC, MSFC and KSC Skylab Program Managers, the MSFC Saturn Program Manager, and the JSC Director of Flight Operations, with the Program Director acting as Chairman and the group relaying decisions to the Flight Director.

Skylab management (March 1973)

NASA Headquarters, Washington DC	
Administrator	James C. Fletcher
Deputy Administrator	George M. Low
Associate Administrator	Homer E. Newell
NASA Associate Administrator for Manned Space Flight (with overall management responsibility)	Dale D. Myers
Program Director	William C. Schneider
Deputy Program Director	John C. Disher
Director of Operations	Robert O. Allen
Director of Experiments	Thomas E. Hanes
Director of Engineering	Melvin Savage
Director of Reliability, Quality and Safety	Haggai Cohen
Director of Budget and Control	J. Pemble Field Jr
Marshall Space Flight Center, Huntsville, Alabama	
Director	Dr Rocco A. Petrone
Deputy Director, Technical	Dr William R. Lucas
Deputy Director, Management	Richard W. Cook
Program Manager	Leland F. Belew
Deputy Program Manager	Stanley R. Reinartz
Chief, Program Engineering and Integration Project	George B. Hardy
Chief, Ground Support Equipment Project	Porter Dunlap
Chief, Experiment Development and Payload Evaluation Project	Jack H. Waite
Chief AM/MDA Project	Floyd M. Drummond
Chief, Saturn Workshop Project	William K. Simmons
Chief, ATM Project	Rein Ise
Manager, Saturn Project	Richard G. Smith
Manager, Mission Operations	H. Fletcher Kurtz
ATM Project Scientist	James B. Dozier
ATM Engineering Manager	Eugene H. Cagle
Johnson Space Center, Houston, Texas	
Director MSC/JSC	Dr Christopher C. Kraft
Manager, Skylab Program Office (SPO)	Kenneth S. Kleinknecht
Deputy Manager, SPO	Arnold D. Aldrich
Manager, Missions Office, SPO	Alfred A. Bishop
Manager, Engineering Officer, SPO	Willis B. Mitchell Jr
Manager, Manufacturing and Test Office, SPO	W. Harry Douglas

Manager, Orbital Assembly Project Office, SPO	Reginald M. Machell
Manager, Earth Resources Project Office	Clifford E. Charlesworth
Director of Flight Crew Operations	Donald K. Slayton
Chief of Astronaut Office (Code CB)	Alan B. Shepard
Chief of Skylab (Astronauts) Branch, CB	Charles Conrad Jr
Director of Life Sciences	Richard S. Johnston
Director of Flight Operations	Howard W. Tindall
Chief, Flight Control Division, Flight Operations (MCC) (Flight Directors: Hutchinson (Silver); Lewis (Bronze); Puddy (Crimson); Shaffer (Purple); Windler (Maroon))	Eugene F. Kranz
CSM Manager	Glynn S. Lunney
Kennedy Space Center, Cape Canaveral, Florida	
Director	Dr Kurt H. Debus
Manager, Apollo/Skylab Program	Robert C. Hock
Manager, Ground Systems Office	W.L. Halcomb
Manager, Program Control Office	J.C. Leeds
Reliability, Quality Assurance and Systems Safety Office	O.L. Duggan
Manager, Skylab Space Vehicle Office	A.R. Raffaelli
Director of Technical Support	P.A. Minderman
Director of Design Engineering	Raymond L. Clark
Director of Launch Operations	Walter J. Kapryan
Director of Launch Vehicle Operations	Dr Hans F. Greene
Director of Spacecraft Operations	John J. Williams

The Flight Control Team (FCT) was the focal point between experiment and mission planning, was in direct control of the mission, and would implement management requirements. Part of this team would track progress toward achieving mission and experiment objectives, as well as tackling system and experiment anomalies, mindful of crew safety and mission success. Direct control of the flight was the responsibility of the flight control personnel in MOCR and SSR. Whichever FCT was on duty, there would be a representative of the FMT in the MOCR to monitor mission status and flight crew activity. The Flight Director had overall responsibility for crew operations, to meet programme and safety objectives.

Spaceflight Tracking and Data Network

The STDN was a worldwide network that was established for all NASA's Earth orbital spacecraft, and consisted of nineteen fixed stations, supplemented by two mobile stations. They were located within a band that stretched around the globe, mostly between the latitudes 50° north and 50° south. The most northerly station was in Fairbanks, Alaska (65° north) and the most southerly was in Orroral Valley, Australia (35° south). For Skylab, twelve ground stations were used: Merritt Island, Florida; Bermuda; Ascension Island; Grand Canaria; Carnarvon, Australia; Guam; Hawaii; Corpus Christie, Texas; Goldstone, California; Canberra, Australia; Madrid, Spain; and Newfoundland, Canada.

Major Skylab contractors

The Boeing Company, New Orleans, Louisiana	S-IC stage of Saturn V
Chrysler Corporation Space Division, New Orleans, Louisiana	S-IB stage of Saturn 1B
General Electric, Huntsville, Alabama	Electrical and ground support equipment
IBM Corporation, Huntsville, Alabama	ATM digital computer system; instrument unit for Saturn V and Saturn 1B
Marshall Space Flight Center, Huntsville, Alabama	Apollo Telescope Mount
Bendix Corporation (under contract to Martin Marietta)	Payload Integration MDA, ATM C&D console
McDonnell Douglas Astronautics, Huntington Beach, California	S-IVB stages for Saturn 1B; S-IVB Orbital Workshop; OWS payload shroud
McDonnell Douglas Astronautics, St Louis, Missouri	Airlock Module; mission operations requirements
North American Rockwell, Downey, California	S-II stage, Saturn V; Apollo CSM
Perkin Elmer, Norwalk, Connecticut	ATM Hα telescopes
Actron Industries, Monrovia, California	Earth terrain cameras
AiResearch Division of Garrett Corporation, Torrence, California	Astronaut life support assembly
Delco Electronics, Santa Barbara, California	Guidance and navigation systems
General Electric, Houston, Texas	Flight garments and crew provisions
ILC Industries, Dover, Delaware	Apollo/Skylab spacesuits
Martin Marietta Aerospace, Denver, Colorado	Payload integration; crew operations and training; flight operations
Massachusetts Institute of Technology	Guidance and navigation systems
Whirlpool Corporation, St Joseph, Michigan	Skylab food system

There was also the instrument ship *Vanguard* (VAN), located for Skylab at the port of Mar Del Plata, Argentina, to ensure additional coverage in the southern hemisphere. The fleet of eight Apollo Range Instrumentation Aircraft (ARIA) was also available to 'fill in the gaps', and the three prime TV reception stations were located at Goldstone, California.

The STDN was managed by NASA's Goddard Space Flight Center (GSFC) in Maryland, and during the Skylab missions it was controlled by the MCC to provide tracking, telemetry, and communication (both voice and TV). This was achieved by a data uplink to the workshop and the VCSM and a data downlink to MCC. Connection between the STDN stations and the MCC was provided by means of the NASA Communications (NASCOM) system. The ground stations had either 30-foot or 85-foot antennae.

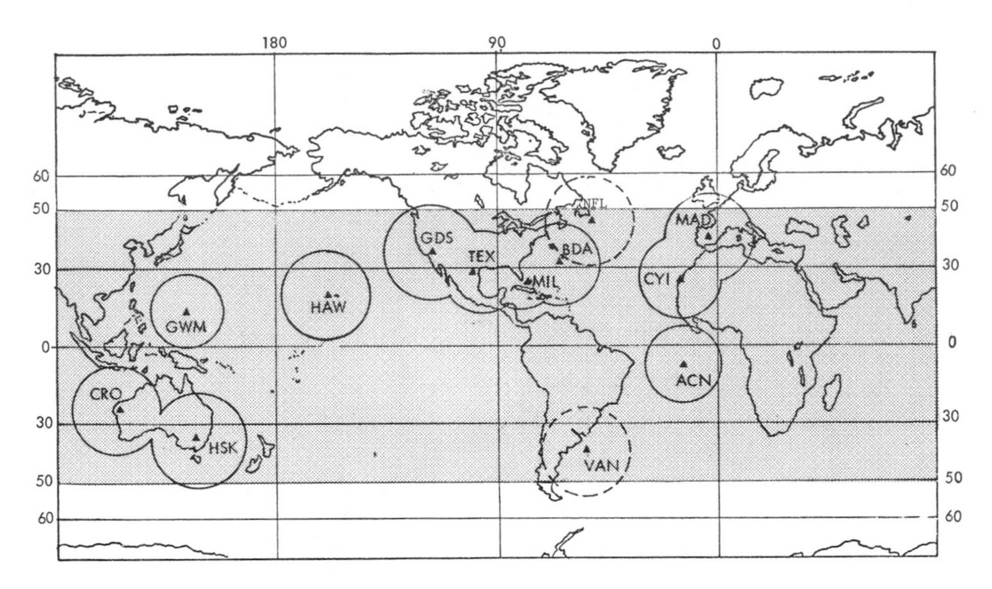

Network configuration for Skylab.

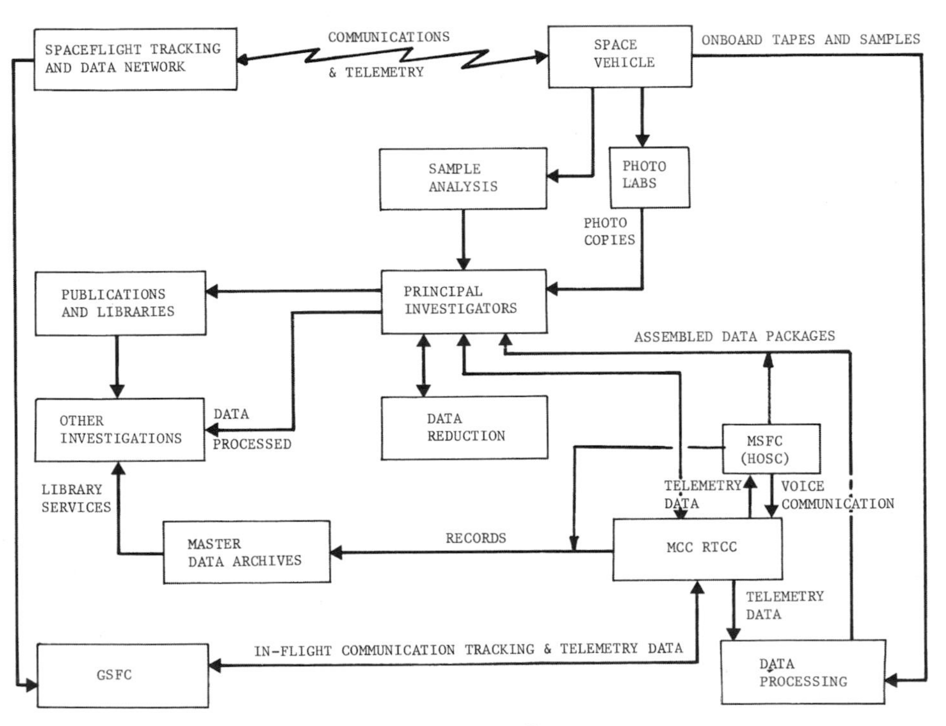

Skylab data flow.

After a record-breaking 28 days in space, Skylab 2 CM splashed down in the Pacific Ocean, approximately 834 miles south-west of San Diego, California, at 09.50 EDT, 22 June 1973.

Recovery operations

Recovery of both the flight crew and the CM was the responsibility of the US Department of Defense (DoD), which had provided support for ocean recovery operations from Project Mercury, and continued through to ASTP in 1975. DoD personnel were trained in NASA's requirements, and were dispatched during the last week of a normal mission to be on station in the primary landing area. Other teams were on standby across the world to support an emergency landing – whether a launch abort or an early mission termination.

The primary recovery area was in the Pacific Ocean, with SL-2 and SL-3 targeted for off the coast of California, west of San Diego, and SL-4 in the mid-Pacific. An aircraft carrier was designated the prime recovery vessel for all three missions. Secondary recovery areas were those after orbital insertion but before normal end of mission (EOM), with at least 24 hours lead time to establish the recovery forces in the Pacific.

For Skylab, the extended duration of the missions meant that full support could not be maintained throughout the mission; but to protect medical data, rapid recovery was required to allow the flight crew to begin post-flight medical examinations after either early or nominal recovery. It was established that for either an early return or the planned end of mission, the time between splashdown and delivery of the flight crew to the Skylab Mobile Laboratories would no be more than an hour.

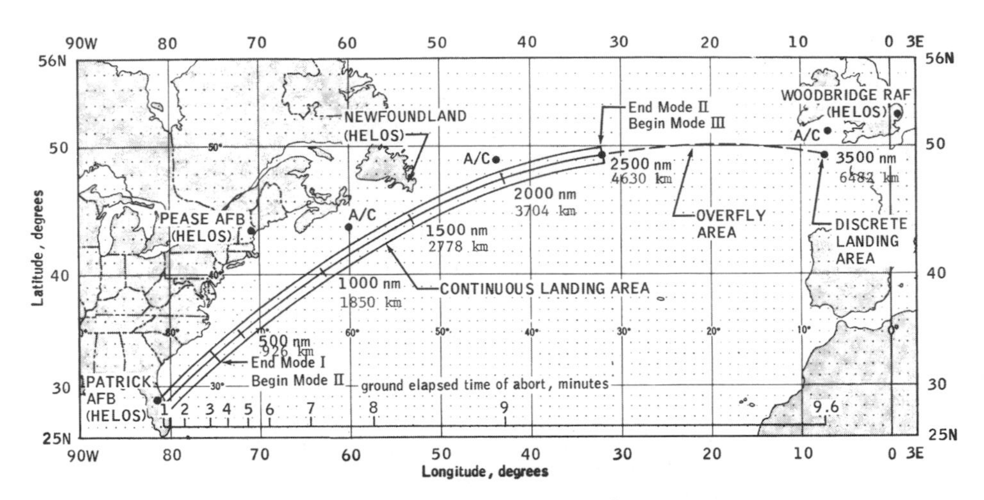

Launch abort area and recovery force posture.

The last Skylab CM floats in the ocean during recovery.

The Skylab Mobile Laboratory was a group of six US Army Medical Unit Self-contained Transportables (MUST), modified to meet post-mission medical requirements of Skylab. They were installed on the prime recovery ship, and the nearest land bases for either nominal or early end of mission transportation to the landing sites.

For launch abort, three modes were available. Mode 1 stretched 500 nautical miles along the launch azimuth (direction) from the Cape, and was covered by helicopters from Patrick AFB. (This covered the first three minutes of powered flight

Skylab 2 CM with the three astronauts still inside is winched onto the prime recovery vessel *USS Ticonderoga* at the end of their marathon 28 days in space. Unlike the Apollo astronauts, who were winched to helicopters after splashdown, the Skylab astronauts remained in their spacecraft for recovery.

using the launch escape system.) Mode II was available if an aborted landing was called in the 500–2,500 nautical mile range across the Atlantic (3–9 minutes GET after LES jettison, non-propulsive using the full lift capability of the CM), and helicopter rescue would be supported from either Loring AFB in Maine, or the Gander International Airport station in Newfoundland. Mode III covered an aborted landing at 2,500 to 3,500 nautical miles (over nine minutes, and using the SPS system towards an area south of Great Britain to prevent a ground landing in Europe, if orbital velocity was not possible), and aircraft and recovery support would be dispatched from RAF Woodbridge, England.

Early entry would be targeted for a secondary landing area near Hawaii, with support from forces located on the islands. The contingency landing area was considered to be anywhere around the globe between 50° north and 50° south should an immediate recovery be required, with capacity for land or sea recovery support.

A global team

With all of these systems and procedures in place, it is obvious that once the programme had been established, preparing for and supporting the flight operations

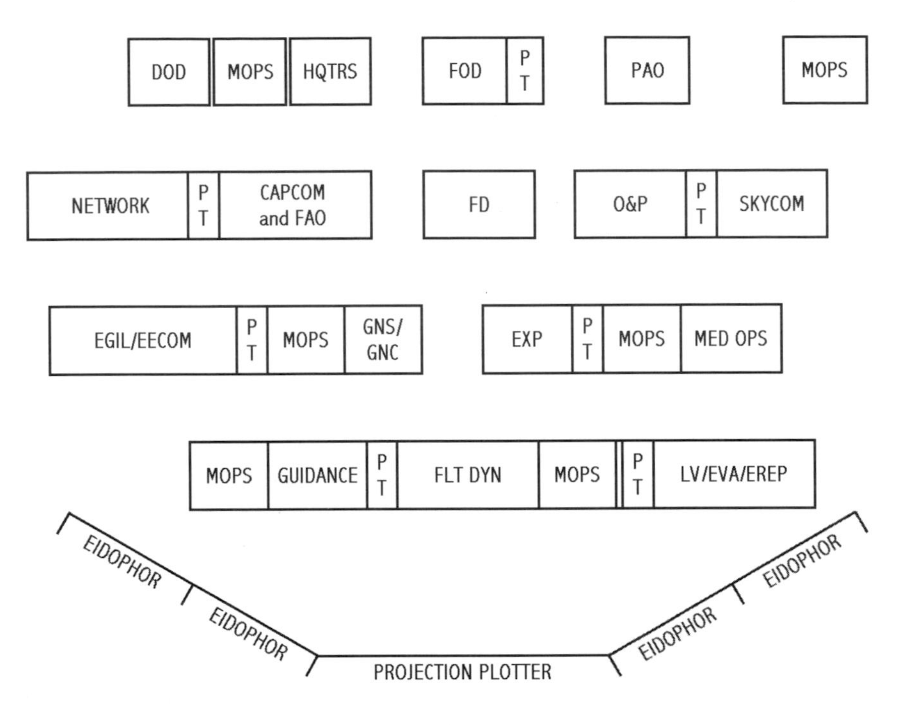

Layout of the Mission Operations Control Room, Houston, from where Skylab was controlled from launch to splashdown.

was a major global effort. Following any mission operations in space masks the huge infrastructure involving thousands of personnel from NASA, contractors, experiment PIs, world-wide tracking facilities and the DoD.

By 1972, in little more than a decade Skylab had progressed from paper studies to the application of Apollo hardware to extended scientific missions, to a fully operable pioneering space station. There remained only the task of flying the missions; and to do this, one more element of the programme was required – the astronauts.

The human element

In just four short years between December 1968 and December 1972, twenty-four American astronauts travelled to the Moon. Three went there twice, and twelve of them landed on and walked across its surface. They were a proud and unique group of explorers, and were applauded across the world.

They were followed by an even smaller group, just as unique, who were equally proud of their achievements and their contributions to the manned exploration of space. These were the nine men who flew the three missions onboard Skylab between May 1973 and February 1974. Two of these were also Apollo 12 Moon-walkers, but the other seven were less well remembered by the public than were their Apollo colleagues or the original seven Mercury astronauts. Their contribution to human space exploration was perhaps as important as the Apollo missions had been, and in pioneering the skills required for living and working away from Earth, their efforts had more long-term importance than going to the Moon.

Prior to 1973, out of 45 missions into space (27 American and 18 Soviet) since 1961, only three had lasted fourteen days or more (Gemini 7 for two weeks, Soyuz 9 for 18 days, and Soyuz 11 for 23 days). Of the rest, all the Soviets flights had lasted for less than a week, and only six American flights had exceeded ten days.

The majority of training accomplished by these crews involved the launch phase, plus techniques for rendezvous, docking and crew transfer, EVA in space (and for the Americans, on the Moon), emergency and contingency scenarios, and re-entry and recovery training.

On all the missions, space habitability – that is, the skill of actually living inside the spacecraft – was an obvious requirement, but in small spacecraft on short missions the facilities were basic although adequate enough for the crew to endure any inconvenience until recovery. On the slightly longer missions, crew time was taken up with additional experiments, observations and research, which produced a scientific return, but was not carried out at the expense of mission success or crew safety. On the missions of more than ten days, the scientific objectives were expanded, but were still restricted in what could be done due to the design of the spacecraft, or the limitations of the mission duration. It was only when the Soyuz 11 crew became the first resident crew to stay on board a space station that useful scientific research from prolonged spaceflights seemed to be a real possibility. However, even on that first space station mission, in-flight hardware failures, an abbreviated training programme, and the tragic loss of the crew during the recovery phase, underlined the fact that any success in long-duration spaceflight would be a challenge.

For the Skylab astronaut team, the first priority was to devise a method of training for missions of a duration that no-one had attempted before and in an environment that was almost impossible to recreate on Earth. This could only be done by experiencing short periods of reduced gravity in aircraft flying parabolic curves to counter the effects of gravity, or under water, each of which could total no more than a few hours.

Each mission was to last several weeks, and because this was the only station that had been budgeted for in the foreseeable future, there was a desire to obtain optimum results from every day that the crew were on board.

How these nine men came to be chosen for the three flights is as intricate as the story of Skylab itself. An early decision at NASA led to the selection of pilot-trained crew-members in the first three astronaut groups in 1959, 1962 and 1963. The selection criteria – initially aimed at test-pilot skills – were relaxed by the third intake, but still demanded several years of jet pilot experience. By the third group, however, several candidates had also gathered academic and scientific research qualifications in addition to their piloting skills. The era of an American astronaut with a scientific or engineering background (rather than a purely piloting career) would finally emerge with the prospect of the scientific research missions that evolved into Skylab.

THE ORIGINAL ASTRONAUTS

By the end of 1963, NASA had selected a total of thirty pilots for its astronaut team to train for and fly on the pioneering Mercury, Gemini and Apollo programmes. From the first American flight into space in May 1961, to the second landing on the Moon in November 1969, every seat on a total of twenty-two missions was occupied by members of those three selections – some of them flying two or three times. It was this group that took up President Kennedy's challenge and personally participated in landing a man on the Moon by the end of the decade.

Group 1; 9 April 1959 (seven selected; known as the 'Original Seven')
Lt Malcolm Scott Carpenter, USN, aged 33 at selection.
Capt Leroy Gordon Cooper Jr, USAF, 32.
Lt-Col John H. Glenn Jr, USMC, 37.
Capt Virgil I. 'Gus' Grissom, USAF, 33.
Lt-Comdr. Walter M. Schirra Jr, USN, 36.
Lt-Comdr. Alan B. Shepard Jr, USN, 35.
Capt Donald K. 'Deke' Slayton, USAF, 35.

All of this selection were educated to bachelor degree level and had an average flying time that exceeded 3,000 hours. Six completed flights in the Mercury programme. On 5 May 1961, Shepard became the first American in space during a 15-minute sub-orbital flight (Freedom 7), followed by Grissom on the second sub-orbital flight (Liberty Bell 7) in July 1961. John Glenn flew America's first orbital mission (Friendship 7, three orbits) in February 1962, followed by Carpenter (Aurora 7, three orbits) in May, and Schirra (Sigma 7, six orbits) in October of that year. The series was completed by Gordon Cooper (Faith 7) flying a 22-orbit mission on 15–16 May 1963. Slayton had been medically grounded in 1962.

Group 2; 17 September 1962 (nine selected; known as the 'Next Nine')
Mr Neil A. Armstrong, aged 32 at selection.
Maj Frank Borman, USAF, 34.
Lt Charles 'Pete' Conrad Jr, USN, 32.
Lt Comdr. James A. Lovell Jr, USN, 34.
Capt James A. McDivitt, USAF, 33.
Mr Elliott M. See Jr, 35.
Capt Thomas P. Stafford, USAF, 32.
Capt Edward H. White II, USAF, 31.
Lt John W. Young, USN, 31.

Again, this group's experience was in operational flying (with an average of 2,900 hours), and they held degrees in aeronautical engineering. Known as perhaps the best selection NASA has ever announced, this group included some of the most well-known American astronauts from the following two decades – from Gemini to the early Space Shuttle programme. One of this group – Conrad – would go on to make his fourth spaceflight on the first Skylab crew.

When the Class of '62 arrived at NASA, it was already generally accepted that even more astronauts would be needed to fulfil the flight requirements leading to the first lunar landings. A decision to select a third group the following year was based on very optimistic planning documents at NASA headquarters that projected 24–30 missions requiring 40–50 crew-members. These plans included:

- Ten manned Gemini flights beginning in 1964 (twenty seats)
- Four manned Apollo Saturn 1 flights starting in 1965 (twelve seats)
- Up to four manned Apollo Saturn 1B flights from 1966 (twelve seats)
- A further six Earth and lunar orbital Saturn V flights, estimated to begin in 1967, leading to the first lunar landing attempts in 1968 or 1969 (eighteen seats).

It was expected that from the first two selections of sixteen astronauts, a couple might fly a second and maybe a third mission. But both Glenn and Carpenter were on other assignments and were not really in the running for a second flight, and Slayton was still medically grounded. No-one really expected that there would be more than forty missions, but at that time, no-one really knew what the result would be, and with a need to provide prime and back-up positions, it was clear that sixteen astronauts would not be sufficient. Therefore, in 1963 the call went out for between ten and twenty new astronauts.

Group 3: 17 October 1963 (fourteen selected; known as 'The Fourteen')
Maj Edwin E. 'Buzz' Aldrin Jr, PhD, USAF, aged 33 at selection.
Capt William A. Anders, USAF, 30.
Capt Charles E. Bassett II, USAF , 31.
Lt Alan L. Bean, USN, 31.
Lt Eugene A. Cernan, USN, 29.
Lt Roger B. Chaffee, USN, 28.
Capt Michael Collins, USAF, 32.
Mr R. Walter Cunningham, 31.
Capt Donn F. Eisele. USAF, 33.
Capt Theodore C. Freeman, USAF, 33.
Lt Cmdr. Richard F. Gordon Jr, USN, 34.
Mr Russell L. 'Rusty' Schweickart, 27.
Capt David R. Scott, USAF, 31.
Capt Clifton C. Williams Jr, USMC, 31.

In order to expand the experience within the Astronaut Office, it was decided to drop the test pilot requirement, although the mandatory jet flying and education to degree standard requirements remained. Advanced degrees or scientific research could also be substituted for operational flying experience. This resulted in a group of experienced pilots with advanced academic skills and experience. In this group, Buzz Aldrin held a PhD in orbital rendezvous, while Walter Cunningham and Rusty Schweickart had performed scientific research, and Bill Anders held a degree in nuclear engineering. Although several of this group worked on the AAP and on Skylab, only Bean would fly a mission to the station.

More scientists – fewer pilots

From the start of the man-in-space programme, there had been calls from the scientific community to include more scientists or doctors, rather than just military jet pilots, in the astronaut corps. For the missions planned at the time of the first selections, the requirement for the crew was to fly the spacecraft as a team and complete the mission objectives. These were mainly in rendezvous and docking, and eventually in landing on the Moon and returning. The Gemini and early Apollo missions were no more than engineering test and development missions to achieve the landing on the Moon, and there was little room for additional scientific experiments, let alone a crew-member trained purely to operate these experiments rather than fly the spacecraft.

The hierarchy of the Astronaut Office in 1964 was based on the military order of peer rating. The Mercury astronauts (Flight A) were at the top of this pyramid, and held the most senior positions in the office. Deke Slayton was in charge of flight crew operations, and Al Shepard – also by then grounded by a medical ailment – was the Chief of the Astronaut Office. Gus Grissom headed the Gemini Branch Office, while Gordon Cooper headed the Apollo Branch Office.

The second selection (Flight B) was preparing for early assignments in the Gemini programme, and most of them were assigned under Grissom. The third branch office of operations and training was headed by one of the top members of the second group – Neil Armstrong.

The members of the third selection (Flight C) were under the direction of either Cooper or Armstrong. Those under Armstrong were to be the first in line for selection to Gemini positions, while those under Cooper were being prepared to fly some of the early Apollo Earth-orbit missions, although several were required to fill later assignments in the Gemini programme.

Assignment to a flight was determined by Slayton (assisted by Shepard), working on a set of criteria that included the candidates background in management, flying experience, and command levels in the military. It also was based on academic experiences and personalities, peer ratings, and performance in training. Slayton also included the factors of future assignment on more challenging missions and attrition from the group. It soon became clear that the 'pecking order' in the Astronaut Office was by group selection into the astronaut programme, military above civilians and test pilots over operational pilots. This was to become known as the world of 'astro-politics'.

Alongside the work on Gemini and the astronaut training was the development of the Apollo spacecraft and Saturn boosters. At Marshall, von Braun's team wanted to methodically test each element of the Saturn rockets with a live first stage, dummy upper stages and non-operational spacecraft. Only when this was successful would a live second stage be added, but still with an unmanned spacecraft. A crew would only be added once all aspects of the launcher were tested and man-rated.

When Headquarters reviewed these plans it was decided that this approach was a waste of time, budget and hardware to fly equipment that was not going to be used operationally. The elimination of several of the staged test flights and opting for 'all-up testing' also led to the cancellation of several of the manifested manned flights – and the available seats for the astronauts.

Scientist-astronauts

As the first astronauts were being trained, it was recognised that pilots were needed to crew the pioneering test flights beyond the atmosphere. But once the theory of spaceflight had been proved, it was argued that flying scientifically trained crew-members would generate greater returns from each mission. A geologist could explore the Moon and perhaps even be a member of the first landing crew, oceanographers and meteorologists could study the Earth from space, astronomers could operate orbiting telescopes, and physician-astronauts could examine the effects of human spaceflight and the adaptation to weightlessness. The military was also evaluating the usefulness of trained observers onboard orbital reconnaissance spacecraft in a range of proposed DoD programmes.

By July 1962, recommendations to NASA by its Space Science Board suggested that 'scientist-astronauts' should be included in the programme, and should even be assigned to the first landing crew. This idea was not well received by the pilot-astronauts, as the priority for the first landing was to get there and return safely.

Science could wait until subsequent landings. The political wheels started turning, however, and over the next few months there was much debate concerning the criteria that the first non-pilot astronauts should fulfil.

Those involved included Bob Voas, who had been a member of the first astronaut selection board in 1959, and who began to liaise between the scientific community and NASA management. At NASA, Robert Gilruth (Director of the MSC in Houston), Joe Shea (Chief of the Apollo Programme Office, MSC) and George Low (Deputy Director, MSC) frequently discussed the idea of bringing scientists into the programme. They gradually convinced Jim Webb (NASA Administrator) and George Mueller (Deputy Administrator) that it would be advantageous to include scientific specialists in a future astronaut selection process.

In August 1964, a selection plan and a list of suitable criteria had been drawn up by NASA's Associate Administrator for Space Science, Homer Newell, and the Head of the Space Science Board at the National Academy of Sciences, Harry Hess. During the same month, the Ad Hoc Astronomy Panel of the Orbiting Research Laboratory Study Group had suggested that while a flight-trained astronaut would be an essential member of the crew of a space laboratory, with the addition of an astronomical observatory it would be beneficial to include a qualified astronomer to direct scientific operations onboard the laboratory.

As a result, on 19 October 1964, NASA and the National Academy of Sciences issued a joint call for astronauts for a fourth group that would be significantly different from the first three groups. A candidate had to be born on or after 1 August 1930, be a citizen of the United States, and stand no taller than 6 feet (to be able to fit inside the Apollo spacecraft). This allowed younger and taller candidates to apply. The significant change was that the candidate must also hold a PhD or equivalent in natural sciences, medicine or engineering. A BS or MS degree was not enough, and flying hours were not required, as all candidates would be trained at a year-long USAF jet pilot training course after selection. However, preference would be given to those with flight experience.

The first scientist and doctor in space

Ironically, this call came just one week after the Soviets had orbited the first of what was thought to be a new family of manned spacecraft – Voskhod. Its crew of three was an Air Force Pilot-Engineer, Vladimir Komarov, an aerospace engineer, Konstantin Feoktistov, and a physician, Boris Yegorov. Neither of the latter two could fly jets. Voskhod was actually a modification of the Vostok that had carried the first solo cosmonauts into orbit, and was not a significant leap in spaceflight technology, as was at first thought. The crew landed after only one day, so their contribution to science on manned spaceflight was limited by duration, the confines and restrictions of the spacecraft, and a short training programme of only a few months. What they *had* managed to demonstrate, however, was that spaceflight was not limited to the realm of the heroic jet pilot. Scientists could also be prepared for, and survive spaceflight, and could fly alongside the pilots.

The first NASA scientist-astronauts

NASA had suggested that for the spaceflights being discussed at that time, two fields of research were beginning to predominate: geologists for the Apollo lunar landing missions, and doctors for the longer-duration missions. It was hoped that about twenty candidates suitably qualified in each of their fields could be short-listed to make the final selection to join the astronaut programme. The Academy of Sciences hinted that this was far to narrow, and suggested expanding the search to include astronomers, physicists and meteorologists. NASA agreed, and the selection process continued.

By 31 December 1964, NASA had received a total of 1,351 applications, and after a preliminary screening, on 10 February 1965 the names of 400 applicants (including four women) were submitted to the National Academy of Sciences for judgement of their academic credentials. Slayton was hoping for ten geologists and ten doctors, out of which three from each field would be selected, aiming to use no more than one from each on early Apollo crews. NASA headquarters had other ideas, and indicated that they were looking for a total of ten to fifteen for the final group. Slayton countered that he needed at least ten in each field to allow for attrition during training, which necessitated a short-list of about fifty scientists!

In reviewing the final 400 applicants, the Academy of Sciences was unable to produce a pool of suitably qualified candidates to meet this demand, and could only propose sixteen candidates. This group was then sent to NASA for medical and psychological testing and interviews. From these, only six finalists – half the number NASA had originally intended – were announced on 28 June 1965:

Dr Owen K. Garriott, PhD, an electrical engineer, aged 34 at selection.
Dr Edward G. Gibson, PhD, a physicist, aged 28.
Dr Duane E. Graveline MD, aged 34.
Lt-Cmdr Joseph P. Kerwin, MD, USN, aged 33.
Dr Frank Curtis Michel, PhD, a physicist, aged 31.
Dr Harrison H. Schmitt, PhD, a geologist, aged 29.

Kerwin and Michel had already qualified as jet pilots during their military service, and on 29 July 1965 the others were sent to Williams AFB in Arizona, to begin flight school. They would not begin astronaut training until the summer of 1966. However, in August 1965, Graveline resigned from the programme due to difficulties in his marriage. Almost immediately, the Scientific Six became known as the Incredible Five.

As the sole geologist in the group, Schmitt seemed destined for an assignment to Apollo, but it would be a long struggle to see that dream realised – as Lunar Module Pilot on Apollo 17 in 1972. Garriott, Gibson and Kerwin would fly to Skylab, while Michel would leave NASA in 1969 without ever having reached into space.

The group was called the Incredible Five so that the other astronauts could distinguish them from 'real' astronauts – the test pilots. There was apparently one other distinct difference between the scientists and the pilot-astronauts. When a scientist-astronaut was told of a new and astounding fact, he would stroke his chin with one hand, place the other behind his back, stare at his shoes, and say 'Why that's incredible, just incredible'. When the same thing was told to a pilot-astronaut, he would look you blandly in the eye, and say 'No shit?'

Apparently, the scientists soon learned of the selection method for assigning crews. The flight order seemed to be test pilots – pilots – secretaries – and then scientists!

ASTRONAUT OFFICE TECHNICAL ASSIGNMENTS

As the new group began flight training, some of the other astronauts were beginning to receive Astronaut Office (CB) technical assignments on future programmes, including early studies in extending the range of Apollo flights to include more scientifically orientated missions.

On 6 August 1964, Astronaut Chief Al Shepard informed the astronaut group of several technical assignments that would be conducted pending assignment to a flight crew. The majority of these were associated with the Gemini series and Apollo lunar missions, but Rusty Schweickart was also assigned to 'future programmes and inflight experiments' within the Apollo branch office, which pointed to the extended Apollo flights then under discussion. Walt Cunningham was also assigned to non-flight (ground) experiments in the same branch office.

Preparing hardware and experiments for any spaceflight takes a considerable amount of time, and involves many evaluations and mock-ups before a design is qualified for assignment to a flight – if at all. The technical tasks for the astronauts were to review the ideas and proposals and evaluate their suitability for flight in areas of operational use and crew safety. Hundreds of plans, proposals and ideas passed through the office, but many were never to reach design configuration, let alone assignment to a flight. This was also at a time that the Gemini and Apollo programmes were taking up so much attention that no-one in the Astronaut Office thought much about the paper studies on extended Apollo missions that were so far in the future – at least, not until after the AAP office was formed in late 1965.

With three of the Group 4 astronauts away at flight school, the remaining two took on technical assignments. On 23 September 1965, Curt Michel was assigned technical duties in experiments and future programmes within the Apollo branch, while Kerwin assumed technical assignments in pressure suits and EVA in the operations and training branch. He would also play a key role in the thermal vacuum test programme of the Apollo Command and Service Modules. Schweickart moved to Guidance and Navigation assignments – the first step in his assignment to an early Apollo crew towards the end of 1965.

Apollo 1 and Apollo 2 assignments

Slayton had been assembling the first Apollo crews during 1965, and produced a plan towards the end of the year:

Mission	*Crew (Commander/Senior Pilot/Pilot)*
SA-204/CSM-012 (Apollo 1)	Prime: Grissom–Eisele–Chaffee
	Back-up: McDivitt–Scott–Schweickart
SA-205/CSM-014 (Apollo 2)	Prime: Schirra–White–Cunningham
	Back-up: Borman–Bassett–Collins

Then, in December, just prior to the official announcement being made public, Eisele dislocated his left shoulder. He was exchanged with White from the Schirra crew, to allow the injured astronaut to heal after corrective surgery.

Subsequent changes to the manifest for Gemini, and delays to the Apollo hardware and flight programme, were reflected in the crewing for these first Apollo flights. In December 1966, Apollo 2 was cancelled as an unnecessary duplication of Apollo 1, and the Schirra crew was reassigned to support Grissom's crew on Apollo 1. The McDivitt and Borman crews moved to form the new second and third Apollo flight crews.

MOL astronaut selection

While NASA had been working on the Gemini flights and preparing the first crews for Apollo, the USAF had progressed with their Manned Orbiting Laboratory programme to the point of selecting of pilots to man the missions. All of the pilots selected for the MOL had to be serving military officers, holders of a BS degree, and graduates of the Aerospace Research Pilot School at Edwards AFB, California. There were originally planned to be five two-man MOL missions, each of thirty days, with the crews flying an adapted Gemini spacecraft. This meant that with a prime and back-up team of four astronauts, twenty seats were available, with perhaps one or two flying twice, and the back-up crews rotating to fly later in the prime positions. To accommodate this, the USAF selected seventeen pilots for the MOL between 1965 and 1967. A further selection was cancelled with the termination of the programme in 1969.

In addition to completing Gemini spacecraft training, the MOL astronauts were to train to operate medical, technical, and military, scientific and strategic experiments during the mission, from inside the MOL research laboratory 'space station'.

The first selection for the MOL occurred on 12 November 1965 (Group 1), and consisted of eight men (six USAF and two USN), of which one would later play a key role in subsequent NASA programmes:

Lt Richard H. Truly, USN, aged 28 at selection.

The next selection, on 30 June 1966 (Group 2), consisted of five candidates (three USAF, one USN and one USMC), all of whom would later transfer to the NASA astronaut programme:

Capt Karol J. Bobko, USAF, aged 28 at selection.
Capt Robert L. Crippen, USN, 29.
Capt Charles G. Fullerton, USAF, 29.
Capt Henry W. Hartsfield Jr, USAF, 32.
Capt Robert F. Overmyer, USMC, 29.

The final selection was made on 30 June 1967 (Group 3), and included just four USAF candidates, only one of whom would transfer to NASA:

Maj Donald H. Peterson, USAF, aged 33 at selection.

What is interesting about these selections is that although NASA was beginning to consider assigning scientist-astronauts (admittedly pilot trained) to future AAP missions, the USAF had selected pilots, and not military scientists, for its MOL programme. Some members of the MOL astronaut group had attained academic qualifications in chemical, electrical and nuclear engineering to supplement the aeronautical engineering requirement, but none of them were pure scientists.

The Original Nineteen

By September 1965, NASA had a total of thirty active astronauts, including the three scientists attending flight school. This was expected to be sufficient for all of Gemini and the first Apollo missions leading to the first lunar landing. However, the Apollo Applications Program Office was beginning to plan the Apollo landing missions after the first successful landing, as well as a programme of research missions in both Earth orbit and lunar orbit.

Although these plans varied constantly, the range encompassed ten lunar landings, three S-IVB wet workshops (launched by Saturn 1Bs), three S-IVB dry workshops (Saturn V), four flights of an Apollo-supported ATM, and several Earth-orbital and lunar-orbital flights. This totalled more than forty manned Apollo missions from 1968, with 120 flight seats available in probably less than a decade!

It soon became clear to Slayton that, allowing for an expected small percentage of attrition, the group of thirty astronauts he had available was totally inadequate. He surmised that even if everyone flew twice, he would still need around 20–24 new astronauts in addition to the thirty that he already had!

Phase 1 of the selection began on 10 September 1965, with a target to recruit at least fifteen astronauts (Group 5). The selection criteria were similar to those of 1963, but with the age limit raised to 36. By 1 December 1965, 351 applications had been received, which was reduced to a short list of 35 finalists. When Slayton was asked how many he needed, he replied, 'as many qualified applicants as we can get!' After further medical tests, evaluations and interviews, nineteen new astronauts were announced on 4 April 1966. In parody, because of their class size over the smaller first selection, they dubbed themselves 'the Original Nineteen':

Vance D. Brand, aged 34 at selection.
Lt John S. Bull, USN, 31.
Maj Gerald P. Carr, USMC, 33.
Capt Charles M. Duke Jr, USAF, 30.
Lt-Cmdr Ronald E. Evans Jr, USN, 32.
Maj Edward G. Givens Jr, USAF, 36.
Mr Fred W. Haise, 32.
Maj James B. Irwin, USAF, 36.
Dr Don L. Lind, 35.

Capt Jack R. Lousma, USMC, 30.
Lt Thomas K. Mattingly II, USN, 30.
Lt Bruce McCandless II, USN, 28
Lt-Cmdr Edgar D. Mitchell, USN, 35.
Maj William R. Pogue, USAF, 36.
Capt Stuart A. Roosa, USAF, 32.
John L. Swigert Jr, 34.
Lt-Cmdr Paul J. Weitz USN, 33.
Capt Alfred M. Worden, USAF, 34.

This class would be combined with the members of Group 4 for academic and survival training, and featured the final members of the crews that would fly American space missions up to 1975, using Apollo-type spacecraft. This included four of the nine who would fly on Skylab (Carr, Lousma, Pogue and Weitz).

Another member of this group played a key role in Skylab, and was equally qualified as a scientist, though he was selected as a pilot. Former USN pilot Don Lind had actually tried twice before to enter the programme. He held a PhD in atmospheric physics, and was employed as a scientist at NASA's Goddard Spaceflight Center. He had applied in 1963, only to be told that he did not have enough flying hours, and that his PhD was not a substitute. In 1965 he was as qualified as Michel and Kerwin in military flying experience, but was 74 days too old. Lind argued that when the other members of the group completed training in a year's time, the 74 days would not mean much, and that as a fully qualified pilot and PhD, he could save NASA time and money in training him. NASA was not impressed, and he was not accepted. In 1966, at his third attempt, he made the cut – which he surmised was to stop him turning up again at the astronaut selection office!

The AAP Branch Office

By August 1965, the development of a series of extended Apollo missions had progressed sufficiently for an Apollo Applications Office to be established at MSC in Houston. It became responsible for crew activities within the programme, and it was not long before the first astronauts were assigned to it.

On 3 February 1966, a CB memo from Al Shepard announced the creation of a new branch office – the Advanced Programs Office – within the CB structure, with former Mercury astronaut Scott Carpenter as the Branch Chief. Carpenter had recently returned from a temporary assignment to the USN Sealab Man-in-the-Sea habitation project (during which he spent thirty days living and working underwater in Sealab II in August 1965). The office also included Kerwin, assigned to pressure suits and EVA development, and Michel on experiments. The memo went on to state that Schmitt, Garriott and Gibson would be joining the branch in September, after their return from jet pilot school. The Advanced Program Office was associated with the CB Apollo Program Office.

Later that month, on 28 February, See and Bassett were killed when their T-38 crashed. They had been in training for Gemini 9, and their loss affected Gemini crewing assignments for the remainder of the programme. The original Gemini 9 back-up crew (Stafford and Cernan) was reassigned to the vacant prime crew, and Slayton brought in Bean and Williams as their new back-ups. Bill Anders joined Neil Armstrong as back-up to Gemini 10. At the completion of those missions, Bean and Williams would be assigned to Apollo Applications assignments.

Early astronaut input from the Advanced Program Office into AAP development was evident in a memo dated 6 May. Kerwin and Slayton expressed CB concerns over the lack of experiment planning and hardware operational safety in the S-IVB wet workshop configurations under consideration. By early August, after completing assignments on Gemini 10, Bean was assigned as the first Chief of the AAP Branch in the Astronaut Office.

On 3 October 1966, Shepard outlined further CB technical assignments in the AAP Branch. Bean headed the group, and assigned under him were Anders, Engle, Lousma, Pogue and Weitz. The Chief of the Experiments Branch was Garriott, and under him were Gibson, Lind, McCandless, Michel and Schmitt. By 5 December, Kerwin and Givens had also joined the Experiments Branch under Garriott. These two offices exchanged members several times until 4 April 1967, when both were merged into the AAP Branch Office, consisting of twelve astronauts under the leadership of Al Bean.

On 20 October 1966, a memo attached to a Mission Requirements document for Saturn Apollo Applications 209 (SAA-209) included a description of the proposed mission. This flight consisted of the (first) S-IVB as a live (wet stage), an airlock, and CSM 105, on a planned 14-day mission that could be extended up to 28 days. In the

During a visit with Bill Pogue to Marshall in February 1967, astronaut Joe Engle examines the area from where he might have to retrieve ATM film canisters. Even without a pressure suit it is a tight squeeze between the LM ascent stage and the ATM rack.

attached memos for this flight, Al Bean listed the first day's time-line, and an EVA to attach an electrical umbilical from the AM to the SM. In an AAP-209 (S-IVB) presentation, technical assignments for this proposed mission were listed as ' Bean–Anders–Lousma', and though not a crew in the official sense of the term, it was the first time that a three-person team had been assigned to technical assignments on a specified AAP mission.

As it was, Anders moved to Borman's Apollo crew in December 1966. (No replacement was found in the search of the documents.) Five years later, after completing their Apollo assignments, Bean and Lousma were reunited in the second Skylab crew, and were joined by Garriott.

After the cancellation of Apollo 2, new crews were required to begin training for the first six Apollo missions, leading to the first landing. Slayton assigned eighteen astronauts to fly and back-up Apollo 1 (204), 2 (205) and 3 (503). The back-up crews for Apollos 1–3 were expected to rotate to the prime positions for Apollos 4–6, although not in their entirety. Some members could expect to be reassigned to other Apollo crews or across to AAP. Flying the Apollo spacecraft required a far more complex training and preparation programme than had been used on Gemini. Because of this, Slayton evolved a third (or 'support') crew for each Apollo mission. They would represent the flight crews at meetings, and help out in tests and simulations whenever the prime crew commander needed them.

The first three support crews, assigned in December 1966, all came from the new fifth astronaut group. Givens was reassigned from the AAP Branch to join the Apollo 1 support team and was replaced by Aldrin, recently back from Gemini 12.

Fire in the spacecraft

On 27 January 1967, the Apollo 1 crew was conducting a pre-flight countdown simulation inside their spacecraft on the launch pad at the Cape, when a flash fire engulfed the capsule. The three astronauts died in a few seconds. The accident was a tragic blow to the programme at a time when the Moon seemed to be within reach. The resulting inquiry, and recovery from the disaster, lasted for eighteen months, and the first Apollo crew (Schirra–Eisele–Cunningham) finally flew in October 1968, as Apollo 7.

From then on, Apollo went from success to success. Apollo 8 achieved the historic first manned lunar orbital mission in December 1968, and in March 1969, Apollo 9 flew the first manned LM in Earth orbit. In May, Apollo 10 tested the LM to within just 9 miles of the Moon, and paved the way for the Apollo 11 landing mission to achieve the goal set by President Kennedy eight years earlier. Four months later, Apollo 12 repeated the feat.

Had the fire not occurred, Apollo would probably not have landed on the Moon any sooner. There were difficulties in man-rating the Saturn V, and the LM was progressing very slowly and would not have been ready to fly with a crew before early 1969.

Working in the shadow of Apollo

While most of the world watched, heard or read about Apollo, those assigned to the AAP were almost forgotten as they continued what is actually the most time consuming part of any astronaut's career – preparing flight hardware and training for spaceflight.

On 31 January 1967 – four days after the Apollo 1 fire – Chuck Matthews suggested that the Group 4 scientist-astronauts should be given priority to visit a leading astronomical centre for training in solar observation, since they had been working on ATM design issues since 1965. As tragic and devastating as the recent fire had been, the general feeling in NASA was that the lost crew would have wished the whole programme to continue.

Between 1967 and 1969, as AAP evolved, so did the complexity of the training required to undertake the missions. With several crews assigned to train for Apollo lunar missions, there was already a demand on simulator time, and with AAP missions planned for both Earth orbit and out to the Moon, there was a need for better facilities.

The facilities and training hardware were spread across the America, and the necessary travel took precious time from the already limited training allocations. It was not long before the astronauts began to voice concern about the time it took to reach the training facilities, and a desire to centrally locate as much of the training equipment as possible in MSC or KSC. In a CB memo to all astronauts, dated 20 March 1967, astronaut Bill Pogue, a member of the AAP office, highlighted the problem: 'Crew training for the first two AAP missions is going to be difficult to support with the facilities now available, and it takes no imagination to anticipate greater difficulties during training for later missions. Scheduling problems have already been identified in planning to share the same trainers between main line Apollo and AAP. Even in the main line Apollo alone there appears to be some problem in providing adequate crew training support.'

Pogue was not criticising the outside facilities, but was pointing out that relocating the major training facilities in the main astronaut centres of Houston and Florida would reduce the time to reach them and return, and therefore increase the actual amount of available training time.

In an attached list of 'suggested areas of action', Pogue cited six focal points for further evaluation, related to both the lunar and Earth-orbital aspects of the AAP. One of these was a larger water immersion (neutral buoyancy) facility (now called a WETF), as the only suitable tank was at MSFC and was too small to accommodate many AAP work areas. One of the other points raised was the suggestion to create a single office or section ('directorate') to assume responsibility for keeping track of training programmes and their evolution, to support impending missions. Prior to 1967, each project had its own training office as required, but now there was a need to train several crews concurrently for different missions, in different programmes, and at different stages. It therefore seemed sensible to have a single, dedicated office group to overview the progress, direction and further requirements of the training programme, so that there were no unnecessary gaps or impediments.

This desire from the recent Group 5 intake of astronauts to develop their training skills quickly and efficiently had evolved from their experiences during their first year at NASA, when they received many classroom presentations from engineers who were not skilled at teaching. The course material they used was difficult to work with or interpret, and many of the astronauts found it difficult to follow both the presentation and what the 'instructor' was actually trying to say. It was difficult enough to negotiate the vast array of new techniques required to become an astronaut, and these inefficient courses also lead to frustration within the group. This led to several of the astronauts giving their own peers extra presentations in subjects that they knew well. For example, Kerwin presented informal 'bull sessions' on space medicine, Lind and Michel offered similar classes on space physics, and Mitchell handled guidance and navigation. The engineer-instructors tried their best, but they were not educators. It become so bad during one presentation that the astronauts stopped the class and telephoned the instructor's company to request a different instructor – and their request was granted.

As the programme progressed, the presentations improved, with more qualified instructors, several specialists in their fields, and potential Principle Investigators (PI) for experiments expected to be flown. One of the leading specialists who instructed the astronauts during this period was oceanographer Robert Stevenson who, little more than over fifteen years later, almost made a flight as a Payload Specialist on the Space Shuttle.

Technical assignments

During May 1967, Deke Slayton and a group of twelve astronauts assigned to the AAP completed a walk through the S-IVB workshop to evaluate OWS stowage and simulated EVA exercises associated with activating the station. This was just one of many visits to AAP contractors, experimenters and NASA field centres as part of their role in developing the flight hardware, systems and procedures. The astronauts also attended countless meetings, and undertook individual technical assignments (for example, Pogue handled instrument panel layout).

The areas assigned to the astronauts in the AAP office for the May review were:

Bean	No specific area listed, but he was then Chief of the AAP Branch.
Engle	IVA equipment (umbilicals, donning, doffing, and so on).
Garriott	Communications.
Gibson	Crew quarters layout and controls.
Kerwin	Food, waste and IVA.
Lousma	Activation, deactivation.
McCandless	Experiments AAP 1 and 2.
Michel	Hand holds, tethers, foot rails.
Pogue	Lighting and photography.
Weitz	Experiments AAP 3 and 4.
Cernan	None listed – probably EVA support.
Aldrin	None listed – probably rendezvous/docking and EVA support.
Slayton	Flight Crew Operations Directorate management representative.

Both Cernan and Aldrin were working dual CB assignments. Cernan had been assigned to the Stafford Apollo crew the previous December, and Aldrin had been assigned to the crew in training for the 2TV-1 thermal vacuum tests of the CM, before joining a new Apollo crew of Armstrong and Lovell in March 1967. In June, Ed Givens, in training for the support crew of the first manned Apollo (Schirra) was killed in an automobile accident, and was replaced by Bill Pogue from the AAP group. Pogue was not replaced in the AAP group, but Joe Engle replaced Aldrin in the 2TV-1 tests, and also continued to work in the AAP.

More scientist-astronauts

The second phase of the expansion of the astronaut team to support later Apollo and AAP missions had continued on 26 September 1966. A joint NASA and National Academy of Sciences news release announced that a 'limited number' of scientist-astronaut candidates would be considered for selection the next year.

This time the criteria were expanded to attract a wider field of applicants. They had to have been born on or after 1 August 1930, stand no taller than 6 feet (to fit inside the Apollo spacecraft) and hold a PhD in natural science, medicine or engineering. Interestingly, the criteria also stated that qualified applicants who were not born in the US, but became citizens as of 15 March 1967, could also apply. This meant that naturalised citizens from outside the US could, for the first time, attempt to become NASA astronauts.

By 8 January 1967, 923 applications had been received by the National Academy of Sciences, which was expecting a total of 20–30 to be selected by NASA. Slayton wanted about 10–20 at most, as by then he had more than enough astronauts. After a two-month evaluation, 69 names were forwarded to NASA for final selection.

During the summer of 1967 the effects of cuts on the NASA budget, the increased expenditure of the Vietnam War, and domestic and social problems across the US, led to a decline in the political support and public interest of space operations. This would entail far fewer missions in Apollo or the AAP than first envisaged just a year or two earlier.

There were already discussions to limit the production of the Saturn 1B and Saturn V, which would restrict its use, and the first missions to go would be on the AAP to protect the main-line Apollo lunar programme. Delays had reduced AAP missions from at least twelve to no more than six, and the requirement for astronauts to crew them decreased dramatically. In light of this, Slayton suggested that only five new scientist-astronauts would be required, but since the selection had started, and because of the results of the Apollo 1 investigation, the final details of the budget cuts and the longer-term effects had yet to take hold, and the selection continued on an optimistic note. On 11 August 1967, NASA announced the second group of eleven scientist-astronauts:

Dr Joseph P. Allen IV, PhD, aged 30 at selection.
Dr Philip K. Chapman, PhD, 32.
Dr Anthony W. England, PhD, 25.
Dr Karl G. Henize, PhD, 40.
Dr Donald L. Holmquest, MD, 28.
Dr William B. Lenoir, PhD, 28.

Dr John A. Llewellyn, PhD, 34.
Dr Franklin S. Musgrave, MD, 31.
Dr Brian T. O'Leary, PhD, 27.
Dr Robert A.R. Parker, PhD, 30.
Dr William E. Thornton, MD, 38.

The group included two physicists (Allen and Chapman), a geologist (England), two astronomers (Henize and Parker), three medical doctors (Thornton, Musgrave and Holmquest), a chemist (Llewellyn) and an electrical engineer (Lenoir). It was a broad scope of experience. The youngest astronauts selected by NASA (England and O'Leary) had only recently received their doctorates, while the oldest ever selected (Henize and Thornton) were over the upper age limit of 35.

Both Henize and Thornton had written to Slayton early in the selection process to determine whether their age would present a problem. They were told that the age limit would be waived for exceptionally qualified applicants, and both men filled that category. Chapman was a naturalised citizen from Australia, and Llewellyn was from Wales – the first two men chosen for astronaut training who were not natural American citizens. All were extremely qualified, and standing out academically above all of them was Musgrave, with no less than five advanced degrees. As qualified as they all were, not one of them could fly anything, and they would all receive USAF pilot training to fly jets.

On their very first day in the Astronaut Office, Slayton informed them that the programme had no room for them, and that if any of them wanted to fly, they would wait a long time to be assigned a seat. Slayton was presenting them with a chance to leave, but none of them did so immediately.

The group knew that landing on the Moon on the mainstream Apollo was out of the question, but maybe the AAP would see them gain their ticket to space. They were then to be called 'module pilots', and trained on operating experiments and performing observations from various missions in the AAP. Some had no desire to fly to the Moon to pick up rocks, but wished to perform investigations in their professional disciplines in space; and that is where some of them would soon become disappointed.

The first requirement was to qualify as a jet pilot and to undergo USAF undergraduate pilot training. Unfortunately, due to the demands of the Vietnam War on USAF combat pilots, this training could not begin until the spring of 1968, so for the first few months the group completed an academic training programme before reporting to flight school.

The fact that the eleven scientists were surplus to requirements was reflected in their chosen group nickname – the Excess Eleven, or XS-11. The flight training took twelve months, with most of the group graduating in March 1969. By then, Llewellyn and O'Leary had dropped out of the programme, having had difficulty in learning to fly jets.

For all of them it was a very long wait for a spaceflight – at least fifteen years – and for Chapman and Holmquest it was too long to endure, as they too left the programme in the early 1970s. Beginning in 1982, each of the remaining members made a spaceflight, but on the Space Shuttle rather than on an Apollo-type vehicle. Musgrave went on to complete six flights before retiring from NASA in 1997.

PROGRAMME CHANGES, 1968–1970

As the new group began training, events in the Apollo mainstream programme began to affect the astronauts assigned to the AAP.

Late in October 1967, Bean was reassigned to the Conrad Apollo crew following the loss of Williams in a flying accident. Many in the astronaut office had thought his assignment to the AAP, after filling a dead-end assignment on the GT10 back-up crew, denoted Slayton's lack of faith in his abilities. There were, however, only so many seats available on Gemini and the early Apollo missions, and a competent figure was needed to run the AAP. Slayton later wrote that when he assigned Bean to Gemini 10 he actually became the first of his group (1963) to be given a command role. As back-up, if anything happened to the prime Commander John Young, Slayton felt confident that Bean could step in and handle the mission.

In the autumn of 1966, Slayton needed an astronaut whom he knew could pull the fledging programme into shape – which is what Bean achieved in the year he was in charge. As his career record shows, Bean was not hindered by the assignment. He went on to become the fourth man on the Moon, and later commanded his own mission – to Skylab!

Ed Gibson performs 'human factors engineering tasks' on an ATM mock-up in the 7-foot diameter by 40 feet deep water tank at Marshall in May 1969. Here Gibson prepares to move a film canister along a guide rail from a mock-up LM ascent stage to the ATM Sun-end structure.

When Bean joined Conrad's crew, Slayton needed a new Chief of AAP, and he turned to fellow Mercury astronaut Gordon Cooper, who had stepped in to fill the dead-end back-up Gemini 12 assignment. Stu Roosa replaced Cooper as Chief of the Booster Branch of the CB in November 1967, and by the end of that year the AAP office included Cooper (Chief), Engle (also working 2TV-1), Lousma, McCandless, Weitz, Garriott, Gibson, Kerwin (also working on the 2TV-1 crew and the Lunar Receiving Laboratory issues), Michel, Lind and Schmitt (who would both soon work on Apollo Lunar Surface Experiment Package (ALSEP) development).

By the beginning of 1968, solar physics training was beginning to become an essential requirement for crew preparation as the ATM programme became more complex. Gibson became the CB point of contact for the ATM Committee, and Garriott, the CB liaison for the group between MSC and MSFC, was also assigned as CB representative for the Astronomy Mission Board Committee at Headquarters. No astronomer-astronaut was available, as those who were qualified were from the 1967 selection and at the time were undergoing pilot training with the USAF.

Science or space?

As the effects of the budget cuts spread through NASA in the late 1960s, reports were surfacing in the media that a few of the scientist-astronauts were not happy in having to sacrifice their research time for the exhausting amount of effort required for astronaut training. Apparently they were disgruntled over assurances (presumably from the National Academy of Sciences) that if selected for the programme, NASA would allow them a third of their time to continue their research. In practice, this emerged as no more than 10–15%, and at times was probably much less.

After being told that there would not be a seat on a flight for many years, having the planned expansion at MSC (to incorporate new state-of-the-art scientific research laboratories) put on hold pending the outcome of the reduced budget did nothing to settle the disappointment. As a result, the Academy of Sciences tried to arrange leave of absence for those who wished to pursue research. Deke Slayton explained: 'It's up to each of [them] to develop their own [scientific] program. They initiate it, and if the facilities are available here at MSC, they'll use those facilities. If they're not, they'll go where those facilities are.'

It soon became clear which astronauts were not happy with the situation, and they soon started to leave the programme. Of those who stayed, Schmitt explained that the chance of landing on the Moon was surely, 'the biggest experiment of the century, probably in history', and was too good an opportunity to miss – which he did not.

Other scientist-astronauts explained that they would not change places with anyone in a laboratory, and if they were on a university campus, private research would be carried out in their own time opportunistically – as was the case at NASA.

Towards Skylab

In November 1968, Apollo 7 crew-member Walt Cunningham was assigned by Slayton to take the lead of the AAP Branch Office. Never a favoured position in the office, an assignment from Apollo to AAP was seen as a backward step in the world of astro-politics and future crew assignments.

The Apollo 7 mission had been declared a 101% success in qualifying the CSM for manned flight, but the crew had suffered from colds on board, and voiced objections to additional assignments put upon them. The objections came mainly from the commander, Walter Schirra, but as first-time fliers, both Eisele and Cunningham dutifully supported their commander – and suffered the consequences in their own careers. Neither of them would again fly in space, but in November 1968, Cunningham assumed that a good job on the AAP was a way back into space.

When Cunningham took over the AAP, Garriott had been heading up the office ever since Cooper's assignment to Apollo training earlier in the year (leading to a position as back-up Commander on Apollo 10). Cunningham had been told that he would have the Commander's seat on the first 28-day mission, and this appeared to be more certain than it had been for Bean, Cooper, or anyone else. The AAP missions were not to fly until after the first Apollo landings, and those were planned for less than a year away.

A thankless role

To become Chief of the AAP Branch of CB was not as prestigious as it should have been in the late 1960s. It entailed working under the shadow of Apollo, in a programme that had no guarantee of reaching flight, but with the certainty of a shrinking budget. Although Bean looked to command the first AAP, mission, his hard work in organising the AAP group for Slayton resulted in his securing a coveted place on an early Apollo landing mission.

When Cooper took over, he brought experience that had its origins in the first days of 'the astronaut office' in 1959, but he too looked towards the Moon. In fact, he often expressed a desire to be the first man on Mars; but the AAP was not going to get him there, and so his heart was never really in the AAP assignment.

In 1968, Garriott replaced Cooper, and had already been working on the programme for over two years as the senior scientist-astronaut. He was a leading contender for one of the science crewman seats on the programme. Unfortunately, his tenure in the AAP office occurred during a time of increased focus on with Apollo after the fire, and achieving the first lunar landing. Minimal support for anything other than Apollo did not help him to push the programme forward, and until Apollo had completed the landing, that is the way that it seemed destined to stay.

By this time, most of the rookie pilot and scientist-astronauts had worked on the AAP at one time or another, while the most senior and flight-experienced astronauts focused on the Apollo landings. It seemed to be a case of letting the rookies fix the problems in the AAP before any of the more senior astronauts would become involved with the programme.

Into the spotlight

In April 1969, Rusty Schweickart transferred to the AAP shortly after his first spaceflight. Slayton would have recycled him to a landing flight as LMP had he not experienced a bout of space sickness during Apollo 9, which surprised and worried many people. At that time, the illness that later became known as Space Adaptation

Syndrome was virtually unknown. It was usually the result of quick movements shortly after entering space as the body settled down to the lack of gravity. In the early small spacecraft this was not a common problem, as movement was restricted. Though cosmonaut Titov first experienced it in 1961, most of the few bouts of sickness had been glossed over. During Apollo 9, the extra room afforded by the CM and LM combination allowed the astronauts to roll, twist and turn as they passed between the two spacecraft through the tunnel, and the extra room for movement could bring on the uncomfortable sensations that were experienced by Schweickart. If sickness occurred during a landing on the Moon it could be disastrous, and so as a precaution, Slayton moved Schweickart to the AAP.

In 1969 there were more than thirty astronauts assigned to Apollo crew training, in prime, back-up, or support roles through to Apollo 12. With others soon to be selected for Apollo 13 to 15, Slayton was short of available astronauts, and reassigned Gibson to a temporary assignment on the support crew of Apollo 12.

During most of the first half of 1969, Gibson was also working (with Kerwin and Weitz) on AAP EVA simulations in the water tank at Marshall, on a full-scale replica of the ATM and OWS, determining procedures for fitting and replacing ATM film canisters. The scientist astronauts from the 1967 selection also returned to Houston, swelling the ranks by nine. O'Leary and Llewellyn had left in 1968, and Michel, from the first scientist group, was in the process of departing that summer.

The nine new astronauts received their first technical assignments from Slayton in April 1969. Allen, Chapman, England, Henize and Parker were assigned to the support crews for Apollo missions from 1970, and this took them out of the running for AAP assignments. Slayton assigned Holmquest, Musgrave, Lenoir and Thornton to work on AAP development issues, and transferred Group 5 astronaut Lind there as it became clear that Apollo 20 would be the first landing to be cancelled from the programme. Lind had widely expected to become the second scientist on the Moon (Schmitt was the leading contender to become the first, around Apollo 18 or 19 if the missions survived that long), and for any astronaut who hoped to fly to the Moon, the realisation that this was not to be was a huge disappointment.

MOL transfers

By the autumn of 1969, the closure of the Saturn production lines, the budget limitations, the cancellation of Apollo 20, and the lack of definite plans for anything with a crew on board beyond 1974, all made the prospects for just one spaceflight very bleak. For the nearly forty astronauts available for flight assignments, there remained the seven (soon to be five) Apollo landing missions left to fly, and the three AAP flights that were planned. Ten missions with three seats (thirty places) meant that at least ten active astronauts would not make a first or return flight into space before the mid-1970s.

The last requirement of the Astronaut Office in August 1969 was more astronauts, but that is what it got with the cancellation of MOL. Of the seventeen astronauts selected, fourteen were available to NASA. Slayton looked at the flight manifest, and at the astronauts that he still had queuing, and said that he did not want them. But Mueller realised that if the Space Shuttle was ever going to be authorised, the USAF was an important ally for military applications for the space-plane, and it would be a sensible move to accept the former MOL trained astronauts. Most of them were interested in such a move, but Slayton limited the group to those under 36 years of age (NASA's upper limit for acceptance to astronaut training) and seven made that cut. The other six were eligible for operational flying duties in the USAF.

On 14 August 1969, the 'Seven' were named as:

Maj Karol J. Bobko, USAF
Lt-Comdr Robert L Crippen USN
Maj Charles G. Fullerton, USAF
Maj Henry W. Hartsfield Jr, USAF
Maj Robert F. Overmyer, USMC
Maj Donald H. Peterson, USAF
Lt-Comdr Richard H. Truly, USN

Having already trained on Gemini and spacecraft systems under MOL, there was no formal training programme for the group, and they were immediately given CB technical assignments. Crippen, Overmyer and Truly went to AAP engineering development assignments, while the other four supported imminent Apollo missions.

The end of Apollo

While Cunningham was in charge of the AAP, the programme emerged from the shadow of Apollo in 1969 and evolved into Skylab, America's next manned space programme, in 1970. At the same time, three Apollo flights were axed, and some of the veteran astronauts were preparing to leave the programme or seek out their last chance of a trip into space.

Cunningham was hoping that his two years' work on Skylab would pay off with the command of the first mission; but when Pete Conrad began to express interest in the programme, Cunningham realised that he might lose the first flight to Conrad, hoping to at least take the second.

By August 1970, Conrad had been named the new Chief of the Skylab Astronaut Office, which had by then grown to include the crews from the cancelled Apollo missions. Potential Commanders of the missions now included Conrad, Cunningham, Bean and Schweickart. Tom Stafford was not interested in Skylab, unless he was assigned the first command, which had gone to Conrad. He did not want to go back to the Moon, as on Apollo 10 he had approached it to within 9 miles, and that was close enough for him. He was now considering retirement from the USAF and NASA, and attempting to run for the Senate. Stafford had showed no enthusiasm for circling the world for a month, and instead, late in 1969, took over Shepard's job as Chief Astronaut while the Mercury astronaut trained for Apollo 14.

Pilot candidates included Weitz and Lousma (both of whom had worked on the AAP for years, and had been in line for assignment to the cancelled Apollo 20), Carr and Pogue (from the cancelled Apollo 19) McCandless and Lind. The scientist candidates were Garriott, Gibson, Kerwin, Musgrave, Lenoir, Thornton and Holmquest, with some of the former MOL astronauts as support crew.

With all the cuts, the Space Science Board asked NASA for assurances that each

Skylab crew would include a pilot commander and *two* scientist-astronauts; but Slayton soon suppressed that idea. Despite the heavier science objectives, Skylab was still a pioneering engineering challenge for extended spaceflight, and he wanted the first crew to include two experienced astronauts who were proven at troubleshooting the problems that would undoubtedly occur. It was also important that there were two fully qualified pilot crew-members on board, as it was felt that there would be a considerable amount of technical and engineering duties that would need to be accomplished to keep the station running. The Skylab crew training programme at Houston was initially orientated towards systems management and malfunction procedures, mainly because the scientific hardware was not ready, but also because this is what needed to be learned first.

From the astronaut's point of view, one of the mysteries of the Astronaut Office was how crews were selected. In the early 1970s, some of the scientist-astronauts approached NASA HQ, and asked how they could work to ensure being assigned to crewing if they did not know what was required of them. Deke Slayton responded typically to this request, at one Astronaut Office meeting: 'Well, Al Shepard and I sit down over a beer and shoot the breeze, and that's how we select the crews!' That did nothing to settle the question, so if 'Slayton and Shepard rolled dice or threw chicken bones on the floor' to arrive at the selection of who flew with whom still remained a mystery. Slayton had openly stated that to him if was far easier to teach pilots to pick up rocks on the Moon than to teach scientists how to fly jets.

The debate concerning science-trained crew that had surfaced in 1968, resurfaced in 1969, and continued into 1971. In October 1969, the active scientist-astronaut group had complained to NASA Headquarters that, even after Apollo 11 had achieved the first landing, all the prime and back-up crew positions were still filled solely by pilot-astronauts. Slayton's reply stressed the hazards of the Apollo missions, and pointed out that a dead geologist was no use to anyone! Acceptance of the importance of science on Apollo by the pilot-astronauts was still an uphill struggle, but with Skylab the opportunities broadened. It would be less hazardous, and the scientist-astronauts therefore endeavoured to put their members on board.

In 1971 the main complaint was that as long as a test pilot (Slayton) picked the crews and a test pilot (Shepard) was in charge of the Astronaut Office, then pilots would get priority over scientists. Even the pilot-qualified astronauts – who were not selected in the 1965 or 1967 scientist-astronaut groups, but had a keen interest or background in science rather than flying – had experienced disadvantages in flight assignments because of that interest.

Homer Newell, the top ranking scientist at NASA, recommended that at least one geologist (Schmitt) should be assigned to a lunar crew, and that the method by which crews were selected for flights should be re-examined for Skylab, to place two scientists on each mission.

Slayton and Gilruth stressed the expected amount of time that system problems would take up, and pointed out that since each one of them would in any case be cross-trained, specific academic experience was not relevant. It would be very difficult to change a training approach that was linked to systems management without seriously impacting upon the launch schedule.

Undeterred, the scientists countered that perhaps two scientists could fly on the third and last mission; but when the three Soyuz 11 cosmonauts were killed during recovery in June, the argument for flying pilot-trained astronauts strengthened.

On 6 July 1971, a recommendation from Dale Myers, Associate Administrator for Manned Space Flight, agreed that priority should be given to operational considerations, and so two pilot-astronauts and one scientist-astronaut would fly on each Skylab mission, adopting Houston's original proposal. It was also confirmed that a physician should be on the first crew, but that exactly who would make up the crews should be left to Houston.

The assignments

The plans for Skylab envisaged at least three manned missions, and possibly a fourth – and in theory, maybe a fifth. This entailed fifteen flight positions for nine secured seats, with the back-up crews for flights 1 and 2 probably flying missions 4 and 5. Their back-ups would come from those flying the first three missions, and so there were no dead-end assignments as a back-up with no flight. The support crew could be filled from the most recent selection – the MOL transfers.

By early 1971 Slayton issued the assignments during a Monday morning pilots' meeting of the CB:

Crew	*Prime*	*Back-up*
1	Conrad–Kerwin–Weitz	Cunningham–Musgrave–McCandless
2	Bean–Garriott–Lousma	Schweickart-Lenoir-Lind
3	Carr–Gibson–Pogue	Schweickart–Lenoir–Lind
4	Cunningham–Musgrave–McCandless	From the crews of 1, 2 and 3
5	Schweickart–Lenoir–Lind	From the crews of 1, 2 and 3

In 1999, Kerwin stated that Slayton named the group in a pilots' meeting in 1970, and that they spent some time in trying to determine who would be on which crew – until they realised that the sequence was first-crew, then back-up, second crew, then back-up, third crew, and soon. Garriott also stated in 1999 that both Slayton and Shepard had worked on the assignments after five years of observations, determining who would form the best pairings, as crew bonding was very important. Kerwin added that the astronauts had been working as 'crews' for five years before they flew – which meant in 1968. Presumably this meant the AAP group and when Cunningham came into the branch in November 1968. It also seems clear that Kerwin had always been intended for the science seat on the first mission

Not everyone in the CB received this proposal favourably. Cunningham had expected to lose the first mission, and maybe the second, but not the third – and to a crew of rookies. With no guarantee that there would be a fourth flight, Cunningham left NASA in August 1971. Conrad recommended Carr as Commander of the third crew because of his sterling work on the Apollo 12 support crew.

The assignments also saw no place for Holmquest who, despite working on the AAP since 1969, attempting to secure a medical assignment on a flight crew, had even been omitted from the back-up crews. On 21 May 1971 he took a leave of absence for a year to pursue teaching and medical research at Baylor University

School of Medicine in Houston. Since August 1970 he had worked part time on his medical studies, and when he returned to the office in 1972 he opted for a second leave of absence before finally resigning from NASA in September 1973.

With Cunningham gone, a new back-up to the first crew was needed, and so Slayton moved Schweickart up and looked to Dick Gordon or Vance Brand coming off Apollo 15 (and having lost Apollo 18) to fill the position as back-up to the second and third missions. Gordon had already indicated his intention to retire after '15', so Brand became the back-up to the second and third missions.

Don Lind was to be disappointed a second time, and could never understand why he was not on the third crew as the science pilot. The scheduling indicated the three main areas of equipment and research field development scheduled for Skylab. In priority order of the science objectives, the first mission was aimed at medical experiments (because of the first American opportunity to extend flights beyond two weeks), and Kerwin was the most logical and senior person to handle them. The second mission was aimed at ATM studies of the Sun, and Garriott's background pointed him towards that assignment. The Earth resources package on the third mission was a late addition, but Lind had been involved with this for some time, and thought that this would lead to his being assigned to the third crew.

Gibson, who was given that assignment, had studied atmospheric physics, as had Lind, and was also becoming a specialist in solar physics, having worked on the ATM design for five years. This dual experience (and perhaps his seniority over Lind, having been selected in Group 4 rather than Group 5) was reflected in the assignment. Lind still hoped to possibly fly on the fifth Skylab visit, or perhaps be assigned to the rescue crew with Brand.

When all of this was being planned, an alternative plan was to have the second crew (Bean) back up the first, and the third crew (Carr) back up the second, and have the three pilots (Schweickart, Lind and Brand) back up the third and be available for the rescue missions as required. The time required for specific mission training, and the fact that the training plan was still under development, made this abbreviated system useless, reverting to Slayton's original plan of three prime crews and the two back-up crews who might be assigned to the fourth and fifth missions.

The reason why two, and not three, back-up crews were assigned was offered in an explanation by Tom Stafford in 2001. Although not totally clear thirty years later, Stafford remembered the need of assigning a physician (Musgrave) to back up Kerwin, and the importance of that first mission to activate the station, and so was formed a specific first mission back-up crew with an experienced Commander. The other two crews were working on the development of the whole training programme, and building on each other's experience and preparations for the last two flights. It was reasoned that the back-up crew for the second mission could easily play the same role for the third mission. There was no discussion of a single back-up crew for all three missions, and due to availability of unflown Group 5 (1966) astronauts, two back-up crews worked better than three.

The naming of the Skylab crew

On 16 January 1972, the official crew assignments for Skylab were announced:

Crew	*Flight*	*Duration*	*Prime*	*Back-up*
1	SL-2	28 days	Conrad–Kerwin–Weitz	Schweickart–Musgrave–McCandless
2	SL-3	56 days	Bean–Garriott–Lousma	Brand–Lenoir–Lind
3	SL-4	56 days	Carr–Gibson–Pogue	Brand–Lenoir–Lind

There was no mention of a fourth or fifth mission, but the option remained open, depending upon the real-time situation. Brand and Lind would train as rescue crew-members for all three missions, and the support crew would consist of Thornton and Henize from Group 6, and Crippen and Hartsfield from Group 7. They would also fill the roles of CapCom for the three missions, along with Truly and members of the back-up crews – Schweickart, Musgrave, McCandless and Lenoir. On 19 February 1973, the final astronaut assignment to Skylab was announced, with Bob Parker becoming the programme scientist for all three manned missions, to ensure that the science requirements during the flights were compatible with the programme requirements. During the mission he would respond to the programme director and the programme managers, to ensure that in-flight science requirements were being met by the flight crews.

SKYLAB TRAINING

When the astronauts were brought together under the Skylab programme in late 1970, there was no single office from which they all worked. The Skylab astronaut 'office' was, in fact, several smaller separate offices within the CB, located on the whole of the third floor of Building 4 at MSC. There were no dedicated offices, however, and some of the group tried to organise the desks into the same smaller rooms.

The programme had an official name, and the designs of the components were advanced to the point of being constructed. There were lists of experiments to be performed, and countless documents of the planned objectives and expanded achievements from the missions. But the astronauts had nothing on which to train.

Apart from physically training for the longest manned missions that the Americans had ever attempted, the actual training programme had to be developed at the same time as the hardware and experiments for the mission. This was not simply a case of assigning a crew who then jumped into simulators to determine how to operate experiments or live on the station. All of this was also under development at the same time.

From late 1970 there was very limited access to simulators. Indeed, the only item that was available was the CM simulator that was still being used by the crews of Apollos 15–17. Conrad's crew was due to be the first to fly to Skylab, and they worked as pathfinders for the training as they evaluated their own mission. Carr and Pogue handled the development of the training protocol, and Bean and Lousma worked with the other four.

A review of the training for Skylab is presented below, but it might be asked how the crews determined which aspect of the training was most important. The crews

Pete Conrad evaluates a check-list during training in the MDA at the material processing facility in the Skylab trainer at JSC.

looked upon the training in two phases. Firstly, there were the systems and operations that were required to make Skylab work, and to allow them to live on board. Secondly, there were the science investigations with their own operating procedures.

For training, the 'pilots' concentrated on the systems and procedures, and the 'scientists' concentrated on the experiment and research fields. The beginning of the operational phase was the Apollo CM, the OWS and its facilities, how they worked and what was needed to make them work. Training for this evolved into a list of steps – an operation checklist – which formed the basic training document that would evolve as the programme developed. It allowed the astronauts to become familiar with the basic workings of the whole system.

With the CSM simulators being mainly occupied by Apollo into 1972, the Skylab crews only managed to obtain occasional sessions in them when they were able. Conrad's crew took priority, as they were first to fly, and Bean used the simulator at the Cape. Once the first crew had launched, the CM simulator became more available to the third crew, but since Pogue had specialised and worked on the CSM since joining NASA in 1966, this was never a major issue for that crew.

The simulator used to train the flight controllers in mission control highlighted the lack of operational training hardware. They had the longest amount of time to

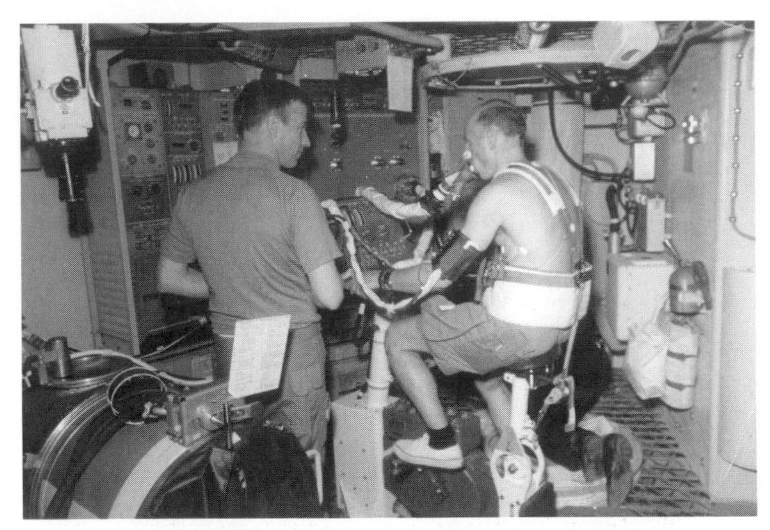

Pete Conrad rides the bicycle ergometer in the Skylab trainer. Paul Weitz monitors his commanders progress as the machine determines Conrad's metabolic effectiveness.

prepare for all aspects of the programme, and Gene Kranz therefore organised a very basic 'simulator' linked to consoles. It was operated by an engineer flipping switches to make lights come on. Skylab's flight controllers needed this mock-up, as the real Mission Control was still occupied with controlling and training for Apollo missions. Sometimes this 'crude but effective' system became available to the Skylab astronauts, who tried to obtain 'sim-time' as often as possible.

The training mock-ups of the OWS, MDA, and AM evolved as the training evolved, and often required updates to the workbook developed to operate the system. The crews quickly became aware of how the flight equipment could soon become different from the training hardware.

On the experimental side, very little was done on crew training before 1971. The scientists frequently went off for a day or two to different parts of the country to work with PIs and vendors, ensuring that the experiment hardware was compatible with the flight equipment and that the operating procedures were actually workable.

An example of this was the hours of effort which Karl Henize devoted to working the stellar experiments into the programme, spending a lot of time liaising with the astronomers in designing the hardware and research programmes. As a professional astronomer, he could understand the objectives, and the requirements of the PIs, and as an astronaut he knew what could or could not be done by the crew. He then talked to the crew, showing them how the hardware worked and why they should carry out the operation in a certain way. Henize also stressed what they were trying to determine looking from the experiment or hardware. The fact that he was providing his own experiment helped him with his explanations for the crews.

Throughout the two years of training, the mock-ups and trainers constantly changed, and it was a challenge to keep up with this to ensure that what they trained on was the latest version that had been selected to fly. At first they trained in groups

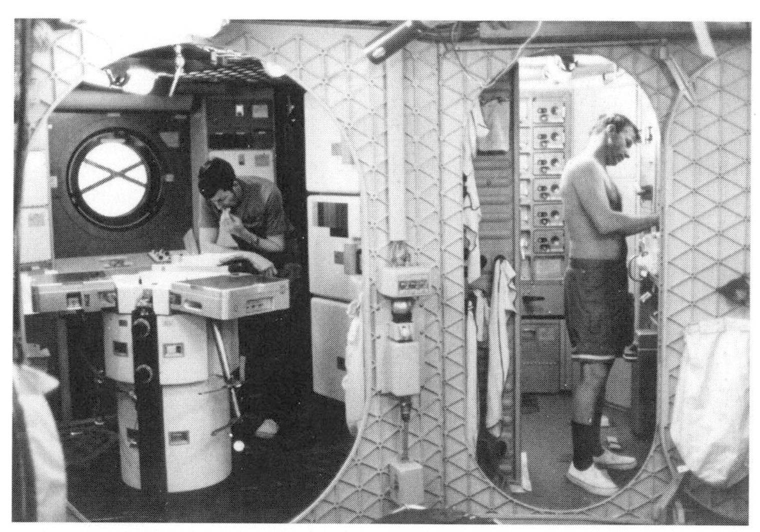

Kerwin takes a snack while he studies the flight plan in the wardroom, while Weitz prepares for personal hygiene in the waste management compartment in the OWS training mock-up at JSC.

In the locker room. Kerwin (*left*) and Weitz assist each other in suiting up in Building 5 at JSC for pre-launch training activities.

– either pilots or scientists, or as prime, back-up, or the whole group. Only later did they form into the three 'crews' to define their own flight procedures.

The missions were originally to last up to 56 days, and a ground simulation of a 56-day mission would help identify and rectify many operational and procedural difficulties. The flight crews were kept up to date by regular daily reports and occasional visits to the simulator where the test was being held.

However, there were no extended simulation runs for any of the three-flight crews. Instead, they operated mini-sims that began in the mock-ups at 06.00 hours (wake-up time) and lasted until lunch or 22.00 hours (bed-time). In these they practised certain procedures or hardware operations, inserting habitability training into a 'working day'. These were problem-free simulations to practise the procedures and become familiar with their operation. Only later, in the integrated simulators with mission control and simulation engineers, were the crews presented with malfunctions and emergency situations.

This was a great deal of hands-on training, rather than book study late into the evening at home. There were scores of lectures, classroom sessions and study on the scientific activities, which were handled by PIs in Houston. This was then translated to the mock-ups in the simulators, to put the theory into practice. Any 'dead time' in the simulations was skipped, to compress the time allocated to training and not leave the crew waiting around for the next step.

A bench review was first conducted prior to simulating the event on training devices. Sometimes the crew worked on flight hardware in white rooms, and never

Pete Conrad operates the T027 photometer experiment in the OWS trainer at JSC. The grid floor was developed as a launch-weight reduction measure, and to assist in air-flow through the station.

saw any training hardware. With them were a PI, and their training officer for that item, procedure, or experiment. Each crew had a main training officer – a Training Co-ordinator – and beneath him was a team dedicated to different hardware or experiments, which liaised between the crew and the PI or vendor, and who also developed a training procedure list.

Initially, scientists were very sceptical whether the astronauts (mainly pilots) could be competent enough to meet experiment objectives, or to understand what they were trying to do. However, they were pleasantly surprised at the success of the flights, although, Bill Pogue could never quite understand what some of the medical experiments were trying to achieve, and Jerry Carr thought that he became a better student of solar physics when he stopped actually trying to be one and simply became a competent observer. The desire to deliver the best performance could sometimes pressure the crews into trying to learn more than they could handle.

This was where the support crews came in to help deal with hundreds of minor developments, procedures and problems that needed tracking and tasking as the complexity of the training evolved.

Conrad organised a weekly group meeting to discuss progress in certain areas and then at the Monday morning CB office meeting each crew would report their progress – or lack of progress – in that area. A factor which later arose was the debriefing from the first crew used as guidance for the second and third crews. The crew debriefing was held with Slayton, Shepard and Kleinknecht, in a locked room. There was a prepared briefing sheet to follow, listing a sequence of important items

The three SL-2 crew-members prepare a meal in the wardroom trainer at JSC during an 11-day simulation of selected mission days of their 28 day mission. (*Left–right*) Kerwin, Weitz and Conrad.

to discuss so that everyone had something to say and nothing was forgotten. Each crew-member was responsible (prime) for a series of systems or hardware, backed up by the other two and trained to that effect. These were the a speciality areas that they worked on for their flight.

Once in flight, the crews could ask questions of the previous crew – basically on where things were located. They were usually told that something had been moved – or they could not remember! Inventory control was a major issue, as everything in training had a place and was put back; but in flight this was not always the case.

At KSC, only the CSM simulator was available, but training for emergency pad evacuation and crew fit and function were accomplished. EVA training followed a similar procedure, with a bench review and an unsuited run using the mock-ups and a platform (or 'cherry picker') to reach the EVA areas. The suited exercises were then completed both in 1 g and in a water tank (at MSFC) to evaluate full procedures.

Review of the training

Success in any manned mission in space is dependent upon the quality of the hardware, the skills of ground support, and the ability of the flight crew to perform assigned functions in both planned and contingency situations. Since the beginning of the space programme, flight crew training has been an integral part of all missions, but with Skylab's extended missions, duration added a new complexity to the training of both the flight crew and the mission control teams. As well as performing the ascent, rendezvous and docking, EVA, and recovery tasks common to many spaceflights, Skylab astronauts now had far more scientific and housekeeping duties to perform.

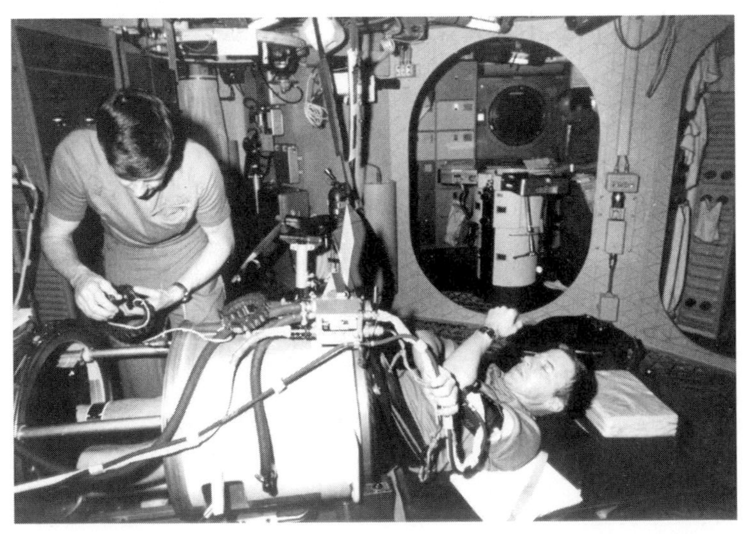

Weitz is inside the lower body negative pressure device in the Skylab trainer at the MSC, Houston, while Kerwin checks the settings of the experiment that places stress on the heart and blood vessels by slight suction on the lower part of the body.

The nine primary and six back-up astronauts underwent identical training, allowing for the replacement of either the complete crew or any individual with minimum delay, should the need arise.

The astronauts each completed approximately 2,150 hours training by the time of launch (see the Table on p. 140). This is equivalent to completing a four-year college degree course in just three years, and was almost double that averaged on the Apollo lunar missions. In addition, there was also an unrecorded amount of time that included personal study, physical exercise, informal briefings, aircraft proficiency training, as well as crew and astronaut office meetings and briefings.

Formal training began with a series of background briefings in November 1970. This provided general information and orientation in spacecraft systems and experiments. The group also participated in spacecraft testing and reviews of flight plans and procedures, and began training in solar physics.

In January 1972, specialised training began in experiment operation, and the simulation of specific elements of the flight, as well as EVA and IVA activities. The following month, integrated crew training began in which either the astronaut crews or the teams of flight controllers used simulators and trainers to gain experience in orbital and flight operations. Then, in November 1972, integrated mission training commenced, training each flight control team alongside the astronauts to gain experience in working together. During this phase, mission simulators were linked with MCC, simulating the conditions of the actual missions.

In an effort to reduce the crew-members' overall workload, specialisation was still an essential element of Skylab crew training. The Commander had overall responsibility for the CSM launch, rendezvous and docking, EVAs, de-orbit and entry, and the success and safety of the whole mission. The Science Pilot was responsible for the ATM, EVA, and medical experiments, and the Pilot was the specialist in the AM, MDA, OWS, and Earth resources experiments. However, each received sufficient cross-training to ensure the successful completion of critical and mission-important tasks or experiments.

The primary elements of Skylab training were:

- Background training (scientific and academic).
- Systems, experiments, operations briefings and reviews.
- Operation procedures training (simulators and mock-ups).

Each crew completed 450 hours of briefings and reviews, with CSM briefing taking up 95 hours, in which the emphasis was on major subsystems and operations. All the astronauts had completed Apollo CSM training as part of their general astronaut training upon entering the programme, but such was the importance of the vehicle that their skills needed to be maintained, even though they would only spend a short time in the CSM during the flight.

The crews completed a programme of six-hour system briefings, concentrating on system anomalies and late modifications. Orbital assembly briefings accounted for another twelve hours, during which the crews gained a working knowledge of the OWS, AM, MDA, and ATM. Briefings on the Saturn 1B lasted eight hours, and

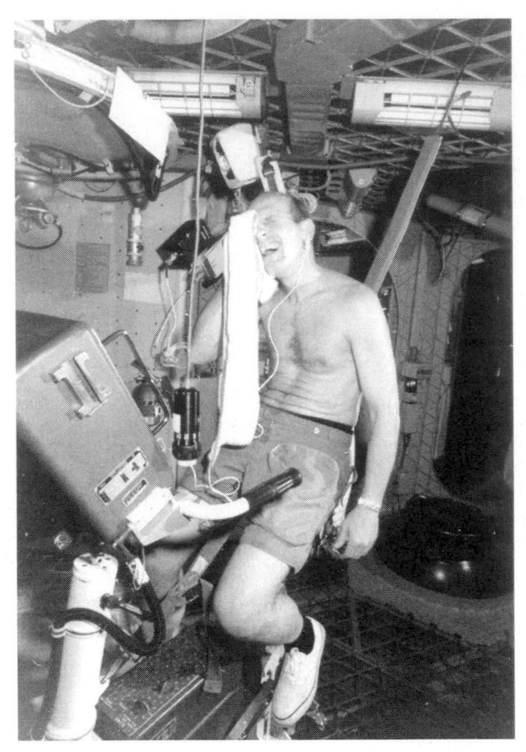

No gain without pain. Conrad takes a break from riding the bicycle ergometer in the Skylab trainer. The device is used for understanding the metabolic effectiveness when measured workloads are placed on the muscles in microgravity. Note the data-recording device clipped to Conrad's ear-lobe.

there were 110 hours of solar physics, 75 hours of flight plans and checklists, and a further 50 hours on mission rules and techniques.

Each crew completed 350 hours in systems tests and selected spacecraft, and experiments tests were conducted to confirm the operational acceptability of crew stations and crew equipment (150 hours). CSM tests were repeated by the three crews (and the back-ups), whereas other component tests were completed just once, although the second and third crew completed 50 hours of spacecraft tests.

Spacecraft testing was undertaken at the primary contractor plants of North American Rockwell, (Downey, California – CSM); Martin Marietta Aerospace (Denver – MDA); and McDonnell Douglas (St Louis – MDA/AM; and California – OWS, MSFC ATM). The crews also periodically inspected stowed equipment onboard the CSM and OWS, and spent some 68 hours minutely checking flight equipment for function and operational suitability.

The crew also spent a total of 40 hours on fire, decompression, and end-of-mission water recovery training. A further 96 hours was allocated to other significant system training, such as briefings on crew systems (food, waste, EMU, hygiene – 46 hours); TV and photography (30 hours); and activation and deactivation of OSS

Skylab 3 crew members (*left–right*) Al Bean, Owen Garriott and Jack Lousma participate in simulated spacecraft check-outs in the Command Module. Note the stowage under the seats.

using mock-up trainers (20 hours). EVA training was completed on EVA/IVA, and took up a further 156 hours.

1-g training varied by crew (108 hours for the first crew, 127 for the second, and 119 for the third), as did the underwater in the Neutral Buoyancy Simulator (NBS) training (48 hours for the first crew, 57 hours for the second, and 42 hours for the third). Full-scale mock-ups of spacecraft modules (trainers) were used for experiment training, procedural and timeline development, stowage exercise, 1-g procedures and EVA and IVA flight tasks.

The NBS at MSFC was used for 48 hours of zero-g training to prepare the crews for performing assigned EVA, including the installation and removal of film magazines, and recovery of samples from outside the vehicle. The 40-foot deep, 75-foot diameter NBS at MSFC was also used, with wire mesh mock-ups of SL modules and full-size replicas of all four major elements of the cluster (ATM, workshop, MDA and airlock) submerged in tank. The crews used weighted suits to remain suspended in a relatively stable position. Vacuum IVA transfer (using contingency activation of the MDA or OWS) and use of the Astronaut Life Support Assembly (ALSA) was also practised. The NBS was also invaluable in developing procedures for solar sail and Sun-shield deployment. Several other astronauts (Gibson in particular) had spent countless hours in the tank developing EVA techniques over many years during the development of Skylab. The 30-second stints of zero-g from the parabolic flights of the KC135 were used in training for eating, drinking, manoeuvring, tumble and spin recovery (esential, given the volume of the OWS) and EVAs to exchange film magazines.

The third crew during training at the Mission Training and Simulation Facility at JSC, Houston, during February 1973. (*Left–right*) Pogue, Gibson and Carr. Carr sits at the simulator representing the control and displays for the AMT which would be located in the MDA on Skylab.

The crews spent 98 hours on medical training, receiving practical training in diagnosing illnesses at an outpatient level. They trained to use the thermometer, stethoscope, sphygmomanometer and opthalmoscope, and to use the microbiology, haematology and urine analysis aspects of the In-flight Medical Support System (IMSS). They also used the incubator for growing microbial cultures. Further medical training included the use of the dental and tachometry kits that would be onboard the station, to enable them to cope with crew illness or accidents, minor injury, and dental care and chemotherapy. They also completed stressed first-aid resuscitation and supportive measures training in case of major illness or injury. The training was sufficient to support the patient until transport to definitive medical care, and included the use and care of the operational Bioinstrumentation System.

Because of the duration of the missions there was a possibility that a tooth would need to be pulled or that minor surgery would be required. To train for this, dentistry work was performed at an Air Force hospital; and one General even rearranged his schedule so that he could say his teeth had been worked on by astronauts! To train for the surgery, the group spent Friday and Saturday night in a leading Accident and Emergency Hospital in Houston, where they conducted simple sewing-up procedures and minor operations, and also went out with the emergency services.

The full mission simulators, part-task simulators, experiment task simulators and various engineering development simulators, used for rehearsal and practice by all crews, took up to 695 hours of each crew's training – about one-third of total crew training.

The CSM Simulator (300 hours) gave the astronauts comprehensive training in all segments of CSM operations, including countdown simulations, operations, launch

Ed Gibson makes notes in a training manual while working at the ATM simulator inside the 1-g trainer for the MDA at JSC in September 1973. The long hours of astronaut training were necessary for the crew to know for certain which switch should be in what position – and when!

orbital insertion, assembly pointing, de-orbit and entry. The simulator could perform independently of, or integrated with, the Skylab simulator (SLS) and MCC.

The Skylab Simulator (300 hours) was an integrated SWS procedures trainer consisting of a functional ATM C&D console, STS C&D panel, oxygen and nitrogen control panel, aft compartment panel, lock compartment control panel, and OWS electrical display and C&W control panel. The astronauts used this simulator to train on operations in the MDA/AM, OWS and ATM systems, plus activation, deactivation and orbital operations. Operational training for ATM experiments was conducted in the SLS. Its highly specific nature necessitated a training syllabus separate from other SLS activities, and was completed on the SLS ATM control panel.

The CM Procedures Simulator (CMPS) (80 hours) was used to develop crew proficiency in rendezvous and entry procedures (both nominal and contingency). Rendezvous training began with CSM orbital insertion, and ended with docking at the OWS.

The Dynamic Crew Procedures Simulator (DCPS) (150 hours) gave the crews additional practise in launch and launch abort familiarisation, normal launch time and launch vehicle failure recovery modes. Follow-on sessions expanded on this to develop crew proficiency in the various abort situations and launch vehicle contingency mode operations.

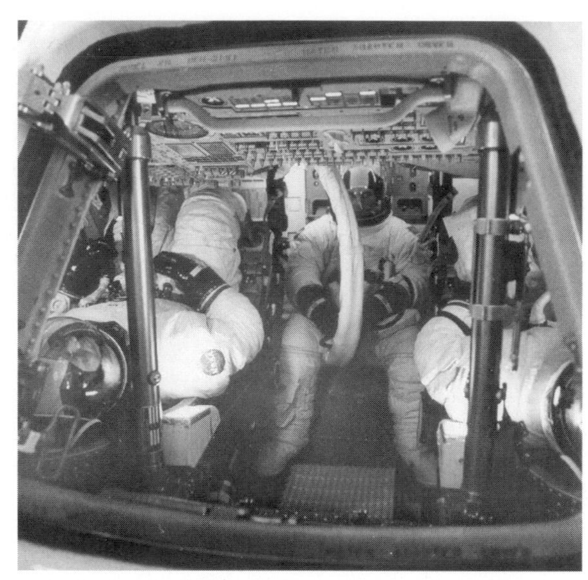

The Skylab 4 crew during altitude chamber test-run training at KSC: Ed Gibson (*centre*), Jerry Carr (*left*) and Bill Pogue (*right*). Although centre couch had been removed, there is still very little space inside the CM.

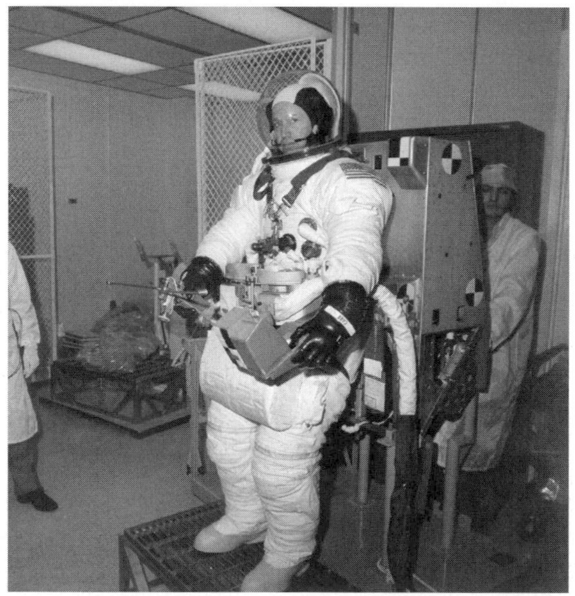

Jerry Carr during a 1972 M509 suited operation of the manoeuvring backpack.

Besides all of this, approximately 430 hours were spent on experiments, consisting of 160 hours of briefings and 270 hours of operational training, primarily involving mock-ups and special simulators. The SL-2 and SL-3 crews spent 134 hours on medical experiments, while the SL-4 crew spent only 98 hours. Training was carried out in mock-up flight-type gear, or in specialised simulators, such as the Six Degrees of Freedom Simulator (M509), the Air Breathing Simulator (M509 and T020), and the EREP simulator (S190–S194).

Finally, rescue training varied from eight to 24 hours (SL-2 used eight hours, SL-3 had used sixteen hours, and SL-4 had used 24 hours), involving briefings, reviews, mock-ups, and simulators as part of a nominal mission if rescue was initiated. This was in addition to training for the rescue crew given prior to the flights.

In preparing for the rescue mission profile, Brand and Lind conducted sea trials in a CM in the Gulf of Mexico. None of the prime crew participated in this training, but one test was for a sea worthiness of a CM with five men in the water rather than three. As the CM was buoyant either upright (Stable 1) or apex down (Stable II) before the self-righting balloons inflated, the tests would confirm post-landing activities for five astronauts in the cramped CM. On the rescue mission, Brand occupied the left and Lind the right seat, with both covering the centre seat instrument panel's activities (where Lenoir would normally have sat). During the return home the Commander of the rescued crew occupied the centre couch, and his two colleagues' couches beneath Brand and Lind.

Brand and Lind had trained for this, and required a simulated 'rescue crew' to practise the five-person post-landing choreography. Two support crew-members took the lower seats, while Lenoir filled the centre couch position. Prior to the test, Lind mentioned that because they would become disorientated in six-foot waves, and flip over several times, they would not then know whether they were right side up or upside down, being strapped tight in the seat. When each of them unbuckled they should grip something to steady themselves from falling until they became aware which way 'up' or 'down' really was. The group looked at him as much as to say that they did not need reminding of such an obvious task.

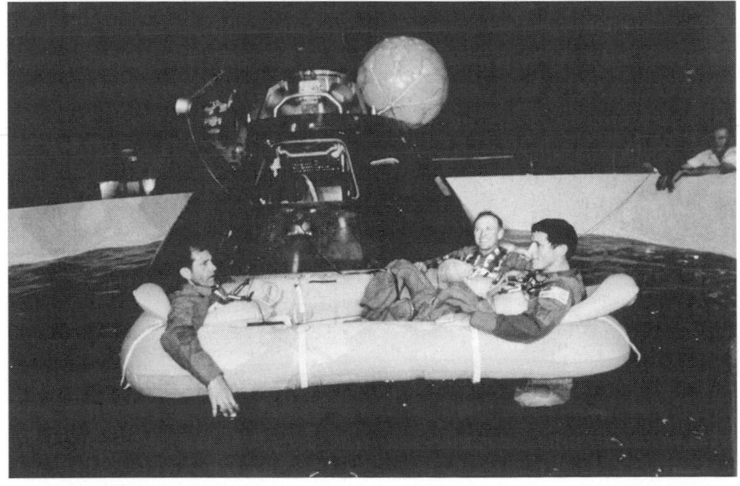

In Building 30 at JSC, Skylab 4 astronauts undergo CM egress training in a water tank.

During the tests, as the CM was again flipped over, the command to crewman 5 (Lenoir) to be the first to unstrap was given. As he did so he fell what he thought was upwards, but was in reality down, into the apex of the Apollo, 3–4 feet way, and slammed into the structure. He looked up at Lind, with that 'don't say a word or I'll hit you' look. Lind really tried not to break into a smile! Of course, the rest of the crew suggested that maybe hanging on was a good idea after all.

The total training for SL-2 amounted to 2,187 hours, for SL-3 it was 2,154 hours, and for SL-4 it was 2,059 hours.

Skylab crew training

Activity	*Hours per crew*
Briefings and reviews	
CSM	95
OWS	112
Launch vehicle	8
Solar physics	110
Flight plan and checklist	75
Mission techniques and rules	50
Systems training	
Crew systems	46
TV and photography	
SWS activate/deactivate	20
Stowage and bench checks	68
Fire and egress	40
Spacecraft test	150 (SL-2); 50 (SL-3 and SL-4)
EVA/IVA	
1-g	108 (SL-2); 127 (SL-3); 119 (SL-4)
NBS	48 57 42
Medical	98
Simulators	
CMS	300
SLS	300
CMPS	80
DCPS	15
Experiments	
Medical	134 (SL-2); 134 (SL-3); 98 (SL-4)
EREP	72
ATM	46
Corollary	178 209 165
Rescue	8 (SL-2); 16 (SL-3); 24 (SL-4)
Totals	2,187 (SL-2); 2,154 (SL-3); 2,059 (SL-4)

SMEAT: THE FORGOTTEN SKYLAB MISSION

While the crews prepared for the three manned missions, a plan emerged to simulate a full-length Skylab mission. On 18 August 1970, a briefing for a ground-based simulation of Skylab medical experiments was held between Dale Myers, the Associate Administrator for Manned Spaceflight, and Robert Gilruth, from MSC. The extent and depth of planned investigations in the biomedical area, and the engineering aspects of such a simulation, led to two actions after this meeting: a costing of such an effort against normal ground-based medical studies; and a study of exactly how the test facility would be designed. The meeting also suggested that a dedicated facility could be designed for use beyond the Skylab missions, as a microbiological investigation facility.

After a review in September, the preliminary test concept proposed that the cost would be cheaper at approximately $8.7 million (1970), against $19.3 million for the much higher fidelity combined Skylab medical experiments and microbiological preventive study. It was also suggested that minimal modifications to the existing chambers at MSC would be preferable to a completely new design for a facility as outlined by Myers. It was argued that although the current configuration was fitted out for Skylab, it was highly flexible with regard to a range of gas mixtures, temperatures, pressures and humidity. Internal equipment and structures could also be readily reconfigured.

There were still many unknowns concerning the health of space crewmen with closed ecological systems in 1970, and it was considered to be better to eliminate as many variables as possible by avoiding new, complicated designs that might never be adopted. It was also much more practical to perform tests in as close an approximation as possible to the design of spacecraft that was to fly.

Many questions would remain unanswered until after Skylab flew, and it was felt that by first addressing the issues on Skylab, and then adjusting the facilities in light of the results from the missions would be a more practical approach, especially given the current budget restrictions and uncertainties as to what would follow Skylab.

A test plan

The objectives of a ground test in of Skylab included:

- Evaluation of flight experiment procedures.
- Functional evaluation of experiments and ancillary hardware.
- Provision of baseline medical experiment data.
- Verification of the analysis, interpretation and reporting plan between a 28-day and a 56-day mission.

A review of potential sites to conduct the test resulted in the selection of the MSC Crew System Division's 20-foot diameter altitude chamber, located in Building 7 at MSC, Houston. The test was originally to include a 28-day test in September 1971, and a 56-day test beginning in November 1971. Both were to feature the operation of the following experiments:

M092	In-flight lower body negative pressure
M093	Vectorcardiogram
M171	Metabolic activity
M071	Mineral balance
M073	Bioassay of body fluids
M112	Man's immunity – *in vitro* aspects
M113	Blood volume and red cell life span
M114	Red blood cell metabolism
M115	Special haematological effects
M074	Specimen mass measurement
M172	Body mass measurement
M131	Vestibule functions
M151	Time and motion

Six test subjects were to be selected – three for the 28-day test and three for the 56-day test. They were to be selected from a pool of candidates (not necessary astronauts), after which they would be subjected to a 21-day pre-test phase, including a controlled environment and the provision of only flight food.

Inside the chamber, conditions would as far as possible duplicate the atmosphere and configuration of the OWS, with sleeping bunks located in an adjacent inner lock. Experiments and crew procedures were to be identical to flight activities on Skylab. In-flight food would be passed in through an airlock, having been stored externally, but would be prepared by crew-members using similar facilities to those on Skylab. The experiment data was to be displayed on the Experiment Support System panel within the chamber, as well as being displayed and recorded externally. The flight waste management system was also to be evaluated in the tests, to maintain conditions as close as possible to those on the flight, and in order to maintain experiment protocol.

At the end of the 28-day run, the crew would be removed from the chamber and would be subjected to a 30-day post-flight test evaluation period. Baseline data evaluated during this period would affect the flight test equipment that would be prepared for the 56-day run. During the second crew's pre-test control phase, recommendations from the first run would be incorporated into their preparations.

The chamber

The chamber had already been man-rated during earlier tests. The main chamber was 20 feet in diameter and 20 feet high, with two floors, and was constructed of stainless steel. It had fifteen viewports and seven penetration bulkheads around its circumference, with two ten-foot diameter, nine-foot long airlocks that also contained viewports and penetration bulkheads. There was a closed circuit TV system and a closed-loop system to cool or heat the atmosphere within. The chamber was designed for used in a 100% oxygen atmosphere, with pressure ranging from 14.7 to 5.0 psia. The vacuum pumping facility had a capacity of 15,000 cubic feet per minute by automatic chamber pressure regulation, with a back-up emergency system that could repressurise from 5.0 to 14.7 psia in six seconds. There were also channels for communications, lighting (with back-up emergency supplies), an automatic oxygen supply system (with emergency back-ups), and a fire detection and water deluge system, plus computerised data recording instrumentation in the chamber, and an adjacent test control room.

Modifying the chamber for the OWS

The illustrations on p. 144 and 145 show the chamber configured for SMEAT. Only one level was configured for OWS simulation, as opposed to the two levels on Skylab itself. This was because the in-flight storage was not required for the test, although the M170 experiment was on the upper deck with additional workstations. Sleep stations were located in the inner lock, with the two inner doors removed, as these were not required. The main chamber and inner lock totalled 380 square feet, which was identical to Skylab's first floor level. The wardroom and waste management compartments were partitioned, as they would be on Skylab, but the rest of the chamber had to be modified to fit in and accomplish the Skylab medical experiments. A small cylindrical transfer lock was used to transfer food and other items into and out of the chamber.

Simulated lighting, caution and warning devices, communications (through a CapCom) and additional TV cameras were also included. The habitability equipment featured as much as possible of what would be on Skylab, including freeze dried, wet pack, and dry pack food, clothing hygienic waste management, recreational activities (board games, cards, and so on) and off-duty activities. Personal hygiene facilities, bedding and clothing were also provided, as close as possible to flight standards. There was nothing to be recycled.

A review of chamber layout, experiments, and crew activity would precede the test, along with a period of crew training running as close as possible to Skylab training. The timescale of the test was dictated by the supply of experiment flight hardware for the test, which was expected to extend preparations beyond 1971 into 1972.

On 2 December 1970, the study was presented at the Office of Manned Spaceflight in Washington DC. It proposed that only a single 56-day test should be attempted prior to SL-1. Any test less than that would only be a result of unplanned events. The design phase continued through to January 1971, culminating in a presentation to the OMSF on 1 February 1971 detailing the elements and objectives of the test.

There were also plans to conduct a post-SL-4 56-day test, using this experiment as a pathfinder for future programmes, to test and implement changes from Skylab and to assist in procuring future programmes, although this did not materialise.

The MSC Skylab programme office was assigned as the field centre responsible for the test on 8 December 1970, and by 4 January 1971, Chris Kraft had complied with instructions to conduct SMEAT under the management of the Medical Research and Operation Directorate, with an MSC steering committee included for management direction.

Preparing for SMEAT

Between 17 and 18 November 1970, representatives from NASA, industry and universities participated in a symposium for a 90-day chamber test. During the

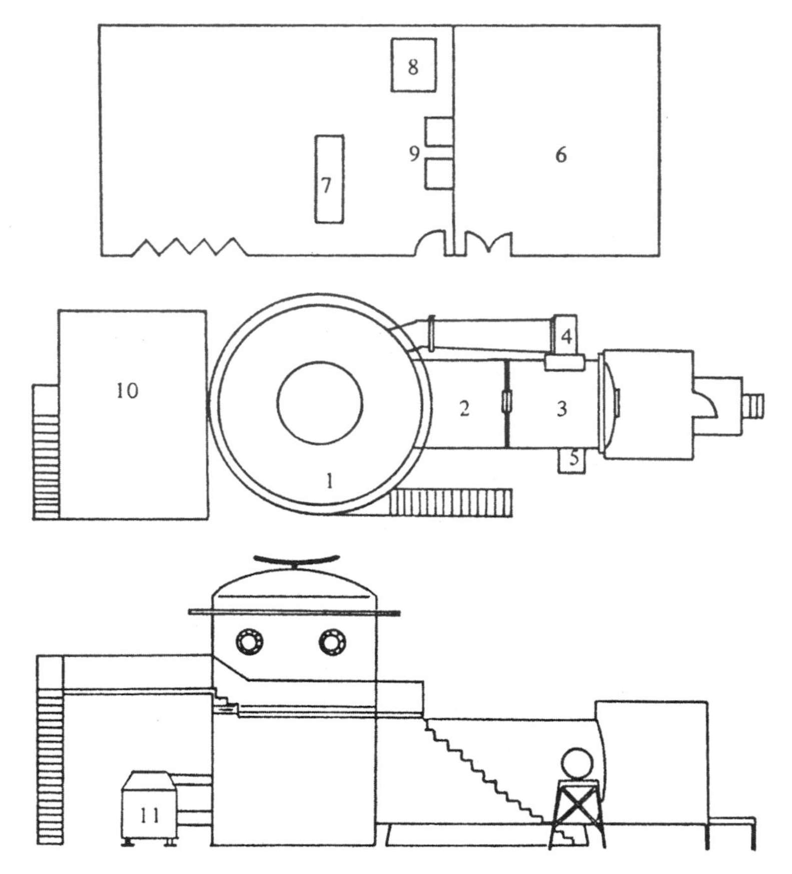

The 20-foot chamber. 1) Main chamber; 2) inner lock; 3) outer lock; 4) decompression chamber; 5) two-foot thermal vacuum chamber; 6) control room; 7) roots Connersville pumps; 8) Beach Russ vacuum pump; 9) Nash vacuum pumps; 10) equipment platform; 11) atmosphere distribution system.

session, papers and discussions focused on water management, atmosphere purification, atmosphere contamination, atmosphere supply, waste management, food management, crew selection and training, habitability and behavioural studies, acoustics and lighting, and medical and physiological aspects. This was an ideal forum for identifying what was needed to support such a test, but it also highlighted the limitations on any hardware or experiments intended for evaluation.

A formal presentation of the proposal for a 'Skylab' medical experiments altitude chamber study, prepared by a combined team from HQ, MSC and Ames Research Center, was received at NASA HQ on 2 December 1970. Following this presentation, MSC was authorised to proceed with planning the design of a 56-day pre-flight chamber programme, to be conducted prior to SL-1 and SL-2, and to be known as the Skylab Medical Experiment Altitude Test (SMEAT).

For the next year, work continued on the configuration of the chamber. On 17

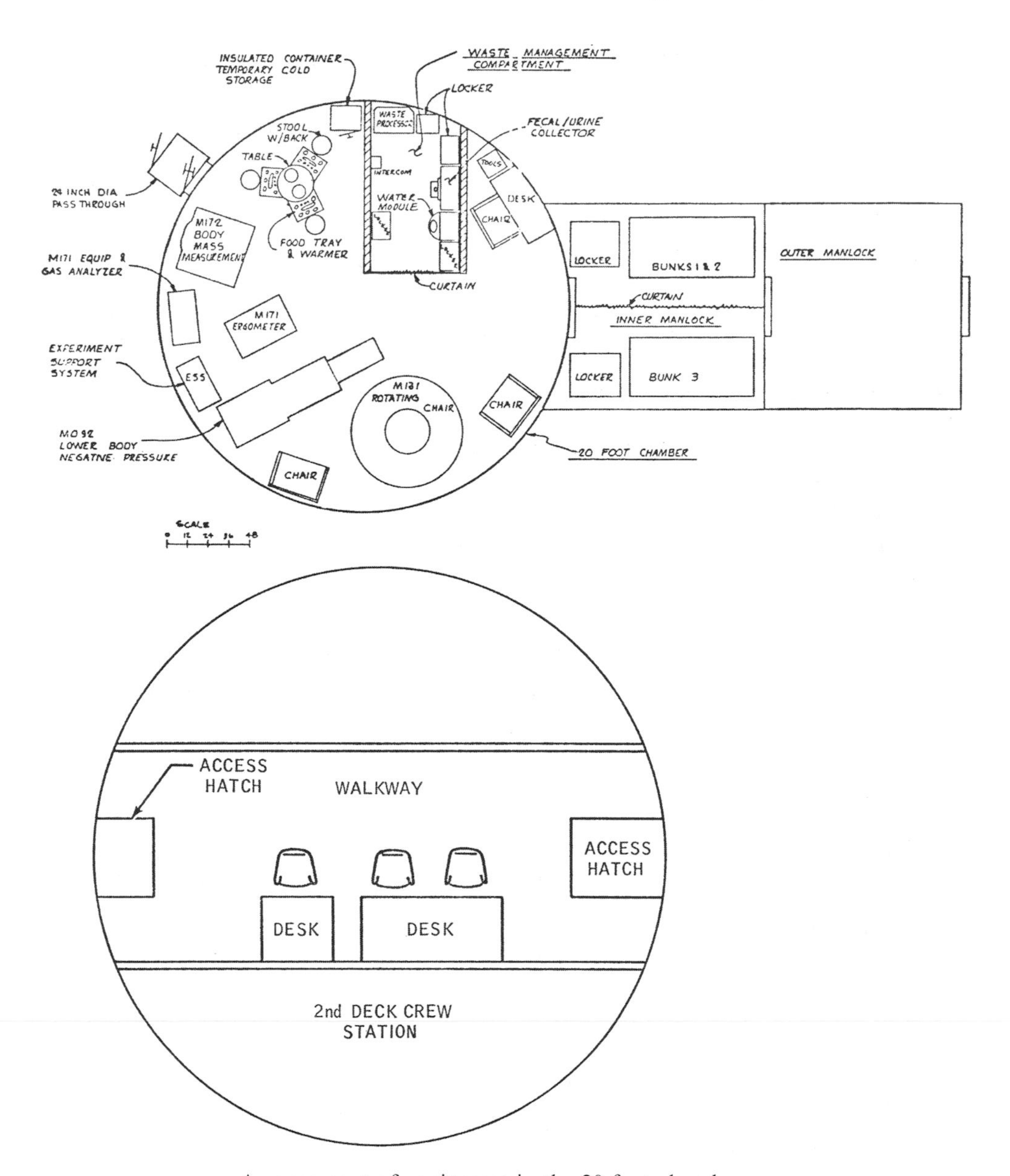

Arrangement of equipment in the 20-foot chamber.

December 1971, a committee was established to conduct an operational readiness inspection of the SMEAT test facility and to certify that the facility would be operational in time for the planned June 1972 start date.

In order to provide a contingency to any serious setback conducted during the test period, on 5 April 1972 the operations management committee decided to review the progress of the test during the actual test period, thus allowing a full assessment of any real-time problems as they occurred. This would also allow them to track problems and implement any corrective actions, to approve and direct changes in test protocol or policy, release reports, meet with media reviews and complete daily reports to the Director of MSC throughout the 56-day duration of the test.

The stated objectives of SMEAT were to evaluate baseline data on medical operations on Skylab's studies of the cardiovascular system, the expenditure of energy to do measured levels of work, and food and nutrient investigations. The test crew would also engage in a full schedule of work, eating, leisure, relaxation, and sleep, comparable to a 56-day Skylab mission in a simulated OWS atmosphere at 5 psia normally found at 27,000 feet.

The SMEAT test was scheduled to begin on 26 July 1972.

Crew selection and training

Very early in the planning for SMEAT, the decision was made to use crewmen selected from those in training at MSC. This was desirable because it would ensure compatibility of general background skill, and motivation between the SMEAT crew and Skylab crews. The team selected to participate in SMEAT consisted of Bob Crippen (Commander), Bill Thornton, MD (Science Pilot) and Karol Bobko (Pilot). Of the three, only Thornton was directly involved in Skylab, although the other two were former USAF MOL pilots who had worked on early Skylab engineering issues since joining to NASA in 1969.

There were no back-up crew-members assigned, as it was felt that it was not time-critical enough for them to be required. It would also save a lot of effort and added complications by not assigning one. The CapComs were to be the chamber operators rather than astronauts, and during the many experiments, the life science personnel would be in direct contact with the test crew to monitor their progress and support any of the added requirements.

The SMEAT astronauts were selected for the programme during June 1971, and began 'crew' participation on 20 July 1971, approximately one year before the test started. During the early stages of their association with the project, each of the men participated in the design of the chamber layout and in operational planning for conducting the test. Crew training began in November 1971, with training actually inside the chamber using test equipment commencing in the spring of 1972. After completion of the test in September 1972, their participation in the follow-on biomedical portion of the experiment continued until 9 October 1972.

The training replicated that of the Skylab flight crews in many respects, with initial briefings on several of the medical experiments becoming more intensive in March 1972 (when the training exercises began for operational procedures) and continuing up to the start of the 'mission'.

The training resembled that for a Skylab mission, and helped to refine that for the flight crews. SMEAT training included medical experiments, briefings on hardware operation practice, briefings on the chamber and system operations, bench checks of storage, crew compartment fit and function checks, maintenance briefings, test procedures, crew flight data file briefings, microbiological training, and emergency procedures. The table below records the planned against actual time spent in preparing for the SMEAT by topic. In planning, 412 hours were allocated; but in reality, each crewman devoted more than 500 hours to the training.

Bench-top briefings and equipment demonstrations followed initial classroom training. This was completed by the crew performing the task in accordance with a checklist procedure in the 1-g trainer.

Planned and actual training times (hours) for crew activity in SMEAT

Item	*Plan*	*CDR*	*SPT*	*PLT*
Experiments				
M092/MO93/M171	62	88.5	100.5	94.5
M074	2	1.0	1.0	1.0
M111-115	3	4.5	6.0	6.0
M133	30	31.0	31.0	31.0
M078A	2	1.0	1.0	1.0
M071/M073	2	–	3.5	3.5
Diet management	7	15.0	15.0	19.0
Waste management	7	11.0	11.0	11.0
M487	3	3.5	2.5	3.5
T003	3	4.0	3.0	4.0
Environment noise	5	6.0	6.0	6.0
Chamber				
ECS/EPS, COMM	6	10.0	10.0	10.0
Crew systems	6	12.5	11.0	11.0
Emergency	12	24.0	24.0	24.0
CCFF/bench check	8	24.0	24.0	24.0
Mission rules	6	3.0	3.0	3.0
ORI/TRRB	10	44.0	4.0	4.0
Dry run test	48	43.0	43.0	43.0
Alt. shakedown test	72	64.0	56.0	56.0
Medical				
Physiology/psychology	4	3.0	3.0	3.0
Medical data system	2	1.75	1.75	1.75
Physical examination	10	12.0	12.0	12.0
Microbiology	4	27.5	27.5	27.5
Inflt. Med Spt Sys	74	77.0	77.0	77.0
Oral hygiene	3	4.0	4.0	4.0
Personal hygiene	1	1.0	1.0	1.0
Miscellaneous				
Data acquisition camera	4	2.0	2.0	2.0
Timeline	4	18.5	15.0	18.5
Flight data file	4	2.5	2.5	4.0
Paper simulation	8	8.0	8.0	8.0
Totals	412	547.25	509.25	514.00

In preparation for SMEAT, changes in the schedule and procedures continued throughout the preparation period. Even when training was scheduled, there were not always firm checklists for operations, and it was more a case of developing the schedule and troubleshooting as they learned the procedure. In some cases procedures were still being developed as they completed the 56-day test, so routine training sessions on prepared procedures became more of a troubleshooting exercise to develop the procedures so they could be used effectively. This was a significant increased workload on the crew, and it also prevented them from completing training in the conventional sense. It often featured 12-hour days and six- or seven-day weeks, especially in the last four weeks prior to the test. But they considered it appropriate, despite the procedural and equipment difficulties.

As with the Skylab astronauts, the SMEAT crew received supplementary medical training off-site from MSC. A three-day course completed at the USAF regional hospital at Sheppard AFB, Texas, included a discussion of the diseases (and limited observation of the symptoms) of the eye, head, cardiovascular, pulmonary, abdominal, and musculo-skeletal systems, plus dermatology. There was also a 40-hour course at the AF School of Health Care Sciences at Wichita Falls, Texas, where medical and emergency care was conducted. This allowed all three astronauts to attempt such treatments as catherterisation of the urinary bladder, nasal gastric intubation, tracheotomy (with a tracheotome) bandaging, splinting, and the administration of a range of medication, for use in dire emergency situations.

In addition to the medical training, the crew also completed a series of walk-throughs to acclimatise to living in the chamber, and tested procedures such as communications, housekeeping, sanitation management, use of hygiene equipment, diet management, and safety and emergency procedures.

The training also encompassed a 16-hour 'wet run' and a three-day shakedown altitude run, prior to the start of the actual test. The 16-hour test was conducted to operate equipment that could not be used at normal pressure levels, or without a crewman inside the chamber. It was during this run that noise measurements were made, with various equipment components running to verify that background noise was within Skylab specifications. During the three-day test, the same protocol was used as would occur during the 56-day test, to provide preliminary evaluations of procedures, medical experiments and certain non-nominal modes of operation.

Activities during SMEAT in 1972 are presented in the next chapter, along with flight activities. By the winter of 1972/1973, the hardware was at the Cape, the astronauts were trained, and the preparations were complete. The Apollo programme was ended and already consigned to the history books. Now Skylab was on the pad and ready to go.

Emblems

The emblems of the three manned missions were approved on 17 November 1972. There was also a programme emblem (the three mission and single programme emblems are on the back cover of this book), an alternative tongue-in-cheek version of one of the manned mission emblems, and also one for the SMEAT crew.

Skylab programme Officially called 'Skylab USA', and taken from the wording on the emblem, this round emblem showed Skylab in Earth orbit with both workshop arrays visible (as it should have appeared) and a large impression of the Sun in the background. The Earth was designed as a cloud-covered planet. The crews would not name each CSM, as they were only for the short trip to and from the station, so for this phase 'Apollo' would be the call-sign. Once the crews were onboard the station, the call-sign 'Skylab' would be used. There was an attempt to use one generic 'name' for all three crews, but it was not successful.

SMEAT emblem This emblem was designed by 'Peanuts' cartoon creator Charles Shultz, and depicted his beagle 'Snoopy'. As he was associated with astronauts and flying, Snoopy wore the leather flyingcap and goggles, and was trailing his scarf. At the time (1972), 'Snoopy' was a favourite of the Apollo crews, and had been the call-sign for the Apollo 10 LM. It was also to become the emblem for the safety award presented by the astronauts to NASA employees – the 'Silver Snoopy Awards'. Originally, the SMEAT crew wanted to put a fish-hook in his mouth, as they thought it appropriate for the kind of medical tests for which they had volunteered. Shultz was not going to put a fish-hook in Snoopy's mouth, however, and instead included the rope which is a little too tight for comfort – just like some of the medical experiments.

Skylab 2 first manned mission Designed by science fiction artist Kelly Freas, it depicted a silhouette of the Skylab against the globe of Earth and eclipsing the Sun, revealing the signet ring pattern the instant before total eclipse occurs. The three crew names completed the emblem. The crew wanted to emphasise that Skylab was for everyone on Earth, and that the work they were doing would benefit all the people of the planet. This was a peaceful mission to begin the real habitation of space, so the planet was depicted simply as a 'big blue marble' with swirling cloud patterns and no distinguishable land masses – an impression of the view from a distance in space.

Skylab 3 second manned mission This emblem features the patriotic red, white and blue colours, and is symbolic of the flight's objectives. Leonardo da Vinci's central figure depicting the human form represents the studies on the human organism, and reflects the work of the Italian scientist, artist, engineer and philosopher. His concepts were centuries ahead of his time, and his studies encompassed astronomy, aerodynamics, hydraulics and anatomy. Behind the figure are two hemispheres depicting the other two main areas of research – the Sun, and Earth resources. The solar disc is represented by a red light, radiating hydrogen atoms and showing surface prominences. The right hemisphere features the study of the Earth's natural resources and environment.

The crew had approached the art department of Rockwell in Los Angeles, McDonnell Douglas in Huntington Beach, and at the Cape, to devise an emblem that covered Earth, Sun and Medical – the three main elements of each Skylab mission. The Huntington Beach design won, although the crew was happy with all the designs offered. The solar prominence was adapted from an image of a prominence on which Garriott had carried out extensive analysis years before, and for the sake of public viewing, the artwork of Leonardo da Vinci's figure was to incorporate 'certain modifications'!

Skylab 3 Wives emblem This was designed without the knowledge of the flight crew, and was first revealed as they opened the storage lockers in their CM. Inside were decals of the emblem, and one reflecting one of the astronauts' strongest mottoes – never lose your send of humour. Reproducing the official Skylab 3 emblem, the crew names of Bean–Garriott–Lousma were replaced by Sue (Bean), Helen-Mary (Garriott), and Gratia (Lousma). The idea originated from French journalist–artist Jacques Tiziou, of Merritt Island, Florida, and the emblem was designed by Ardis Shanks Settle of Houston. The central design was, of course, adopted to feature a female figure, to make it 'girl-rated'. The emblem has become one of the most sought after by collectors, as it was never commercially available, and only 320 were produced.

Skylab 4 third manned emblem Again the crew wanted to emphasise the three main elements of their research on the mission, and all three developed the idea, with Pogue putting together the overall ideas so that MSC graphic artist Barbara Matelski could do the artwork.

In the design, the tree emphasises the natural environment and the Earth resources studies, while the hydrogen atom – a basic building block of the universe – represents man's exploration of his physical world, his application of knowledge, and his development of technology. It also represents the hydrogen in the Sun – another major area of study on the mission. The human silhouette was a representation of human capacity to direct technology with wisdom, tempered by a regard for the natural environment, and was also related to the medical experiments that the crew were to perform on themselves.

The rainbow symbolised the promise made to man from the biblical story of the flood. Taken as a whole, the emblem reflects man's pivotal role in the 'reconciliation of technology with nature by a humanistic application of our scientific knowledge'. The three-sided rounded triangle and the figure '3' represented these three elements, the three crew-members, and the third manned visit to the station. Some of the souvenir emblems of this flight also featured an impression of Comet Kohoutek, which the crew were to try to observe during the mission.

Skylab astronaut assignments, 1972–1974

Flight	*Position*	*Prime*	*Back-up*	*Support*	*CapCom*
SMEAT	Commander Science Pilot Pilot	Crippen Thornton Bobko	None Assigned	None Assigned	Test Chamber Staff
Skylab 1	Unmanned space station				
Skylab 2	Commander Science Pilot Pilot	Conrad Kerwin Weitz	Schweickart Musgrave McCandless	Crippen Henize Thornton Hartsfield	Truly Crippen Thornton Hartsfield Henize Parker
Skylab 3	Commander Science Pilot Pilot	Bean Garriott Lousma	Brand Lenoir Lind	Crippen Henize Thornton Hartsfield	Truly Crippen Thornton Hartsfield Henize Parker Musgrave McCandless
Skylab 4	Commander Science Pilot Pilot	Carr Gibson Pogue	Brand Lenoir Lind	Crippen Henize Thornton Hartsfield	Truly Crippen Thornton Hartsfield Henize Parker Musgrave McCandless Lenoir Schweickart
SL Rescue/5 (not flown)	Commander Pilot	Brand Lind	Schweickart? McCandless?	Crippen? Hartsfield?	As above, as required

Flight operations

As launch preparations were under way to prepare the final mission to the Moon in the summer of 1972, so the hardware for the Skylab programme that was to follow Apollo 17 began arriving at the Cape.

The first Skylab flight components – the Command and Service Modules that were to take the Skylab 2 astronauts to the workshop – arrived at the Cape by air transport from Rockwell's facilities in California on 19 July 1972, and underwent a series of tests and inspections that would be fitted in around the Apollo 17 schedule. The following month the two propulsion stages of the Saturn V and the two for the first Saturn 1B were at the Cape and were also undergoing tests and inspections before being stacked on their respective launch platforms in the VAB.

The elements of the workshop began arriving in Florida during September. The Orbital Workshop and the Payload Shroud completed a two-week sea trip by a USN logistics transport vessel from McDonnell Douglas in California, and was followed by the ATM structure, by air, a few days later.

With the elements in various stages of launch processing, work on the actual flight hardware and not on test and simulation equipment initially progressed quite smoothly. By October the first problem occurred during the deployment test on the micrometeoroid shield. The structure was placed around the workshop, and was likened to 'fastening a corset around an elephant', as thirty-two technicians tightened trunnion bolts to draw the 1,200-lb shield towards the skin of the workshop.

Examination of the structure indicated that several bulges remained, and despite several attempts to loosen and retighten the bolts in sequence, the bulges remained. Ultrasonic examination revealed that only 62% of the shield was actually touching the workshop. It was decided to pressurise the workshop, and the area was then rescanned to reveal that 92% contact was now recorded. It was established that in orbit the difference in pressure from inside the workshop and outside in vacuum would be substantially higher than on the ground, and the situation was accepted.

Even during the actual shield testing, the latch failed, and some of the bolts seemed to be over torqued. This resulted in delays to the process, with the two solar arrays not being attached until November. Over the next six months, various elements of the Skylab payload were tested and checked out, with several minor problems that lingered throughout the whole eight-month launch-processing period. Some of the systems testing was rather rudimentary: for example, several electrical connection problems were checked by two rather basic systems tests, either by wriggling a cable or by probing it with a connector pin.

During the processing for launch, about 61% of the experiment hardware located in Skylab had been removed and replaced due to failures in the test procedure or late changes in design, and this resulted in the requirement to re-check interfaces once the replacement units had been reinstalled. As many of the Skylab components were state-of-the-art, they were highly susceptible to such failures and upgrades, and it was later determined that if such re-testing had not been completed, about one third of the experiments flown on Skylab could have failed in flight.

In February 1973, a ten-day Integrated Systems Test of the workshop on its launch vehicle was conducted, with only minor problems revealed. Later the same month, the first Apollo spacecraft was taken to the VAB for mating with the Saturn 1B, and in March the launch team completed a simulated countdown and lift-off of the Saturn V during a Flight Readiness Test – one of the final milestones prior to moving the vehicle to the pad.

Launch dates had been set for the Saturn V as 14 May, and for the crew, 15 May. Skylab was soon to move into flight operations with the launch of the unmanned station, followed by the first of the three manned missions. Before these are discussed, it is worth recalling the other Skylab 'mission' that is often forgotten, but was equally as important as the flown missions. It lasted for 56 days, but never left the ground.

SMEAT: 56 DAYS IN A CAN

'Skylab has a much better chance of success because of what we learnt in SMEAT,' reported Bob Crippen during the post-test press conference in September 1972. Along with 'Dr Bill' Thornton and 'Bo' Bobko, 'Crip' had recently completed the 56-day test in the 20-foot chamber, in which, the astronaut recalled, the preparations for the test were actually worse than the test itself.

A planned 21-day pre-test diet and waste collection procedure became 28 days when the test was delayed for a week. All three astronauts followed a diet that consisted of the Skylab-type food apart from a few fresh items, that they would eat on the test. The water they drank was specially prepared and constantly measured, and even when they travelled they carried a specially designed suitcase which contained their special diet.

On the day that the test would begin – 26 July 1972 – a medical examination was conducted on the three men, who then completed a required pre-breathing period – which was quite boring, as there was nothing much to do apart from sitting and waiting for the purging of nitrogen from the blood.

Having completed a training session inside the chamber the previous month, the initial sealing inside the facility was no real surprise as the crew settled down to their two-month stay. Their initial observation in the 5-psia atmosphere was that sound

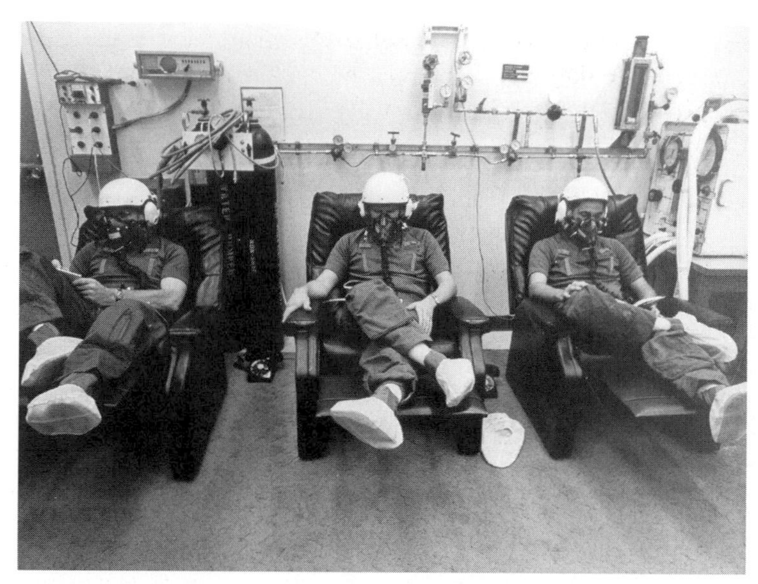

All dressed up and waiting to go. The three Skylab SMEAT crew-members conduct the required pre-breathe period prior to entering the chamber for the 56-day test run. (*Left–right*) Thornton, Bobko and Crippen. Typically, Thornton works on his notes as Bobko and Crippen catch up on their reading. The latter two studied Russian as part of their training to support the joint Russian mission in 1975.

seemed to be further away and somewhat softer, so for a few days they became aware of shouting to each other and becoming hoarse. Another consequence of the reduced pressure was an inability to whistle, and a sneeze was far more milder than expected. One of the other problems was an increase in abdominal gas that gradually demised after two weeks but which never disappeared completely. The astronauts surmised that this was a combination of the reduced atmosphere and the gas produced in the food as it was consumed.

Despite the reduction in sound transmission, the noise levels remained, and the crew were very thankful that noisy items of equipment had been insulated before the test, as it would have been very difficult to live with this for 56 days. One of the effects of sound transmission in the medical tests was the K-sounds used in determining whether the blood pressure was reduced along with heart and lung sounds in a normal environment. Taking measurements were a little more difficult, but the crew soon became accustomed to it, and after a few days they did not notice the difference as they became adjusted.

Once the crew had settled into the facility, a daily pattern featured 07.00 wake-up, and after breakfast they started work around 09.00 through to lunch at 13.00, followed by a second work period from 14.00 until 19.00, when they ate dinner. A 30-minute review of daily activities and a 30-minute period of housekeeping was completed by 21.00, followed by two hours of personal recreation before the sleep period commenced at 23.00 each night.

Some of the recreational activities included a Russian language course for Bobko

The three SMEAT crew-members enter the airlock to the 20-foot chamber, fitted out to resemble the Skylab workshop wardroom for the test. Note the 'Snoopy' emblem crew patch on the wall of the chamber.

and Crippen, as part of their Apollo–Soyuz support crew training, a Skylab CSM course via closed circuit TV, an electronics course, a solar physics course, and a commercial pilots study course. The three astronauts decided upon taking these supplementary activities in order to fill their spare time in a useful way, and to ensure that their workload was representative of actual days onboard the Skylab. Despite a concern that there would be too much spare time, actually became so busy that they had to neglect some of the supplementary activities to complete the more important activities.

Once a week there was a day off, of which half was filled with preparing the weekly report, while the rest featured leisure activities. Each morning the planning for the following day was decided, and was presented to the crew at noon, with the final version sent in a message that evening.

During their confinement in the chamber, all three astronauts were keen to maintain physical exercise, and had intended to use the bicycle ergometer for this purpose as well as in connection with medical experiments. Unfortunately, the design of the exercise machine was seriously lacking in structural integrity, and it soon broke down, requiring it to be passed out through the main airlock for repair.

Baseline testing for the ergometer was to have begun six months before the test, but other hardware failures prevented anything but minimum baseline data collection just six weeks prior to the start of the test period. Both Crippen and Thornton, who had worked on the ergometer prototypes for some time prior to SMEAT, were still unable to test their full work capacity when the machine was returned to the chamber after repair. It was still not sufficient to test the full capacity of the astronauts, and in sheer frustration Thornton got on the machine 'and rode it into the ground' until it seized up a second time.

The crew found the habitation of the chamber 'reasonably comfortable', though sparse in comfort. The most luxurious item they had were garden furniture Sun-loungers that they unfolded in the evenings. Although the bedding, food, eating facilities and hygiene facilities were a marked improvement over the Apollo equipment, none of the crew expressed a desire to radically change their own homes to incorporate the Skylab facilities!.

Throughout the test, one of the more challenging aspects of housekeeping was in maintaining the cleanliness of the chamber. The regular cleaning of the wardroom, the toilet and the experiment work area was given the most attention. In addition it was found that the floor of the chamber needed constant cleaning by using their personal soap and washcloths, as there was nothing else to use.

The crew used a simulated Skylab shower, and found it delightful to once a week have a hot shower. The Skylab astronauts would also find it enjoyable. By washing after each period of exercise they found that they felt personally clean throughout the test. They also found the upper level desks a worthwhile place to study and isolate themselves from the main activities of the chamber.

The meals during SMEAT used the Skylab-type food, which was sealed in small cans that fitted in food trays that included heating elements. Their days meals were preplanned and repeated ever six days in a diet that contained approximately sixty items, divided into freeze-dried food to be reconstituted (dried mashed potatoes or dried soups), frozen meats and ice creams, and thermostabilised foods such as lemon pudding and peaches. It required a great deal of crew input prior to the test to arrive at suitable menus that not only supplied the mineral, proteins and calories required each day for personal health and medical experiments, but were also appetising. By the end of the test, which included the pre- and post-test periods, the astronauts had been eating the Skylab food for more than 100 days, and despite their comments that it was 'adequate', it was a delight to try real food again after the test period ended.

As the eating of food was based on Skylab flight protocol, so too was the collection of waste. Urine was collected via the Skylab system, and was collected, measured and sampled each day. Only one of the crew was required to use this system to evaluate it for the flights, which was quite fortunate, because it often broke down or leaked. This resulted in lost data and a rather messy clean-up period. The other two crew-members collected urine in covered cans and then passed them out through the transfer lock each day.

Solid waste was collected in a 'small potty' – a replica of the Skylab waste collection system (less one step of drying), with air drawn through it, and stored in bags . All other items of dry waste, including used clothing, was stored in bags and passed outside in a manner similar to using the trash airlock on Skylab.

The main purpose of the SMEAT, in addition to habitability, was of course to gain experience and baseline data for the medical experiments. The main experiments were aimed at the cardiovascular system, and consisted of the lower body negative pressure device, the Vectorocardiogram and the Metabolic Activity Experiments. They were completed in a series of runs by two crew-members – one of them serving

Inside the SMEAT chamber in 1972, astronaut Bobko is seen on the bicycle ergometer while Bill Thornton looks on. The 56-day test evaluated systems, experiments and protocol planned for the three manned missions the following year.

as the test subject, and the other as an observer/assistant. The runs took about 2 hours 30 minutes per man, and total participation in the series as a test subject, and then as the observer for the second test run, lasted about five hours and was conducted every third day.

The operation of the experiments provided not only important data for the investigator, but also established a timeline in which to perform the experiments that was valuable in devising a crew activity plan for the missions to Skylab. The SMEAT astronauts uncovered several data-handling and procedural problems that had not been evident during the shorter training runs and during the course of the test. The crew and test conductors re-evaluated these problems so that by the end of the test period they were all solved, and data began to flow much more smoothly. The crew pointed out in their post flight report that it was lucky that there were no serious changes in any of the crew during the early stages of the tests, as the determination of the changes by faulty hardware or procedures would have produced inaccurate data.

Strict recording of all body intakes and the recording of wastes and blood samples were followed with no deviation during the crew participation. The three astronauts evaluated different procedural methods of data collection to evaluate the most suitable for flight operations. Some of their evaluations ranged from capturing samples of exhaled breath for exterior analysis, to inboard measuring devices, to attachment of electrodes on the body. In addition, more basic evaluation dealt with the fixing of experiment harnesses to the body, which threatened to ruin experiment data runs because of the incorrect fit or the extra time it took to secure the devices correctly before beginning the data collection.

The crew evaluated that the LBNP device (M092) would be enhanced with the addition of a back-up waste seal flown on the missions, as the test runs showed increased leakage rates despite all crew efforts, and that even when a new seal had been passed in it still resulted in a failed zipper, and a difficulty in climbing the device, as it was more contoured to the body. In fact, the crew did not notice much difference between the old faulty seal and the new ones.

The Velcro fastener restraint strap on the device for the legs was located at the knees, and tended to remove body hair as it was released, which was repeatedly uncomfortable. Neither would it hold the legs in place, and it needed careful redesign before flight. The Blood Pressure Measuring System resulted in small failures and variances in data recording, which raised doubts about the whole operation – so much so that the 'outside world' asked whether the unit was in error or whether the crew was using it incorrectly. The crew report indicated that 'from our standpoint the BPMS used in SMEAT was obviously in error, and [was] unsatisfactory for monitoring the health and well being of the test subjects.' The crew found that the observer could handle back-up procedures for single failures, but when two occurred in the equipment, then the third crew-member had to assist in the experiment, which divert his time from his own activities.

On Experiment M093 – the Vetrocardiogram – they found that the tattooed marks for electrode sites marked in March 1972 were adequate, but more than four months later were not very conspicuous. The tapes used to stick the electrodes to the skin cause irritation, the electrode sponges were too thin for the electrode cups, and were too dry, The harness was also found to be too small, and despite adjustments it was found to be difficult to use for the full eight weeks.

The M171 Metabolic Activities was a major Skylab experiment, and the difficulties of the bicycle affected not only this experiment but also the crew health programme. The design and integrity of the unit was a frustrating challenge for the crew during the test, and during the missions this could have had serious implications for crew health and data-gathering.

The most serious problem found on SMEAT was in the leaking Urine Volume Measuring System. It was found that the four-pint collection bags were too small. One crewman exceeded six pints of urine a day, while the other two reached the limit of the bags, which leaked; and on six occasions the sample bags had been torn when sampling, resulting in more than two pints of urine splashing over the collection system, the chamber floor and the crewman. It was an uncomfortable situation, and concern about urine in direct contact with the skin was added to the problem. On SMEAT, the hand-washing system was not sufficient to handle this type of repeated leak, and the amount of disinfectant onboard the chamber was insufficient, leaving the chamber not totally antiseptically clean. Simply installing larger bags was not an option, as the space for the bags was limited by the internal workings of the system, and without major redesign to the whole system, larger bags would be difficult to secure.

Pete Conrad expressed a lack of confidence in the whole system, and in zero-g, a floating globule of urine was not something to relish, as it could float around the station, into the crew quarters, or behind electrical circuits, and cause serious problems. Conrad opted for the upgraded Apollo rubber cuff (condom-type) system, as the designers and crew representatives argued for and against the Skylab system.

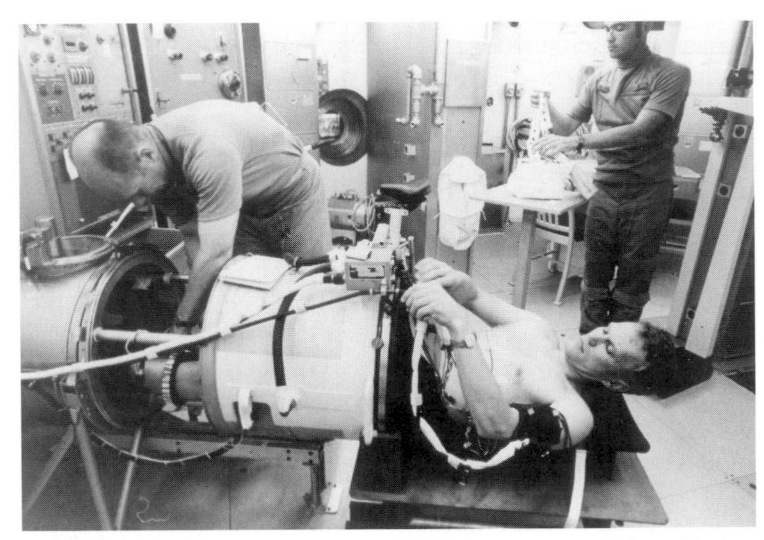

All three SMEAT crew-members work in the chamber mock-up of the orbital workshop wardroom and lower experiment area. Bobko is seen lying in the LBNP chamber, while Thornton adjusts the device prior to a test run. Behind is crew Commander Bob Crippen.

At the same time, other design faults pointed to a complete redesign or rethinking of the medical requirements, or both, and with only six months to launch, the options were limited. The problem was solved by offering a choice of the Skylab device with an eight-pint capacity bag, with either centrifuge collection or the Apollo roll-on cuff. But it was still involved another four months of redesign until an acceptable system was ready to fly.

The crew recognised that one of the most important lessons on SMEAT was learning about themselves in such a close environment for such a long time. There had been concerns that since there was no official screening for crew selection this could cause a problem in crew compatibility but in reality this was never the case, as the crew successfully with each other from the very beginning.

In each of the crew-members was a strong personality, mutual respect for each other's capabilities, and acceptance of each giving 100% effort to the test. Although there were disagreements, there were never arguments, and this was strengthened by the trio working as a team from the beginning of their assignment to the test programme. Their general approach to the crew's input was that 'our motivation to complete the test successfully overshadowed any disagreements between us.'

What did frustrate them on more than one occasion was the ground support personnel, but this was allowed to continue, and so the crew could vent their emotion outside rather than at each other. The 'us and them' syndrome was carefully monitored, and it worked to unite the crew even more, eventually resulting in constant 'leg-pulling' to add to the release mechanism during the test.

One of these 'leg-pulling' episodes that unfortunately did not materialise beyond discussion concerned the crew's clothing, chicken bones and a cat! The crew had found that the Skylab clothing shed a considerable amount of lint, and the vacuum could not cope with the influx of material. The crew therefore redesigned the pick-up brush to cope with the problem. The collected balls of lint also gave rise to an idea for a practical joke that the crew wanted to play on the test conductors outside. The serious manner of the controlled experiments and the medical protocol of Skylab, measuring intake and output of all food-stuff, was a little tiresome on a daily routine, and the crew thought that it would be a break in this strict regime if they gave the impression that they were not alone in the chamber! They were in contact with their families once a week, during the private telephone conversations, and they planned to ask one of the members of the family to smuggle in some fast-food meat-free chicken bones to a friend in the chamber, who would pass them through the airlock without discovery. The crew were then to place, on a used meal-tray, the chicken bones and a couple of hairballs from the suits, with a note saying how much the 'cat' enjoyed the fresh meat! Unfortunately the crew were so busy that they never had the opportunity to include this supplementary experiment in their programme. The look of surprise on the test-conductors faces would have been one to behold!

As the SMEAT progressed, the crew experienced the feeling that perhaps there was more to do than they had time to complete, and as the test neared conclusion, so they were worried that they would be unable to collect those last samples that had been troublesome during the test. The crew never felt despondent as they set out to do a good job for 56 – days and that is exactly what they did.

On the final day in the chamber – 20 September 1972 – the crew were woken at 05.45 with a US Navy call to 'Man Your Mops'. As the trio prepared to exit the chamber that had gradually equalised the pressure to the outside world, Crippen tried out some of his Russian with 'Ochen khorosho, otkroite dver!' ('Very well, open the door!'). He was already in training to support the proposed US–USSR joint flight, and had supplemented his conversational Russian course via closed-circuit TV, and a move-a-day chess game with Ken Snyder, of Crew System Division, during his off-duty time. The result was: outsiders one game; insiders, one game.

At 07.00 the door was opened, and the three of them – with Bobko and Crippen sporting beards – emerged to the outside world and were welcomed by members of their families and Centre officials. Crippen exclaimed 'Man, is it nice to get out?' but added 'The last 56 days were not as bad a we thought they would be'. On 11 September he had celebrated his 35th birthday in the chamber, and did not want to mark the event with a special celebration: 'Like most guys who get to 35, I'd just soon ignore it'. One event that he could not ignore, however, was the news that the family boxer dog had delivered thirteen new puppies. Crippen said that his wife had not thought this to be a problem, but *he* certainly did! After the glamour of an astronaut's assignment in testing space hardware, it was time to return to more Earthly domestic challenges.

Thornton suggested: 'You'd better pick the company you will be with on Skylab. I couldn't have made a better choice,' although he had missed his gourmet meals during the test, and had been looking forward to the experience of eating out after the eighteen days of medical tests as part of the post-test period. He looked forward to a big meal of his own choosing, although he added, 'I am afraid it will be gluttonous, and not gourmet.'

Bobko suggested that, based on their findings, the Skylab crews would find Skylab quite liveable, and later wrote that as crew-members the SMEAT trio had participated in the design, planning, training and operation of the test over a period of eighteen months which they found a valuable experience. Each of their days were fully occupied with experiments evaluation and supplemental activities, offering them a challenge from start to finish, but not wavering in their belief in a successful conclusion. They found that there were no detrimental effects in living in a 'Skylab environment', and that the operational evaluation of the hardware, experiments and procedures that revealed several discrepancies in hardware and procedures and their correction 'should result in a better Skylab flight' to which SMEAT contributed.

Frustration with Salyut

Following the tragic loss of the three Soyuz 11 cosmonauts in 1971, the Soviets halted their space station programme while the cause of the death was established, while discussions continued with the Americans concerning possible joint missions. By the summer of 1972 the next Salyut was ready for launch, and clearly the intention was to have cosmonauts onboard a space station well before Skylab reached orbit. Unfortunately it was not to be, and the ensuing twelve months were as frustrating as the previous year had been traumatic.

On 29 July 1972 the second Salyut was launched, on top of a three-stage Proton rocket, from the Tyuratam cosmodrome. It was intended to support two two-man crews, but just 162 seconds into the flight the second stage control systems launch vehicle failed, resulting in a shut-down of one of the four second-stage engines and a launch abort, destroying the unmanned vehicle.

Parallel to the development of the scientific Salyut stations was the military orientated Almaz series of stations, with similar objectives to the cancelled USAF MOL programme. By 1973 both an Almaz and a third Salyut were readied to fly in orbit at the same time, enabling the Soviets to upstage Skylab by having not one but two space stations in orbit.

On 3 April 1973 – a little over a month before Skylab was planned to be launched – the first Almaz was launched; and in an attempt to mask its true military nature it was called Salyut 2 upon reaching orbit. Initially all appeared to be satisfactory, and the first crew prepared for launch; but on the thirteenth day of the Salyut automated flight, telemetry indicated a 50% drop in pressure inside the main hull, which imparted a slight trajectory change in the orbital path. It was clearly a major malfunction, and the attempt to launch the cosmonaut crew had to be cancelled. Apparently the main engine of the station had a serious design fault that caused it to puncture the main hull when it ignited. Other reports indicated an electrical fire and rupture of the hull that was so violent that the solar array and rendezvous antennae had been ripped off, leaving the station to tumble out of control.

Less than a month later, and just three days before Skylab was launched, the Soviets tried again with their third civilian Salyut, which was launched on 11 May. This time the problems began as soon as the vehicle arrived on orbit, with irregular firing of the attitude control rockets quickly depleting the fuel reserves due to a failed ion sensor on the station. The difficulties had occurred so quickly that the station did not even receive a Salyut name. It was therefore designated Cosmos 557, and again the cosmonaut's remained grounded.

In September 1973, while waiting for the next space station, Soviet cosmonauts participated in a two-day test flight of a new version of Soyuz, but it would be twelve months before the next station was launched, six months after the last Skylab crew returned home. With Skylab, the Americans was set to finally surpass all Soviet space endurance records.

SKYLAB 1: THE LAUNCH OF THE ORBITAL WORKSHOP

The first vehicle to reach the pad was the Saturn 1B for Conrad's flight on 26 February 1973, when it was positioned on Pad 39B. On 16 April this was followed by the Skylab 1 stack, which was rolled out of the VAB to Pad 39A for a series of tests and simulated count-downs.

There were mixed emotions as the final Saturn V was prepared for flight. It was the end of an era, but was also the beginning of a new, short series of flights to create what was hoped would be America's first space station. All depended on this first launch. If Skylab was lost there was a back-up workshop, but the delay could

Skylab Flight Directors Don Puddy (*foreground*) and Phil Shaeffer seated at the flight director console in Mission Operations Control Room (MOCR) at JSC during the early hours of the Skylab 1 mission.

threaten the whole programme. With rain-clouds, storms and lightning threatening during the first ten days of May, actual launch preparations progressed smoothly toward on-time launch on 14 May. As a last-minute addition, and in fitting tribute to the decade of work put into the programme, technicians attached a Stars and Stripes flag to the docking adapter on the cluster inside the launch shroud on top of the last Saturn V. The first United States space station was at last ready to fly.

As the clock ticked towards zero, astronauts Pete Conrad, Paul Weitz and Joe Kerwin watched with personal interest the end of the countdown of the thirteenth launch of a Saturn V. It had been only five-and-a-half years since the first Saturn V had thundered off the pad at the Kennedy Space Center, and now, on 14 May 1973, this was the last of the line. Among a crowd of 25,000, the three astronauts were on hand to witness the historic event. They were only spectators at the very beginning of Skylab flight operations, but if all went as planned, twenty-four hours later, they would launch on a smaller Saturn 1B to rendezvous and dock with the station to begin a month's residency.

A storm front was threatening to move into the Cape area, and with clouds gathering, the 3 hour 30 minute launch window seemed to be under threat. However, exactly on time the last Saturn V prepared to leave the pad.

Kennedy Launch Control: 'T–20 seconds and the countdown continues to go smoothly. Guidance release. T–13... 12... 11... 10... 9... 8, we have ignition, the sequence has started, 6... 5... 4... 3... 2... 1... ZERO. We have a lift off. The Skylab lifting-off the pad now, moving up, Skylab has cleared the tower ...'

At this point, Mission Control in Houston took over the control of the Saturn as the five F-1 engines powered the vehicle into the Florida skies. Lift-off occurred at 13.30 EDT, and with it the flight operations of Skylab had at long last really begun.

The Skylab mission begins with the launch of the Skylab 1 OWS on the very last Saturn V, on 14 May 1973. Seconds after this photograph was taken the the entire mission was threatened when Skylab lost its protective micrometeroid shield and one solar wing during its ascent through the atmosphere.

On the ground, the Earth shook as the Saturn V rose off the pad three miles away from the onlookers – the shortest safe distance. Conrad had ridden one of these monsters through rain clouds on the way to the Moon in 1969, and now he waited for and then felt the sound-wave wash over him as the Saturn gained speed and rode a pillar of flame skywards, straight into the cloud cover.

There was nothing like riding the vehicle from the CSM perched at the top of 'the stack', but witnessing the launch from three miles away came a pretty close second place. Unfortunately this was the last time that anyone would see the spectacle, and the sheer power of such a machine was unlikely ever to be witnessed again. The era of huge one-shot rockets was coming to an end.

Onboard the Saturn, initial operations were performed normally, with the vehicle exceeding the speed of sound in just 60 seconds. Ten seconds later it passed through the period of maximum dynamic pressure (Max Q) on the vehicle, when the forces of the atmosphere were at their most severe.

It was just seconds after passing Max Q that ground telemetry indicated an abnormal situation on the climbing rocket. The data, which went almost unnoticed, indicated a premature deployment of the protective micrometeoroid shield and the number two workshop solar array. At first thought to be an erroneous signal, the vehicle continued its preprogrammed ascent as the first stage separated from the vehicle and was replaced by the ignition and burn of the five J-2 engines of the second stage. Another telemetry signal indicated that the inter-stage had failed to separate, and so the second-stage engines were programmed to burn a little longer than normal to compensate for the extra weight and drag.

At 9 minutes 49 seconds after launch, the second-stage engines were shut down, and three seconds later the stage separated from the Skylab payload. With Skylab at 274.6 miles altitude and 1,118.5 miles downrange and over the Atlantic Ocean, heading away from Newfoundland, the instrument unit on top of the third-stage workshop began a preprogrammed sequence to configure the station for orbital operations. After more than a decade of planning Skylab was at last in orbit – but in what state had it arrived?

First orbits

Initial telemetry from Skylab had been monitored via the fleet of ARIA aircraft off the coast of Greece and the island of Mahe in the Seychelles. As the cluster flew into range of the Canarvon tracking station, the expectation that the data would show the large OWS solar arrays deploy, to supply 12.4 kW, was replaced with surprise that the power levels reached only 25 watts!. As the first orbits was completed, data received at Goldstone and Texas confirmed the reading, and added to the problem by indicating a short in the pyrotechnical relay used to deploy the wings. Furthermore, data received during the second orbit revealed that the micrometeroid shield had not deployed at the planned time. Just three hours after launch, and after evading the storm clouds, Skylab was in a far more serious predicament in orbit.

During the initial post-launch briefing, the Director of Launch Operations, Walter Kapryan, reported a normal count-down and launch sequence with no concern about the weather during the launch. The Saturn V performed normally, taking the Skylab to orbital insertion only 1 second later than planned, at an orbital velocity of 25,096.6 feet/sec. The payload shroud had jettisoned at 15 minutes 25 seconds into the flight, and the ATM was deployed six minutes later with the four ATM solar arrays unfolding five minutes after deployment. Kapryan stated that the ground control had not been able to confirm the deployment of the large OWS solar arrays.

The press attending the conference pressed him as to whether he considered that there was a problem or not. Kapryan simply replied that they just did not know at this early stage. The ground had to wait for telemetry downlink from the next orbital pass over the United States – telemetry data which would reveal the levels of power and the state of onboard batteries. If the solar arrays were not deployed, the full 28-day mission of Skylab 2 could not be flown, but the decision to fly a reduced mission could come later, once the data from the OWS indicated the state of the vehicle.

Asked why good data were not forthcoming, Kapryan indicated that the quality of the data received was poor due to other telemetry 'noise' being received as the vehicle climbed and began its deployment sequence. He also stated that if the wings were not deployed it was doubtful whether the astronauts would be launched the next day. Asked if the crew could carry out a 'fix-it-yourself' type of repair, Kapryan replied: 'We're not set up for an EVA to do that [manual solar array deployment] type of work. They [the solar wings] either go out or don't go out. That is, the astronauts can't do anything about it.'

Skylab management convened to review the situation, which seemed to be worsening by the hour. Without the main solar arrays that supplied 60% of power to the station, there was a serious threat to the planned 140-day manned occupation. If the CSM was launched and docked to Skylab, power could be routed from fuel cells in the Service Module – but only 1.4 kW, and only for fourteen days until the reactants depleted. The experiment programme would be serious hampered with very little power and the entire purpose of Skylab as a scientific research platform would be seriously threatened. Within hours, the 15 May launch of Skylab 2 was cancelled and was rescheduled for 20 May, and the three-man crew flew back to Houston to review a new flight plan.

The plan was now to launch Conrad's crew for a 17-day nominal mission, followed by a further 11 days of habitation and reduced activity to provide medical data for a full 28 days. As the new flight plan emerged, a far more serious problem was developing up on the workshop. Within 24 hours of launch, internal temperatures of the workshop were recorded at 38° C and rising! Studies of the launch data indicated that at just over a minute after leaving the pad, aerodynamic forces had ripped the micrometeoroid shield from the side of the workshop after a premature deployment, and that the solar wings had for some reason been unable to deploy. In orbit, with no protection to reflect solar heat, the temperature rose to threaten the thermal balance inside the workshop, producing not only temperatures excessive for any crew that might enter the vehicle, but also risking the stored onboard consumables and delicate equipment.

It soon became evident that what was needed was a shade to protect the exposed areas. Despite no contingency plans to achieve this or to deploy the solar arrays, teams began working around the clock to devise solutions to work the problem and provide answers to resolve the situation, in a race to evolve procedures, fabricate material, train the crews, and launch before Skylab was beyond help.

Initially it seemed that there was little the astronauts could do, but Pete Conrad was not prepared to abandon Skylab without a fight, and from orbit.

The cause of the failure

At 63 seconds into the flight there was an indication that the shield had prematurely deployed to stand a few inches from the hull of the workshop, and in doing so, aerodynamic forces during the ascent would have ripped off the lightweight metal shield like the skin of a banana. This incident had a mechanical effect on the deployment mechanism of the solar arrays. Initial indications were that the main arrays had only partially deployed, due to the power level being so low.

On 30 July, the boards investigating the Skylab 1 mishap reported that the failure of the micrometeoroid shield 63 seconds into the flight caused the breaking of the Solar Array System (SAS), and at 593 seconds into the flight the second-stage retro-rocket plume exhaust resulted in the ripping off of the array.

The inquiry determined that the most probable cause of the failure of the shield was internal pressurisation of its auxiliary tunnel that forced the forward end away from the OWS and out into the supersonic air stream, where aerodynamic forces tore

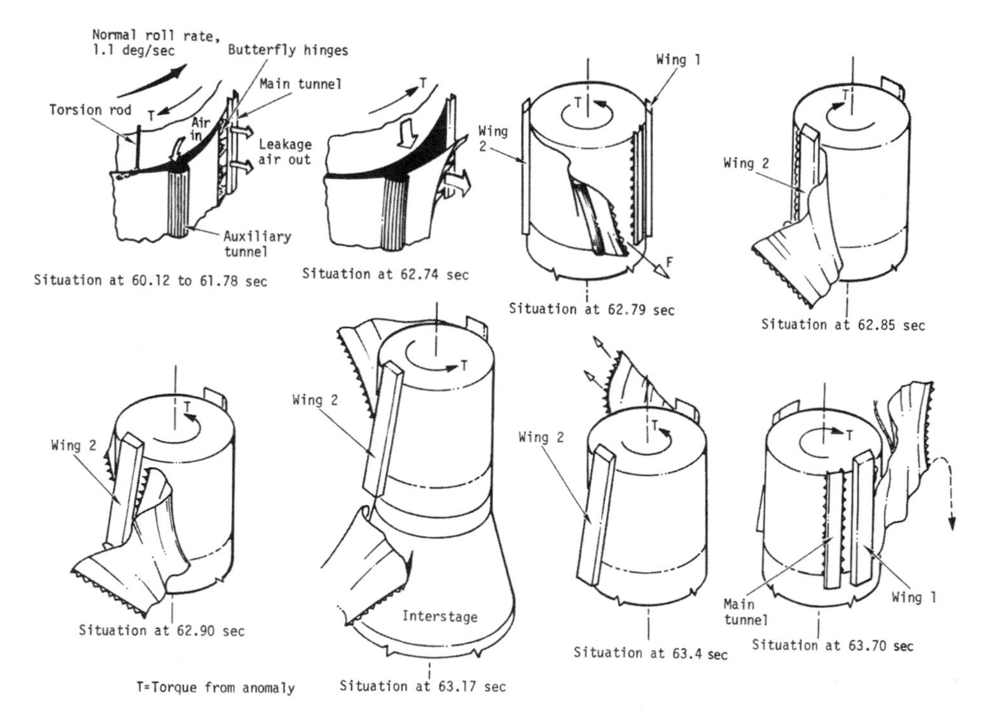

Micrometeoroid shield dynamics, revealing how the shield was ripped from the workshop.

the shield from the OWS. The pressurisation of the auxiliary tunnel was the result of high-pressure air entering several openings on the aft end due to imperfect fittings and seals on the structure. Moreover, the venting analysis for the tunnel was based on a complete seal on the aft end, and the opening on the end of the tunnel resulted from a lack of communication between the aerodynamic, structural design and manufacturing personnel teams. The workmanship was sound, and so the failures were due to inadequate communications between the different teams over a period of time, in 'an absence of sound engineering judgement and alert engineering leadership'. Corrective action was applied to the back-up OWS 'if it is to be flown in the future'.

Power or temperature?

Overnight, ground teams continued to analyse the data, and the following day they presented a statement and briefing on the current position of the station – and indeed, the whole programme.

Bill Schneider reported on the activities in the NASA centres across the country, primarily at Marshall, where the main core of the programme engineering was located. While the team at Houston handled the operational activities of the ailing station, Huntsville was trying to find a solution to the problem.

When the micrometeoroid shield was lost, it took with it the thermal paint pattern designed to balance thermal loads across the Sun-exposed surface. This exposed the bare skin of the S-IVB workshop to the full forces of solar heating. As the controllers witnessed the temperatures rise, it was decided to terminate the pressurisation inside the station until the temperature levelled out. The outside skin was covered with a gold foil layer, and the type of aluminium used in the construction of the stage walls would lose structural strength as temperatures rose. Not wishing to add to the problem with the possibility of internal pressure rupturing a weak point in the hull, the pressurisation was terminated. Recorded temperature levels on the Sun-facing side of the OWS were levelling out at 225° F.

During the night of 14/15 May, the OWS was manoeuvred in various attitudes to provide a better understanding of how it responded from thermal loads and electrical power requirements. This presented NASA controllers with a quandary. To supply the vehicle with maximum electrical power, the ATM solar arrays had to point directly to the Sun. This, of course, also subjected the weakened hull to increased thermal loads. Reducing the thermal loads by pointing the smallest end of the station forward toward the Sun also resulted in loss of solar array alignment with the Sun. Finding the optimum balance of time and temperature between these two attitudes was evaluated as teams of controllers tried to stabilise the power and temperature levels, while others tried to conserve limited TACS reserves.

Heat concerns

Skylab had been launched with most of the consumables onboard for all three crews, with a margin of contingency. There were no planned resupply missions as became a standard operation on Soviet space stations, and the CM on each manned mission only could carry a minimum of fresh supplies to the station. The rapid rise in internal temperature to 54° C caused concern about the provisions stored in the OWS, but tests on the types of canned food stored in the station indicated that they could withstand sustained heat. Plans to carry replacement foods to Skylab were dropped after it was determined that the food would not be altered in mineral content or taste by the heat. However, the astronauts were all provided with added instruction on food inspection.

Other concerns about heat included the onboard medical supplies, and a stripped down resupply list was urgently ordered from the contract pharmaceutical companies to add to the Skylab 2 manifest. Film storage for use on Earth resources photography was also a concern, as the heating and drying of the emulsion on the film would render it useless. Kodak engineers informed NASA that it might be possible to restore the film by rehumidifying the vaults by adding new salt packs to provide moisture, but this could take 20 of the planned 28 days. In addition, therefore new film would be loaded into the CM on the pad for the first mission.

While all efforts in Mission Control and the contract support team focused on lowering temperatures and providing extra time to prepare for the launch of the first crew, Marshall concentrated on developing replacement surface protection, which would be carried on the CM and installed by the crew. In the weeks after the launch, scores of ideas were offered from both inside NASA and from contractors across America. Some of the more fanciful ideas ranged from spray paint and wallpaper to window curtains, deployable weather balloons and extendible metal panels. From the huge list, ten were shortlisted for evaluation and partial demonstration, within the guidelines of 'being light, and the deployment relatively simple', and having the ability to fit inside the CM.

Rescue plans

With input from the Marshall and Johnson engineering directorates and astronauts, three solutions emerged as favourites for further development:

- From a long pole attached to the telescope mount, a shield would be extended across the exposed skin of the workshop. This required more EVA training than the time allowed, although the astronauts had trained for EVA work at the ATM, and in doing so had found that with suitable hand-holds and portable foot-restraints they could easily face the proposed working area.
- A shade would be deployed from the hatch of the CM flying alongside the workshop. This would be attempted as soon as the vehicle arrived near the station, and the crew would visually inspect the damage before they attempted docking. The problem with this least complex design was actually maintaining station keeping during the stand-up EVA.
- The deployment of a shade 'device' through the OWS scientific airlock on the solar side. This was by far the simplest option, as the crew could work from the inside the pressurised – but hot – OWS. The problem with this option was in the design of the shade that had to be packaged to fit through the opening just eight inches square, but then unfold to cover an area 23 feet square. In addition there were no data available to show that the airlock would be free of debris to allow deployment through it before the astronauts could see it.

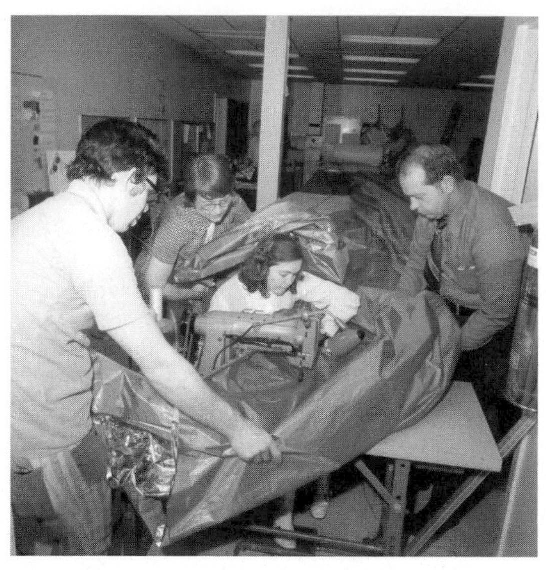

How to make your very own space station sunshade. Take one seamstress and a double needle Singer sewing machine, three layers of material comprised of a top layer of aluminised Mylar, the middle layer of laminated nylon ripstrip, and a bottom layer of thin nylon, and three helpers to feed the material through the machine.

It was therefore decided to pursue the first two options for SL-2, and possibly fly an airlock shade on the second crew once the station had been visually evaluated by the first crew and the temperature hopefully controlled to a more comfortable level.

To evaluate the problems and develop the proposals, a team at Johnson handled the Apollo CM deployment method, and a team headed by veteran spacecraft designer Cadwell Johnson constructed the sail to be deployed. The problems with Skylab had made national headlines since the launch, and while Building 30, housing the centrifuges at JSC, was adequate for the fabrication and testing of the solar sails – it was also on the regular public tour route. The team felt as if they were working in a goldfish bowl as their activities were watched not only by the rest of NASA and members of congress (who paid the bills), but also by hundreds of tourists and the worlds media.

For several days the team worked to meet the launch deadline of 20 May. Seamstresses sewed the sail material, and then parachute packers carefully folded and packaged the device, handing it over to designer engineers to attach various fasteners. The major problem facing the team was in obtaining adequate data on the hardware in orbit, as they found that engineering drawings were not current, and there were, of course, no photographs from space to show the configuration that they had to repair.

The plan for deployment of this shield would see one astronaut stand in the open hatch of the CM, being helped by a second crew-member from inside, while the third flew the spacecraft. This was termed Stand-Up EVA (SUEVA, pronounced 'Seeva') and was tentatively designated the primary shade for deployment. After attaching

Astronaut Bill Lenoir evaluates the deployment of the parasol sunshade device at JSC. Designed to fit in the T027 experiment photometer canister, the canopy, fully deployed, measured 22 × 24 feet.

one part of the shade to the aft end of the OWS, the CM would be then manoeuvred across the rear of the station to allow attachment to a second point. The CM would then be flown across the exposed area of the workshop, deploying the solar array as it went, until, at the ATM end, it would be secured off, allowing the crew to then dock normally at the forward port.

Meanwhile, at Marshall, teams had been evaluating an EVA from the ATM structure. On the evening of 15 May Rusty Schweickart (back-up Commander of SL-2) and Joe Kerwin (SL-2 Science Pilot) entered the huge water tank at Marshall, which housed the Skylab mock-up which the crews used for ATM EVA training. The two astronauts evaluated several devices, including a favoured window shade device, and determined what an astronaut could see from the ATM and from the restricted field of view from inside the EVA helmet.

By 1973, American astronauts had accumulated more than 170 man-hours on EVA since 1965. However only about twenty hours of this was in Earth orbit or in deep space on the way back from the Moon. The remainder had been on the 1/6 gravity of the Moon, and the difficulties encountered during the Gemini EVAs from lack of adequate restraints and additional efforts that put a strain on the life support systems were well remembered. Sending out astronauts so early on a mission for the first time in Earth orbit since Apollo 9 in March 1969, and to perform the most rigorous operation since 1966, with only an abbreviated training programme, was not favoured by JSC.

The parasol sunshade device undergoes deployment tests in the Technical Service shop, Building 10, at JSC.

The post-EVA simulation press conference at Marshall reviewed the experiences of the two astronauts, and concluded that the design needed further evaluation. Within hours of the underwater test, a new design featured two 55.7-foot poles assembled from eleven smaller sections (to allow them to fit in the CM), and then cantilevered from the telescope mounting. Running along the length of the poles through eyelet's was a rope to which the shade was attached. By pulling on this, the deployment would be like hoisting a ship's sail (a resemblance not lost on the all-Navy SL-2 crew) which, it was hoped, would cover the affected area.

Which method?

During a multi-centre management teleconference on the 19 May, the main topic of conversation was the task of choosing the primary method of solar shield deployment. One of the first agreements was the decision to delay the launch of the first crew until the 25th, allowing more time for sail development. There was also an agreement to eliminate the SUEVA option, as it would be at the end of a 22-hour day for the crew, and the contamination effects of firing the CSM reaction control thrusters across the ATM solar panels and telescope were not known. Medical officials were also against attempting to deploy twin-pole assembly early in the mission and before the crew became accustomed to weightlessness, as the adaptation to the environment was still an unknown factor and, indeed, a major objective for the first crew on which to obtain data.

Meanwhile, the development of the scientific airlock deployment had gained momentum. Tests of the system proved that a combination of coiled springs and telescopic rods could fit inside the standard airlock canister, fit into the airlock, and be deployed smoothly.

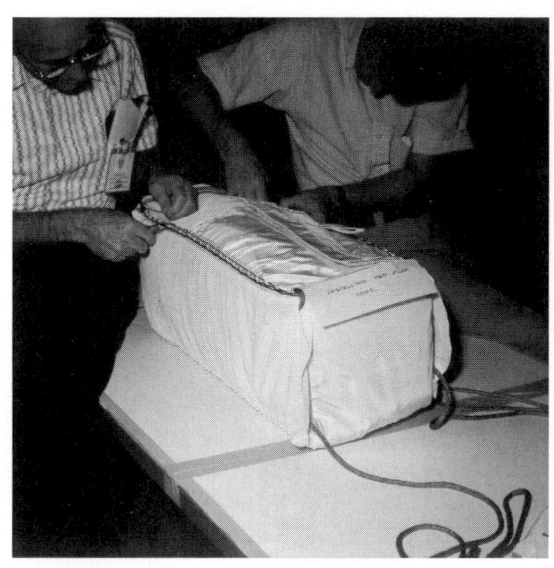

Two technicians package the JSC SEVA 'sail' during pre-flight evaluations prior to SL-2. When unfurled, the 'sail' measured 28 feet in length, and varied in width from 20 feet at the narrow end to 26 feet at the widest point.

Jack Kinzer (a close friend of Conrad), of the Houston Technical Services Division at JSC, developed the system, drawing on his experiences in Apollo CM weight and size stowage limitations. Kinzer jury-rigged his idea from a parachute canopy attached to telescoping glass-fibre fishing rods fitted in hub-mounted springs. Using a canister similar to those used to place experiments in the airlock, and deploying the device by pulling strings tied to the telescoping rods, he clearly demonstrated that as the poles extended and locked, the parachute formed a smooth canopy over the area.

In the teleconference, support for the airlock deployment came from Deke Slayton, Director of Flight Crew Operations, and from Conrad himself as Commander of the mission that would attempt to deploy it. The decision came down to: Marshall twin-pole assembly (still 55 lbs overweight) against the airlock deployment devised at JSC, with the SUEVA in back-up third place.

Airlock deployment selected

It was decided to go with the airlock deployment as the simplest, safest and quickest method, needing less training and use of consumables. However, as an added option the development of the twin-pole system would proceed, to be available for the second crew if necessary.

Development of aluminium rods and stronger springs for the airlock system was complicated when it was found that the airlock on Skylab was actually off-centre from the area to be shaded. This was solved by designing all rods the same length to easily deploy symmetrically, but with the canopy attached off-centre on two of the deploying rods.

Marshall Space Engineer Charles Cooper, assisted by a team of safety divers, works with a 'shepherd's hook' during underwater simulations of techniques which the Skylab 2 crew could use in releasing the trapped solar array on the workshop. As well as devising methods to free the array, and the tools to achieve it, the simulations evaluated designs of restraints and hand-holds in an area not designed for EVA.

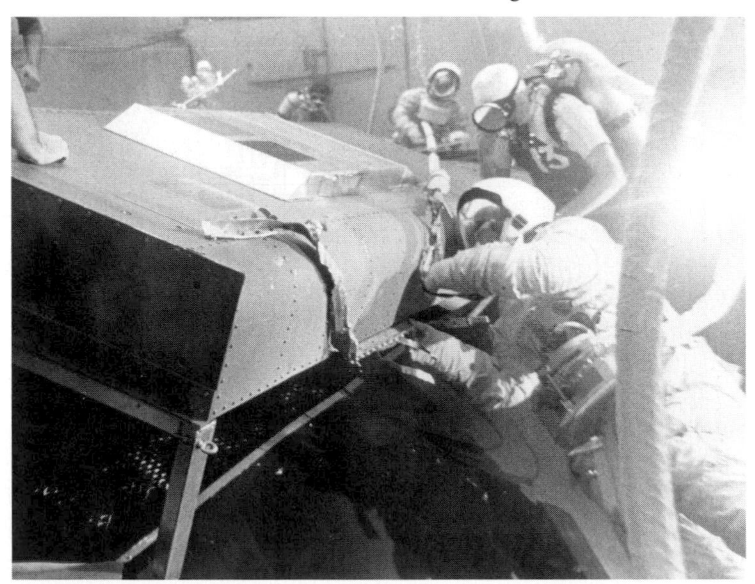

Once the engineers had devised possible methods of performing the EVA, the astronauts tried the procedures themselves. Here, Skylab 2 back-up Commander Rusty Schweickart uses tools under consideration to free the trapped array in the Marshall Neutral Buoyancy Simulator (NBS) (now termed the Water Emersion Test Facility – WETF).

Once the decision had been made to use the airlock shade, the next task was to develop the actual insulation material to be used. The shield had to be lightweight, compact and deployable, non-contaminating, and capable of withstanding a wide range of temperatures, as well as not tending to pull back towards its stowed configuration once deployed. This seemed easier than it actually proved, and for the designers it was a race against time.

It was eventually decided to use spacesuit material less than ½ inch thick, comprising of layers of nylon, mylar and aluminium. The nylon had a tendency to deteriorate under ultraviolet rays, and although it could be coated in thermal paint this would add extra thickness to the structure – a margin that did not exist, due to the already tight confines of the airlock. After further evaluation the decision was made to fly the canopy without thermal covering, as it was expected to last over the 28 days of the mission, and the second crew could replace it if required. In the final hours of spacecraft closeout on the 24th, the parasol and extra equipment was stowed in the cramped SL-2 CM on the pad at the Cape.

As the count-down clock ticked towards zero and the crew prepared for launch, final evaluation of the status onboard the station revealed that power supplies would be adequate for the 28-day mission if the demand was reduced significantly, but would fall short of requirements for the two 56-day missions to follow. A plan was devised to launch a 'solar wing module' to dock with the side port on the MDA to help alleviate this short-fall, but this would also prevent any rescue CSM from docking with the station in the event of a systems failure on the prime CSM. The solar wing module was being devised by Rockwell International, and could be attached to the ASTP docking module and fixed to the radial port, which would still allow a docking by the CSM in an emergency.

Another option revealed itself in further analysis of data from the stricken Skylab. What was left of the thermal shield still held one of the large solar wings in place. If the first crew could clear debris on an early SUEVA, an attempt to unfold the restricted solar wing could hopefully be attempted by the first crew, but more probably by the second, rendering the solar wing module redundant.

To assist the crew in freeing the solar array, a selection of off-the-shelf tools was evaluated, and tree-loppers were found to be the most adequate. A call to a company in Missouri which made tools for power and telephone companies, produced a heavy-duty cable cutter and a universal tool with prongs for prying and pulling. Because of the separation distance of the CM hatch and work areas, these were modified for attachment to a ten-foot pole.

In underwater evaluation on the 22nd SL-2 Pilot Paul Weitz demonstrated that he could free a mock-up solar wing; but the pointed cutters were a hazard to the spacesuits in the confines of the CM, and were replaced by blunt-ended cutters, shipped to the Cape. At 03.00 EDT on the morning of 25th May, just prior to the crew climbing into their CM, the final stowage was completed and the final preparations for launch continued.

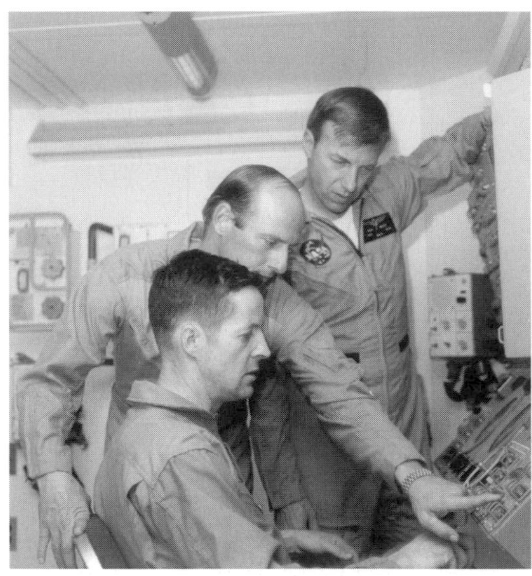

The Skylab 2 crew during training in 1972 at the Skylab mock-ups and trainers located in the mission Simulation and Training Facility at MSC. (*Foreground*) Science Pilot Joe Kerwin, (*centre*) Commander Pete Conrad, and (*background*) Pilot Paul Weitz.

SKYLAB 2: 'WE CAN FIX ANYTHING!'

The launch of Skylab 2 was flawless, and was highlighted by Conrad's comment: 'We can fix anything!', as the Saturn 1B lifted off the pad. By mid-afternoon of the same day, the crew had got their first close-up view of the OWS: Conrad: 'Tally-ho the Skylab! We got her in daylight, 1.5 miles, 29 feet per second. As suspected, solar wing two is gone completely off the bird ... solar wing one is ... partially deployed. There is a bulge of micrometeroid shield underneath it in the middle, and it looks to be holding it down.'

With the astronaut's commentary, the true extent of the damage to Skylab became clear. Not only had the protective shield been lost during the launch, but it had also ripped off one of the power-generating solar arrays, and seriously restricted the deployment of the second. While Weitz filmed the damage, Conrad completed the CM fly-around, and expressed conviction in the crew's ability to free the array during the SUEVA. They also reported the blackening of the gold foil on the station's surface due to the intense solar heat, and that the scientific airlock planned to be used to deploy the solar shield was free of debris.

In order to save station-keeping fuel they docked the CSM to the front port of the OWS while they ate dinner and prepared for the SUEVA. For space veteran Conrad on his fourth spaceflight, the sight of the huge Skylab docked in front of him brought back memories of earlier missions: 'Boy, I've had some big things on my nose in space before, but this is by far the biggest. It sure beats the Agena [Gemini 11] or lunar module [Apollo 12].'

Joe Kerwin – Science Pilot on the first manned mission – undergoes suiting up in the Manned Spacecraft Operations Building at KSC.

Stand-up EVA

After dinner, and clad in full pressure suits, the crew undocked from Skylab and depressurised the CM, opening the side hatch. Paul Weitz worked the tools while 'standing' in the open hatch, and Conrad flew the CM as close as two feet from Skylab. Restrained only by Kerwin, who held his ankles and feet, Weitz later recalled: 'It was a very frustrating experience all in all.'

As a precaution against firing the forward thrusters on the Service Module (those facing Weitz' back), Quad A FWD was turned off so as not to inadvertently 'barbeque' the pilot 'extra-crispy' variety! Conrad had the other fifteen out of sixteen thrusters to maintain attitude. Several times Weitz pulled at the debris, disturbing the stability of the station and causing the gyros to compensate. As the oscillations threatened to pull him out of the hatch, they were approaching the night-side pass, where the attempt would have to be abandoned.

Forty minutes later, as the spacecraft flew in range of the California tracking station, the obvious difficult being experienced in freeing the array was expressed in no uncertain terms by a string of four-letter words. After CapCom reminded the crew they were on VOX (open microphone) Conrad proceeded to give a gloomy report. The problem was a ½-inch wide strip of metal that had become wrapped across the Solar Array System (SAS) during the tearing of the shield. Its metal bolts

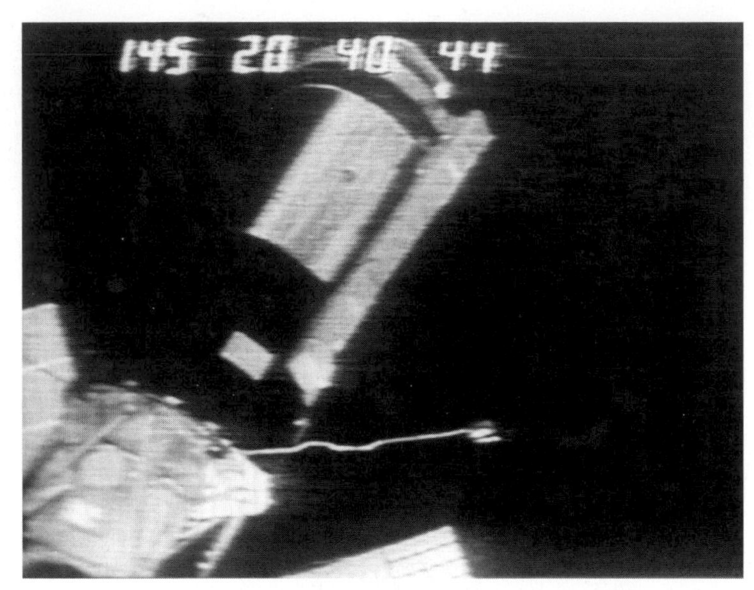

A TV view of the Skylab 1 OWS from the Skylab 2 CM during a fly-around inspection of the cluster. The large rectangular white area is the missing piece of the micrometeoroid shield which was ripped off during the ascent. The figures on the frame are the Skylab 1 ground elapsed time (GMT: days, hours, minutes, seconds).

tangled themselves in the array, and despite Weitz' strenuous attempts to free it, the solar array remained fixed solid: 'We couldn't get it [the SAS panel] out right now. We're all trying to break it [the strap] loose. It's only half an inch strap, but man, is it riveted on!?' Conrad: 'We ain't going to do it with the tools we got.'

They were told to abandon the attempt and dock with the workshop. Trying to get back into the capsule proved as difficult, with the long cutters and boat-hook hitting Conrad twice on the helmet and forcing him to seek shelter until Weitz was back inside. Kerwin too, was kicked in the helmet (as the Pilot snagged console switches) muttering unscientific words of abuse, to again be reminded by CapCom that they were still on VOX.

Docking frustrations

Giving up the attempt, the crew closed the hatch and prepared for docking. When Conrad attempted this, the probe did not attach to the drogue, which added to their frustrations. Three further attempts failed, resulting in the crew depressing the CM and removing the forward tunnel hatch to bypass some of the electrical connections, and then using the SM thrusters to push the two spacecraft together, firing the twelve capsule latches. Conrad later reported that he 'poured the coals' (burned the RCS engines) to achieve docking, and in doing so he actually created a dent in the forward drogue assembly of the MDA.

At the end of a very long and tiring 22-hour day, the crew had finally docked to Skylab and they then enjoyed a well-earned rest in the CM. The workshop would be

Taken during the 7 June EVA, this close-up shows the partially deployed solar array and the aluminium strapping that prevented its deployment. It was this strapping that was later cut by the crew during the EVA to help free the array.

entered the following day, after their sleep period. Safely docked to the front of Skylab, what was waiting for them on the other side of the hatch? They knew the pressure inside was stable, so they would not need their pressure suits; but could they stand the heat, and would toxic gases from overheated insulation be mixed into the atmosphere?

Entering Skylab

The next day, wearing gas masks they opened the hatched into the station and entered the workshop. Weitz was the first inside, and reported a dry heat resembling a desert. He found no evidence of toxic gases and pronounced the atmosphere safe, although at 130° F it was 'rather warm' ! They found, however ,that the humidity was quite low, and that they could stay inside the workshop for up to five hours.

One of their first tasks was to deploy the parasol to reduce the temperature. They proceeded to the airlock area on the Sun-facing side of the workshop and began to assemble the canister containing the parasol. The intense heat was stifling as they worked, and they had to retreat to the relative comfort of the CM from time to time to take a break from the workshop 'oven'.

Preparations for moving the parasol outside took some time, with the crew spending about 6–7 man-hours in assembling the equipment. Conrad alone spent 90 minutes assembling elements of the solar parasol to deploy through the airlock. The extra complication of zero g added to their delays in preparing the equipment for deployment.

They eventually extended the parasol, and the folded arms swung out, spreading the fabric. Visual inspection by the crew, from the CM windows, confirmed deployment, but disappointment in their voices revealed that the shield had not correctly deployed, with only two-thirds deployment due to the crinkling of the parasol. Loud cheers could be heard in Mission Control, however, and NASA expected the shield to unfold correctly during solar passes as it heated up.

Over the next three days the interior of the workshop cooled considerably, and the crew found the limits of the parasol by running their hands across the inner wall of the workshop, feeling the significant changes in temperature. Overnight, the internal temperature dropped to a more comfortable 90° F (close to Houston temperatures), but for the first few nights the crew chose to sleep in the docking adapter, where it was a pleasant 68° F.

Rescue EVA

After the crew indicated they would be able to release the stuck solar array, plans were put in motion to achieve this during an EVA, planned for 7 June. During Mission Day 3 (MD3) the crew talked to the ground about the EVA plans, in place of a planned news conference. As the details of the EVA was defined, the crew configured the spacecraft from launch configuration to preparing the scientific experiments which began in full on MD5, 29 May. While the crew were busy with unpacking the OWS, performing what experiments they could and preparing for the EVA, on the ground a team led by Rusty Schweickart evaluated the problem of releasing the solar array.

Information and pictures from the crew helped Huntsville assemble a reasonable simulation of the actual hardware in space. What Schweickart's team developed was a two-person EVA during which the astronauts would move to the antenna boom at the end of the workshop. They would attach a 26-foot cable cutter on a pole to the debris, and then one of the astronauts would use the pole as an improvised EVA handrail to reach the trapped solar array. Once there, he would attach a nylon rope with hooks at both ends, between the solar wing and the airlock shroud, used to replace the frozen hydraulic damper on the array once the debris had been removed. EVA simulation by Schweickart and SL-4 Pilot Gibson had revealed that despite the

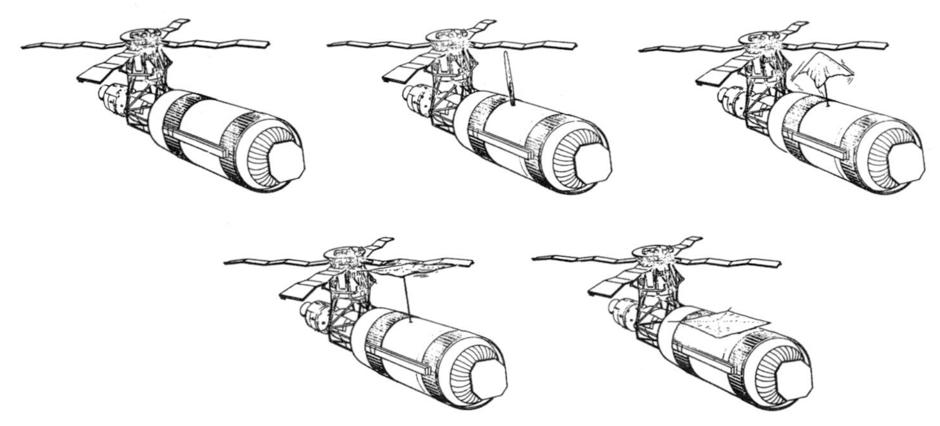

A sequence of drawings illustrating how the parasol sunshade was deployed from the scientific airlock during the first mission.

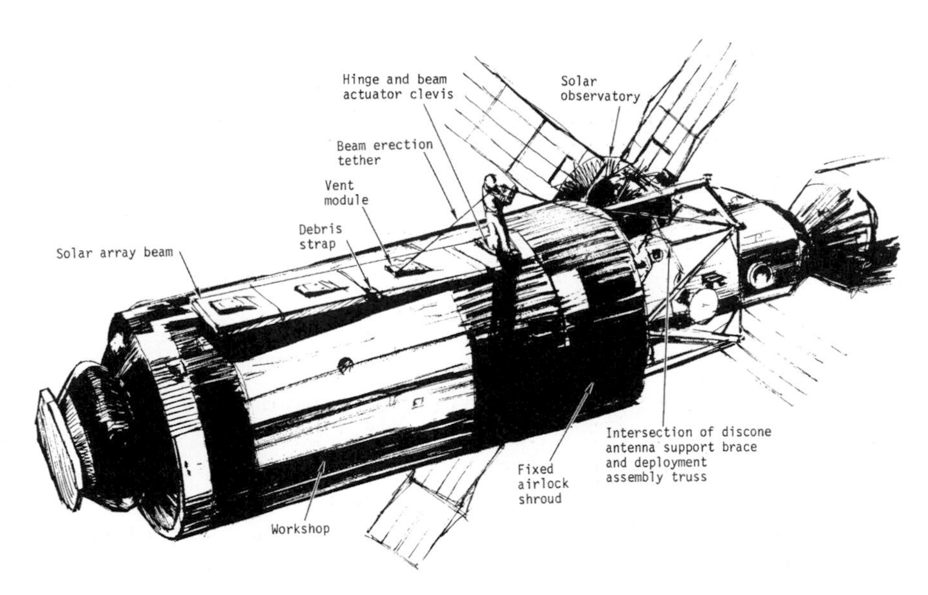

Solar array beam deployment technique.

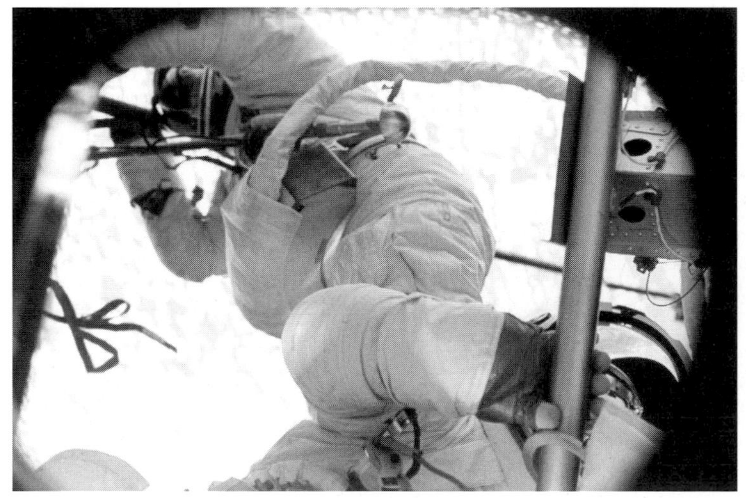

Conrad (*background*) and Kerwin (*foreground*) during the 7 June EVA. Conrad carries cable-cutters to sever the aluminium strapping that prevented the solar array from deploying. Weitz took the picture through one of the optical windows inside the airlock module.

lack of foot-holds in the work area, the task could be completed in the water tank – and what was completed underwater was normally possible in space.

In preparing for the EVA, Conrad was to wear the protective outer work-glove previously worn by Weitz on the SUEVA. These were worn over the integrated pressure-suit glove. Conrad found that the gloves fitted, but the thumb was an inch

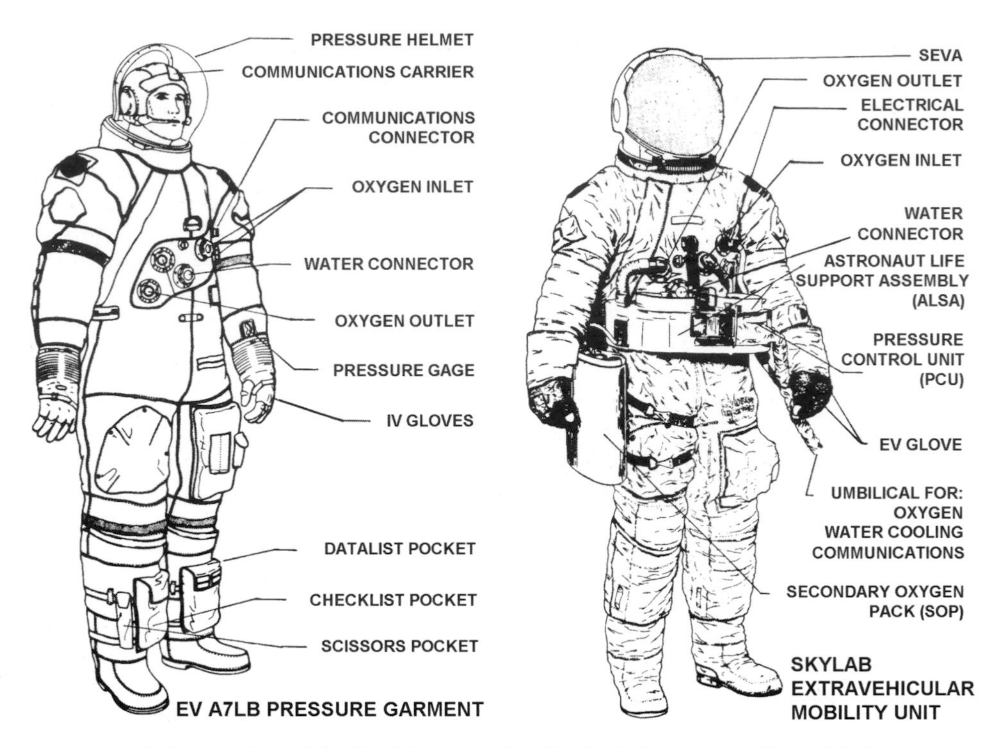

An artist's impression of the Skylab spacesuits. On the left is the Apollo A7LB IVA suit worn in the Command Module for launch, and at right is the EVA configuration of the same suit with additional EVA provisions.

too long. During EVA the thumb was the most important digit to hold things in a pressurised, gloved hand. Conrad cut the end of the thumbs of the outer glove to fit better, but his hand movements were still restricted, as the gloves were not made for his hands. He therefore decided not to use the outer gloves during the EVA.

The Skylab EVA suit was an Apollo A7LB lunar pressure garment modified for use in Earth orbit. There were fewer thermal layers, because the environment was not so hazardous as that on the lunar surface, and the lunar over-boot was not required. Instead, a rigid sole was fitted with a restraint devised to allow use in various foot restraints on the outside of Skylab.

The suit consisted of a liquid-cooling garment, made up of a network of coolant tubes routing cold water around the suit and returned to the workshop via umbilical connection for heat rejection. The Pressure Garment Assembly (PGA) was worn over this, and was attached to the gloves and helmet with visors for solar-ray protection. The suit was used for CSM operations at launch, as well as for EVAs.

The PGA provided an oxygen environment for not only breathing but also for ventilation and pressurisation, and was provided with bioinstrumentation and communications electrical provisions. The suits were stored in the forward compartment of the OWS in foot restraints, and were attached to the drying station, where air was blown through the suits to dry them after each use.

The Life Support Umbilical supplied water, oxygen and electrical power during EVAs. It was 60 feet long, marked at 5-foot intervals, and visually tracked as the crewman exited the airlock. The LSU supplied the Astronaut Life Support Assembly (ALSA) in the suit, where it was regulated and distributed throughout the suit by a Pressure Control Unit (PCU). An additional (back-up) oxygen supply was available by a Secondary Oxygen Pack (SOP) which was usually attached to the thigh of the wearer.

On 7 June, following conversations with Schweickart (who would act as the EVA CapCom), and practise using the tools inside the station the previous day, the EVA crew of Conrad and Kerwin exited the airlock in the darkness of a night-side pass. At first they were disorientated, as they had no visual references other than the Skylab structure: Conrad: 'Where the hell's the world anyway?' CapCom (Schweickart): 'Skylab, Houston, we're right here. We're listening in loud and clear.' Conrad: 'Oh! I didn't mean the *world* world, I meant the clouds and the Earth and sea world underneath.'

Conrad assembled the 25-foot long pole under the floodlights of the airlock shroud, and attached the cutters at one end. He then traversed along the pole to ensure that the cutters were against the debris. This confirmed that he made his way back to Kerwin, who would do the actual cutting.

When Kerwin tried to close the cutters against the debris, he found that they kept slipping because he was not able to secure a firm position for himself. For more than thirty minutes he tried in vain to close the cutters and reported that with one hand wrapped around the pole to restrain himself, using the other free hand to close the cutters was useless.

It was the old problem encountered during the Gemini EVAs: the lack of firm footholds. Putting pressure on the working area (in this case, the cutter) simply produced momentum in the opposite direction, and he floated away. On several occasions he moved his hand away from the pole to open the cutters, only to have them move away from the debris as he tried to use them. With his pulse racing to 150 beats per minute, he passed an area of no communications from Houston, and evaluated his situation. He realised that he needed a firmer position against the edge of the workshop, and so shortened his tether, doubling the line. When communications were restored with Houston, a happy astronaut reported that the cutters were securely fastened to the debris.

Kerwin pulled on the lanyard to operate the cutters: 'Man, am I pulling?' he told ground control – but nothing happened. Seeing this, Conrad then made his was back along the beam to examine the jaws of the cutters. Just as he reached the cutter end, the jaws snapped shut, freeing some of the metal strap in a sudden movement that catapulted Conrad into space. His safety tether restrained his sudden unexpected movement away from the station, after a big surprise for both Conrad and Kerwin. The wing had opened some 20°, and their next job was to extend it to its full 90°.

The partial deployment had been anticipated, but the catapulting of Conrad was unexpected. Before cutting the debris, Conrad had hooked a tether to the vent module relief hole in the boom, and secured the other end to one of the solar observatory support trusses. He had some difficulty attaching the hooks to the array, as the holes on the actual array beam were a little smaller than on the ground simulator! After struggling in the pressure gloves, he decided on only one attachment instead of the planned two on the beam.

When the cutters freed the debris, the frozen damper still resisted normal deployment. Both astronauts heaved on the tether, but to no avail, and so Conrad placed his feet on the hinge, stooped to fit the tether over his shoulder, and then stood up. Kerwin again pulled on the tether, and this time the solar wing suddenly released and pranged open, catapulting both astronauts into space, although again they were saved by their safety tethers.

When asked by Houston about the status of the array, Conrad, laughing, commented: 'I'm sorry you asked that question. I was facing away from it, heaving with all my might, and Joe was also heaving with all his might when it let go and we both took off. By the time we got steeled down and looked at it, those panels were out as far as they were going to go at the time.'

After 3 hours 25 minutes, two tired but very happy astronauts re-entered Skylab after a very successful demonstration of man's usefulness in space – as on-orbit repairmen.

Back to the flight plan

By the following day, solar heating had fully extended the arrays, leading to the generation of almost 7 kW of much-needed power. At last the Skylab crew, ground controllers, scientists and programme managers could look forward to real scientific

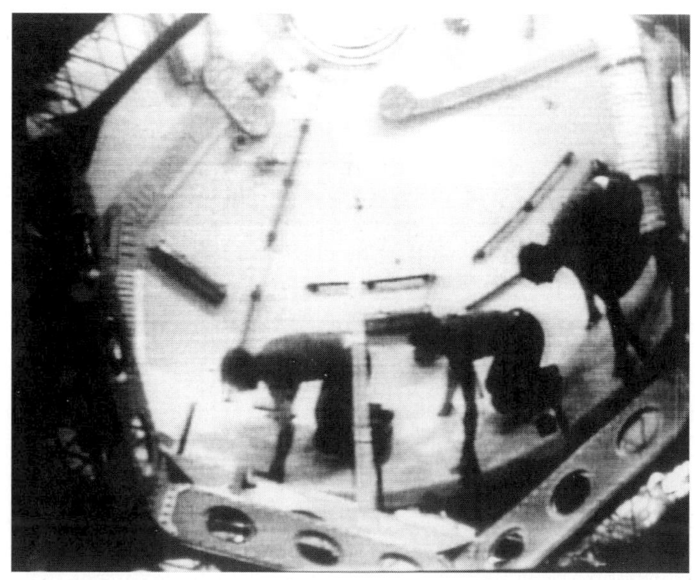

(*Left–right*) Astronauts Weitz, Conrad and Kerwin line up in Conrad's 'Indy Start' for a televised 'run around the dome' on Skylab in the afternoon of Friday, 1 June 1973.

return from the first crew, and plan with confidence for the second and third visits to full duration. Careful monitoring of onboard systems and consumables would continue, and the option of flying the twin-pole array on the second mission remained; but on 7 June 1973, astronauts Conrad, Kerwin and Weitz had saved Skylab from the brink of failure in a dramatic and impressive way.

With the drama of the first two weeks behind them, the crew could settle down to completing their scientific programme for which they had trained for more than two years. An added bonus was a clear demonstration of the usefulness and resourcefulness of humans in space. If Skylab had been a fully automated station, then only limited scientific data would have been gathered before the batteries ran out and the station overheated.

Skylab was saved by the deployment of the protective solar shield and the stuck solar wing, combined with the preventive thermal control manoeuvres and power-saving actions by ground control and the onboard crew. With the station operating in a more controlled and habitable condition, it was time for the crew to proceed full time with their delayed scientific programme and some serious work.

According to Paul Weitz: 'There was a concern in getting the array deployed so that we could get on with our reason for being there. The whole activity package was put together in pieces that could be arranged in various ways, so to accommodate delays due to limited power and other activities associated with wing deployment were not that difficult.'

Science after a rescue

The first few days of activity inside the workshop centred on activating the ATM and EREP experiments ready for data gathering when possible, due to the low power supplies, conducting medical experiments, and preparing for the 7 June EVA. Using the ATM solar arrays for power offered 4.6 kW of electrical power, but 3.6 kW was diverted to systems management across the station, leaving just 1 kW for experiment operations. If the arrays on the OWS had been operating the full power from the ATM, the arrays would have supplied 0.7 kW for ATM operations, 0.45 kW for EREP experiments, and 0.4 kW for medical equipment.

The medical experiments began with runs on the LBNP device and the ergometer. It required a couple of sessions for them to realise that they needed to release the seat and harness. However, even in this unseated position they still had the triangular cleats fitted to pedal, and the handlebars were too low – an adjustment made on the later flights. Experiments with the mass measurement device were completed, but were terminated early due to the heat inside the OWS. One exercise that they had devised was to 'run' around the ring of water canisters in the OWS dome.

The first experiment runs on the ATM and EREP began during MD5 and MD6, although power levels necessitated close management of these and the medical experiments. From MD5, the crew began to sleep in the sleep compartments in the wardroom. By the end of the first week in space, they were becoming efficient in conserving electrical power by turning off lights and fans to supply more power to the scientific equipment, rotating between the ATM solar runs, EREP passes, medical experiments and stellar observations.

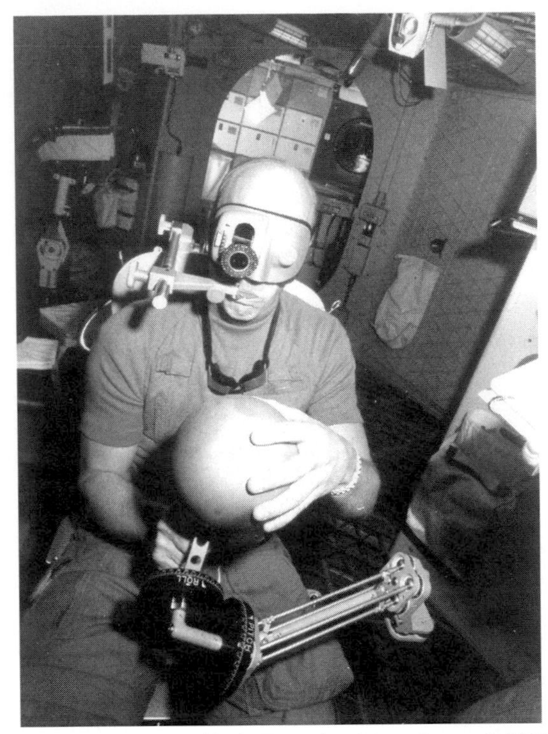

Pete Conrad checks the Human Vestibule Function Experiment (M131) in the work and experiment area of the crew quarters of the OWS trainer at JSC.

Paul Weitz mans the ATM console during the first 28-day manned period. During the mission the crew completed more than 117 hours of solar observations, and gathered 28,739 frames of data.

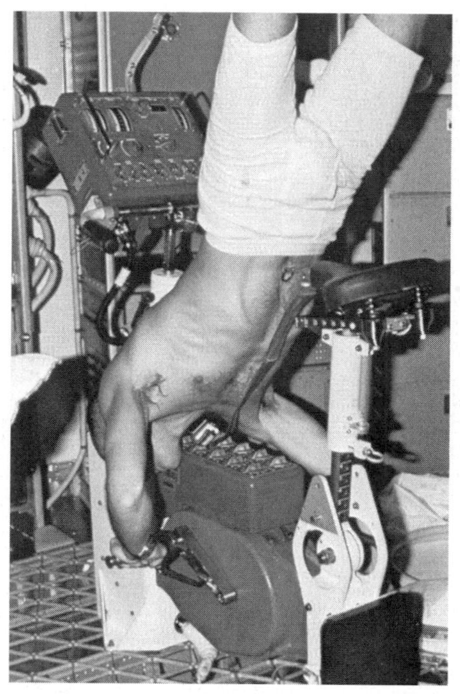

Pete Conrad uses a novel exercise technique on the bicycle ergometer.

Most of the experiments were completed as planned, although six experiments could not be performed using the +Z scientific airlock that was occupied by the parasol. However three of these were performed using modified procedures through the –Z airlock. During the EVAs two more were mounted on the solar observatory truss, and the sixth was mounted to the solar observatory Sun-shield, so that the crew were able to replan the operations of the lost six experiments.

On 2 June Conrad celebrated his 43rd birthday, and during the evening he spoke to his wife and four sons, who were visiting Mission Control in Houston. Mindful of the importance of the EVA five days later, Conrad requested that the next two rest days be deleted to compensate for the preparations for, and activity of, the 7 June EVA. The crew were already beginning to develop an efficient pace of work that reflected how they felt in both achieving the docking and saving the station. The first two weeks were uncertain until the array had been deployed and the power levels rose and the temperature fell. The result was that they worked even harder in the last two weeks to achieve all of the science objectives to ensure that they handed over a good and efficient ship to Bean and his crew.

An overview of science on all three Skylab missions is presented in Chapter 5, but the efforts of Conrad and his crew in rescuing Skylab and obtaining science results was highlighted on MD24, when President Richard Nixon called to compliment Conrad on the excellent performance of his crew. For Conrad, the day was also highlighted with a call from his family on the occasion of his 20th wedding anniversary.

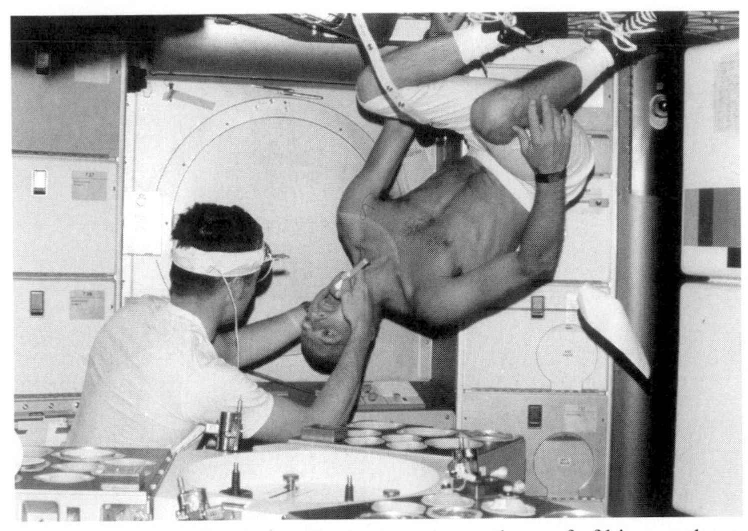

If the patient can turn upside down, the doctor can see the roof of his mouth much more easily. Conrad obliges Kerwin during an oral examination conducted in the wardroom of the workshop.

In the final week before returning home, work on the experiments was interspersed with several re-entry simulations and preparation for EVA on MD26 to retrieve and replace ATM film canisters. There was some discussion in having Conrad's crew deploy the A-frame solar shield, but this was deferred to the second crew. Instead Conrad would deploy a sample of parasol material from the A-frame sail so that half would be in direct sunlight and half in shadow. This would be recovered during mission two to discover the long-term effects of the exposure of the solar-shade material.

The day before the EVA (MD25), the crew were informed that they had surpassed the Russian space endurance record set by Soyuz 11 on Salyut 1 in 1971 at 23 days 18 hours 22 minutes, and that most of the planned experiments for the mission had been accomplished, with 81% of the ATM observations achieved, and 88% of the EREP data runs completed, which gained 60% of the data logged. In addition, the crew had logged more than 90% of all medical investigations, corollary experiments, and the planned five student experiments.

What was even more remarkable was that all the primary objectives had been completed. Most of the Detailed Test Objectives were met, out of 44 planned telecasts they managed 28, plus three unplanned – and they rescued Skylab from what seemed to be certain failure! (The achievements of the Conrad crew can be found in the Appendix on p. 352.)

ATM EVA

In the original flight planning, the first crew were to perform one 2 hour 30 minute EVA (by Conrad and Kerwin) on MD26, to retrieve and replace ATM film cassettes. However, as this was the third of the mission, it gave Paul Weitz a second chance of

Joe Kerwin wears a sleep monitoring experiment cap.

EVA. During the EVA, Conrad and Weitz worked from the 'base camp' – the Fixed Airlock Shroud (FAS) station that was located next to the airlock hatch. It was from here that they began the ascent to the ATM between five workstations along a Deployment Assembly route (commonly known as an 'EVA trail'). In order for them to move across the route, single and double handrails were positioned for translation. The double handrails resembled a ladder, but without the rungs and the astronauts found that moving across them was very easy. Painted blue for visibility, they actually faded in bright sunlight, and the alphanumeric designs – called 'road signs' by the astronauts – were difficult to see in the sunlight. 'Moving across the handrails was as easy as driving down a freeway,' one of them commented.

Moving the ATM film canisters from the airlock to the ATM was actuated by a 'film tree', which was moved by way of three extendible booms located in the EVA bay near the FAS. The controls for these motorised booms were located next to the EVA hatch and were able to be operated manually or by using a backup pulley –type system aptly termed the 'clothesline'.

During EVA, Conrad replaced canisters of film in the six experiments, cleaned the optics of the white-light coronograph with a fine brush to reduce the glare, retrieved exposed materials samples from the exterior of the workshop, that had been exposed since launch six weeks earlier, and then returned to the airlock. The EVA crew also reported on the condition of the parasol, the ATM thermal coatings, and those on the CSM, which had already been in space twice as long as any previous vehicle of the type. Observations of this kind were important to the lifetime evaluation for the SL-3 and SL-4 CSM vehicles that were to remain in space at least twice as long as had the SL-2 CSM.

One final task of Conrad was in his role of acknowledged space repairman, when he attempted to free a stuck electrical relay on Circuit Breaker Relay Module 15. Acting on instructions from the ground, and in an untypical NASA low-tech way, Conrad hit the device with a hammer and freed the relay to supply electricity into the stations power system.

At the end of the EVA the crew had logged 5 hours 21 minutes in three EVAs – twice as much as originally planned.

Home comforts

In addition to the experiments and operational activities of the Conrad crew, the three men were the first to live onboard the station and use the facilities provided for personal hygiene, crew comfort and habitation.

Sleep Three sleep compartments were provided for the crew-members, and each contained a sleep restraint, a light, stowage for clothing and personal items, an intercom, towel holders, and a privacy curtain. The Science Pilot's sleep station was also provided with equipment that was associated with the sleep monitoring experiment. The crew-member adjusted the compartment's noise level, ventilation and personal comfort as he required.

Complete body restraint was provided by adjustable fittings to suit the individual. The assembly featured a welded tubular frame, a thermal back assembly, a comfort restraint, top and bottom blankets, a headrest, and cover and body restraints; and there were additional blankets and headrest covers for exchange throughout the missions every fourteen days. Although adequately restraining the crew-member during sleep, the restraints allowed for reasonably quick egress in an emergency situation.

Personal hygiene This facility was located in the waste management compartment, and was used to maintain health and personal grooming. There was a water module

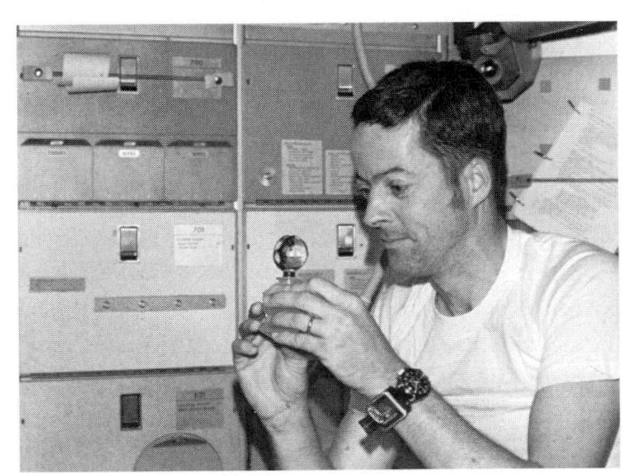

Joe Kerwin monitors the release of a globule of liquid during a science demonstration exercise, and for his own entertainment.

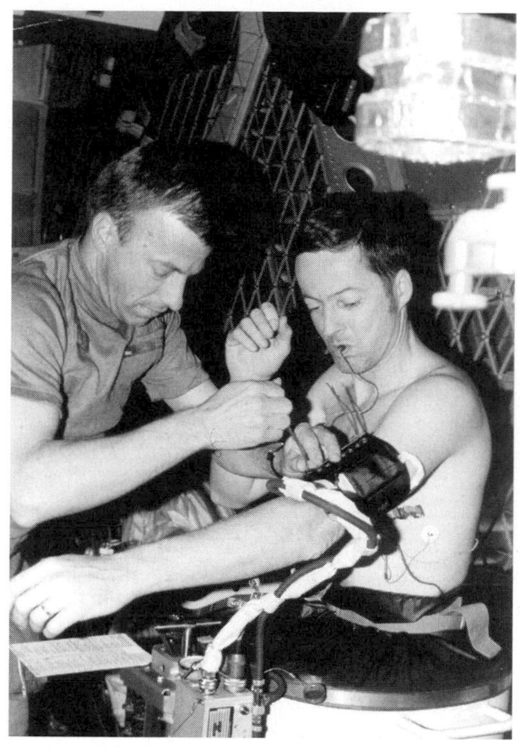

Taken during the mission, this time Kerwin is the test subject in the LBNP chamber, while Weitz assists with a blood-pressure cuff on the Science Pilot's arm. The data collected assisted in the understanding of his period of cardiovascular adaptation during the flight, and provided information on his response to returning to gravity after the flight. The wall behind Kerwin is the grid floor.

that allowed partial body-cleaning, consisting of a hot water dispenser with a wash cloth squeezer. The crew dispensed water into the cloth, then placed it in the squeezer to squeeze out excess water into a bag that was drained through a filer into the waste tank through a vacuum dump system, allowing the crewman to use the wet wash cloth as required.

The shower featured a collapsible enclosure that utilised a constant airflow as a gravity substitute to move water over the occupant. A 6-lb water capacity bottle was filled from the waste management compartment water heater, pressurised with nitrogen, and attached to the showers ceiling location. Using a manually operated hand-held spray nozzle, the nitrogen gas pressure expelled the water from the bottle through a transfer hose to the nozzle and on to the user. A soap dispenser was fixed by Velcro to the ceiling, and supplied a quarter of a fluid ounce of liquid soap for each shower. Removal of water from both the astronaut and the inside of the shower was by means of a suction head that featured an air blower that pulled air through a filter that protected the blower, and was connected by hoses to a centrifugal separator that deposited waste water in a collection bag.

The mirrors were made of unbreakable polished stainless steel. Personal hygiene equipment for each crew-member included a shaving razor featuring blades, or a wind-up razor, a tooth-brush and toothpaste, combs, brushes, nail clippers, files, scissors, shaving cream, hand cream, and body deodoriser.

Drying stations provided an opportunity to dry towels and wash-cloths, and

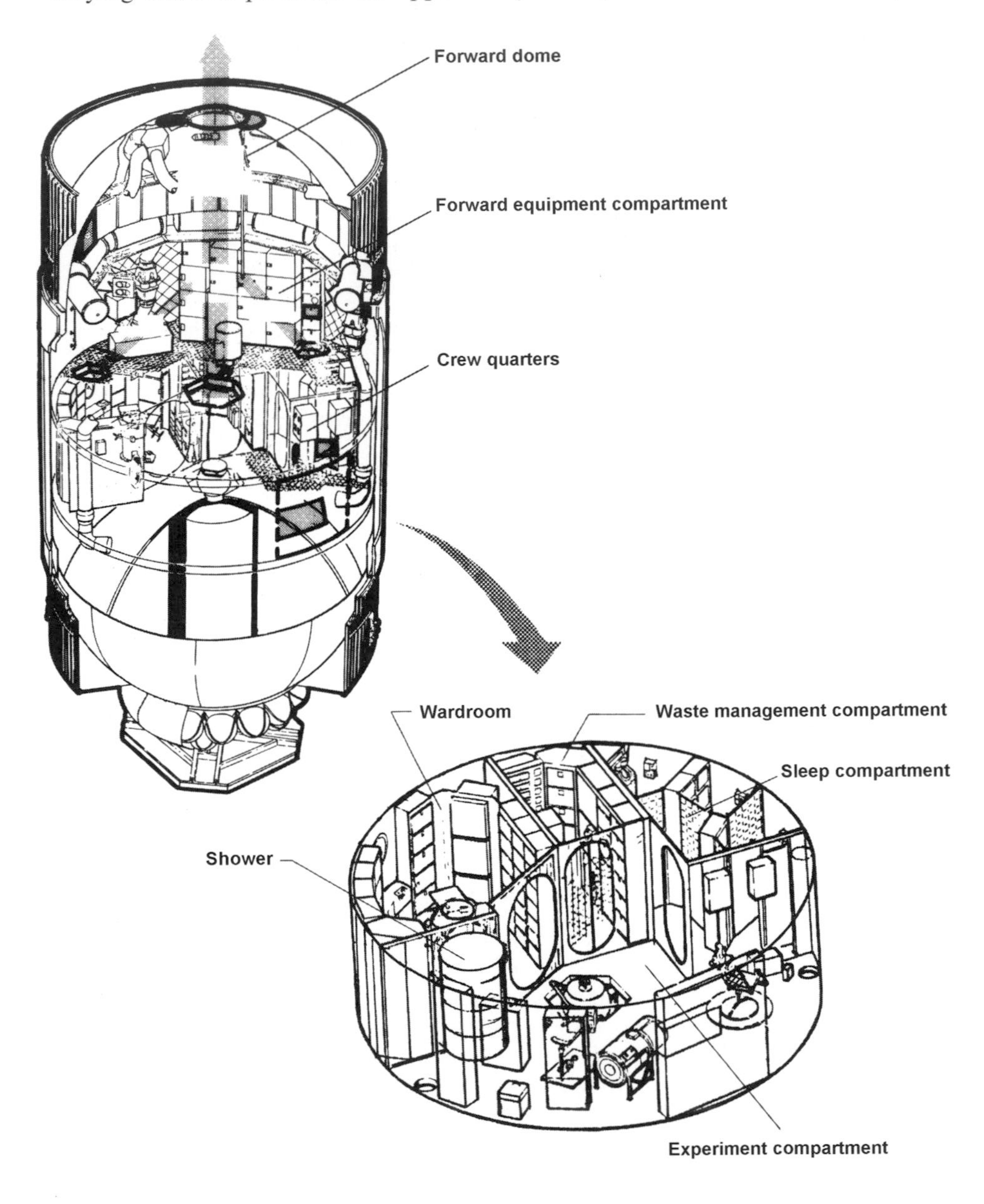

Crew quarters in the OWS.

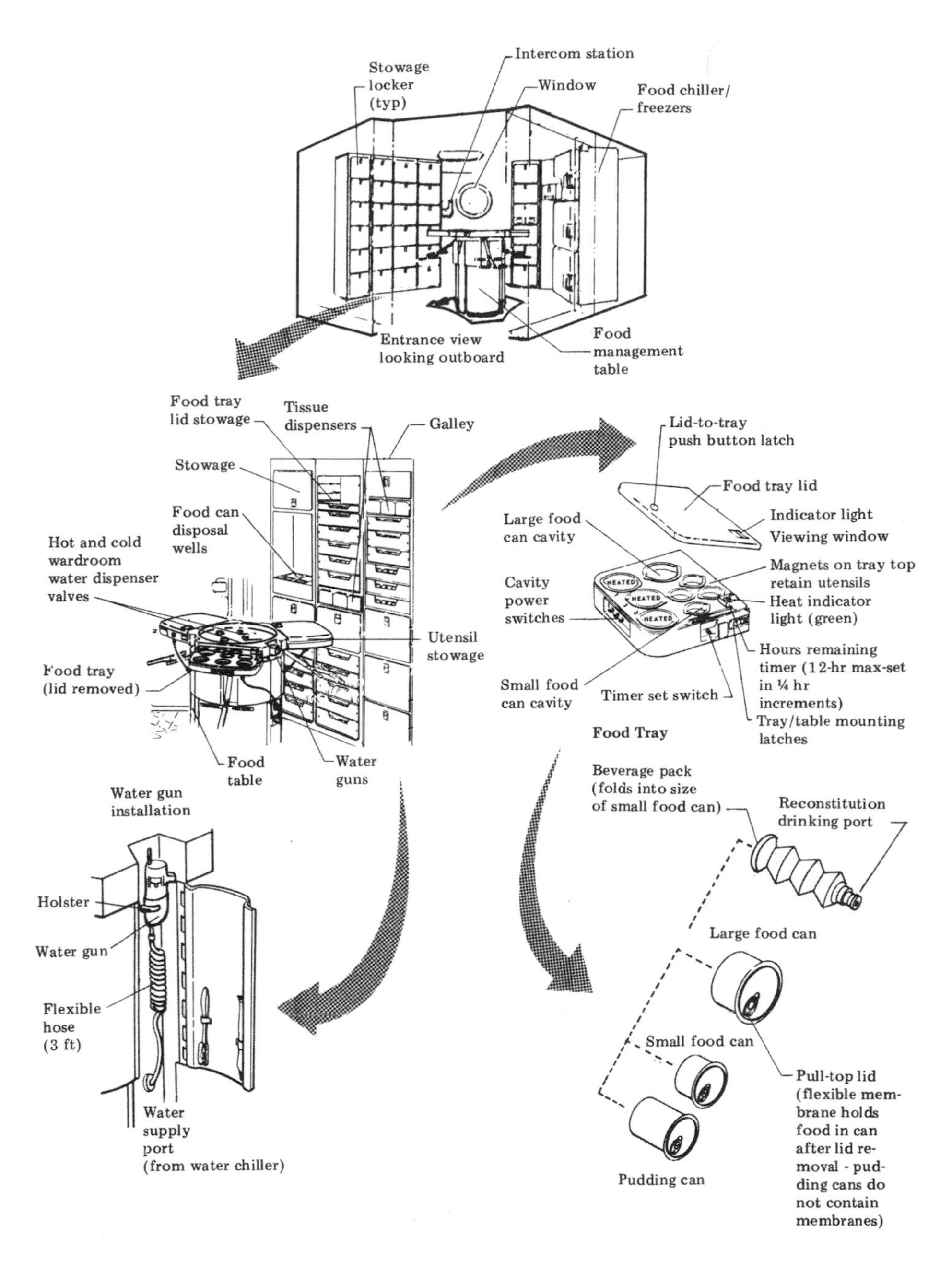

Wardroom eating facilities.

there were 840 reusable 12-inch square wash-cloths, and 420 reusable 14 × 32-inch towels for skin drying. Wash cloths were located in three dispensers, one for each man, holding fourteen days' supply of two per day. Nine lockers held the replacement wash cloths. Towel dispenser modules in the wardroom held eighteen towels each, providing a six-day supply of one towel per man per day. They were all made from rayon polynosic terrycloth, and included coloured stitching for individual crewman colour-code identification.

Crew identification for the hygiene kits and dispensers featured the 'Snoopy' emblem, with colour-coded backgrounds of red for the Commander, white for the Science Pilot, and blue for the Pilot

Waste management The management of body waste of nine men, and the daily trash from what became more than 170 days of manned operations, was a challenge to the designers. To complicate the design further, samples from the body waste were required for some of the medical experiments – a requirement that had been known since 1968. NASA examined the two available systems – one devised for the USAF MOL programme, and the other for the Biosatellite programme. In 1969, and during the following two years, the development of a suitable system for Skylab became a challenging problem of how to collect and store sample waste products. A restriction in being able only to test the design in parabolic flight resulted in a plan to fly the Skylab faecal collector on the Apollo 14 lunar mission, although this met with a mixed reaction. The MSC Skylab Office supported the test, while the Astronaut Office was indifferent to it; but when Al Shepard, the Commander of the flight, learnt of the plan, he simply vetoed it. He was adamant that he would not include a faecal collector system in his mission objectives!

On Skylab, the waste management compartment performed the same function as a toilet on Earth. The user sat on a seat, but faced the floor!. The Skylab toilet was on the wall of the compartment – which is not a problem in zero g, although to restrain the user on the seat, there was a lap belt and hand-holds. A pair of foot loops on the floor allowed standing use of the urine collection device, and a blower unit provided a gravity substitute airflow (suction) to draw the waste towards either the faecal or urine collection systems.

Solid waste was deposited in a bag that was removed after each use and replaced with a new bag. The sealed bag was then weighed on a mass measuring device, vacuum dried, and stored in the WMC. Liquid waste was drawn into one of three storage areas (drawers) by means of the astronaut's individual collection sheaths. The urine was then stored for 24 hours. It was, however, difficult to collect the urine, separate the air, and withdraw a daily 4 fluid ounce sample. A refrigeration coolant system cooled the urine separator and the sample bag, which were stored until the end of the mission. Samples of crew vomit were also retained for post-mission analysis.

The remaining urine was deposited, with other 'biologically active waste' (clothing, filters, food cans, sleep restraints, tissues, wipes, towels, washcloths, and other items) in trash collection bags, through the trash airlock, into the oxygen tank below the wardroom floor. The bags were collected from eight locations around the

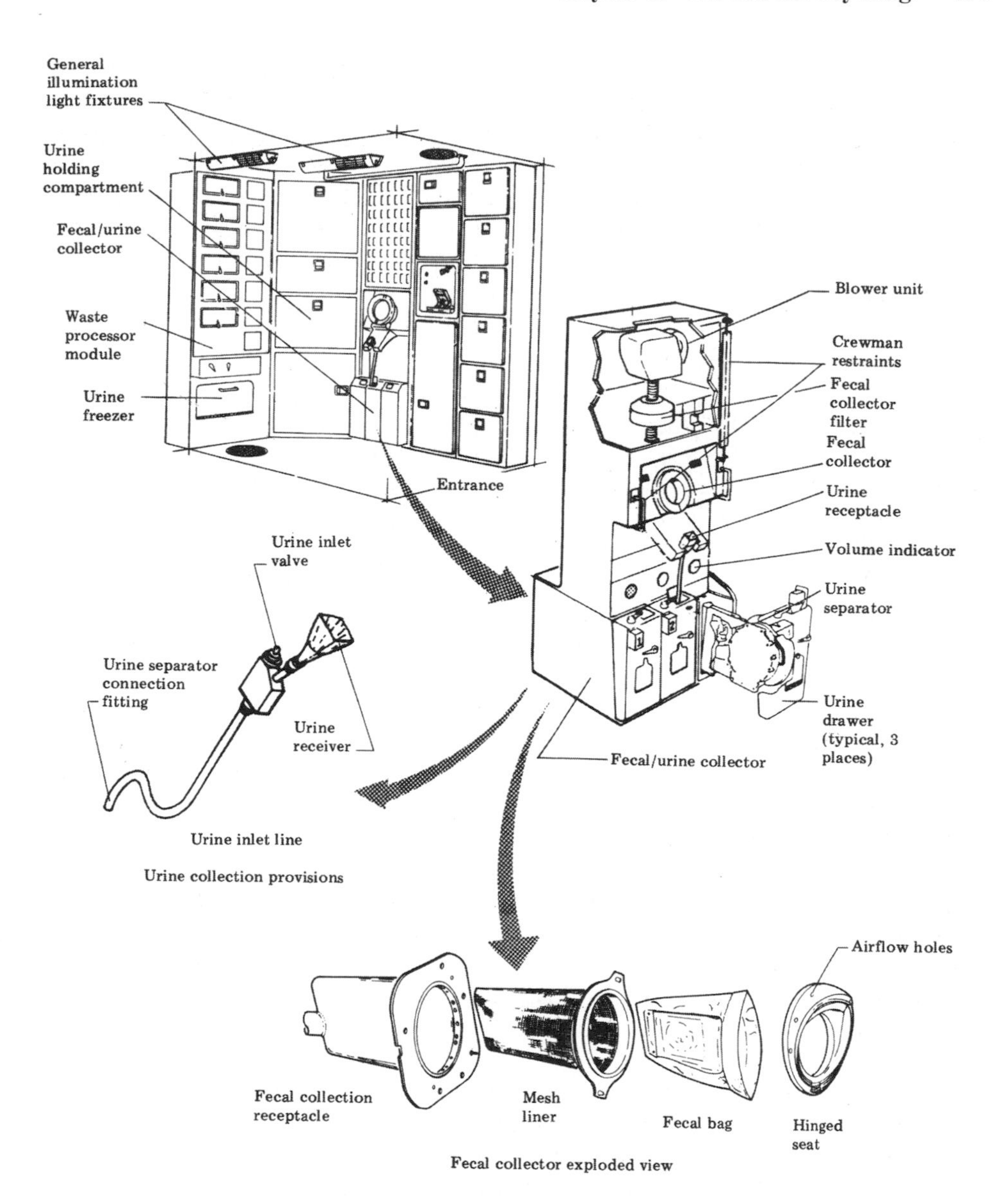

The waste management compartment.

station – usually daily, sometimes weekly – and were placed in the collector, sealed, and passed through the airlock into the oxygen tank.

Meal preparation The Skylab diet offered the energy requirements for each of the crew-members, based on their own body weight and age. Recommended daily intake of calcium, phosphorous, sodium, magnesium and protein nutrients were managed to within 2%. Each diet would follow National Academy of Science dietary

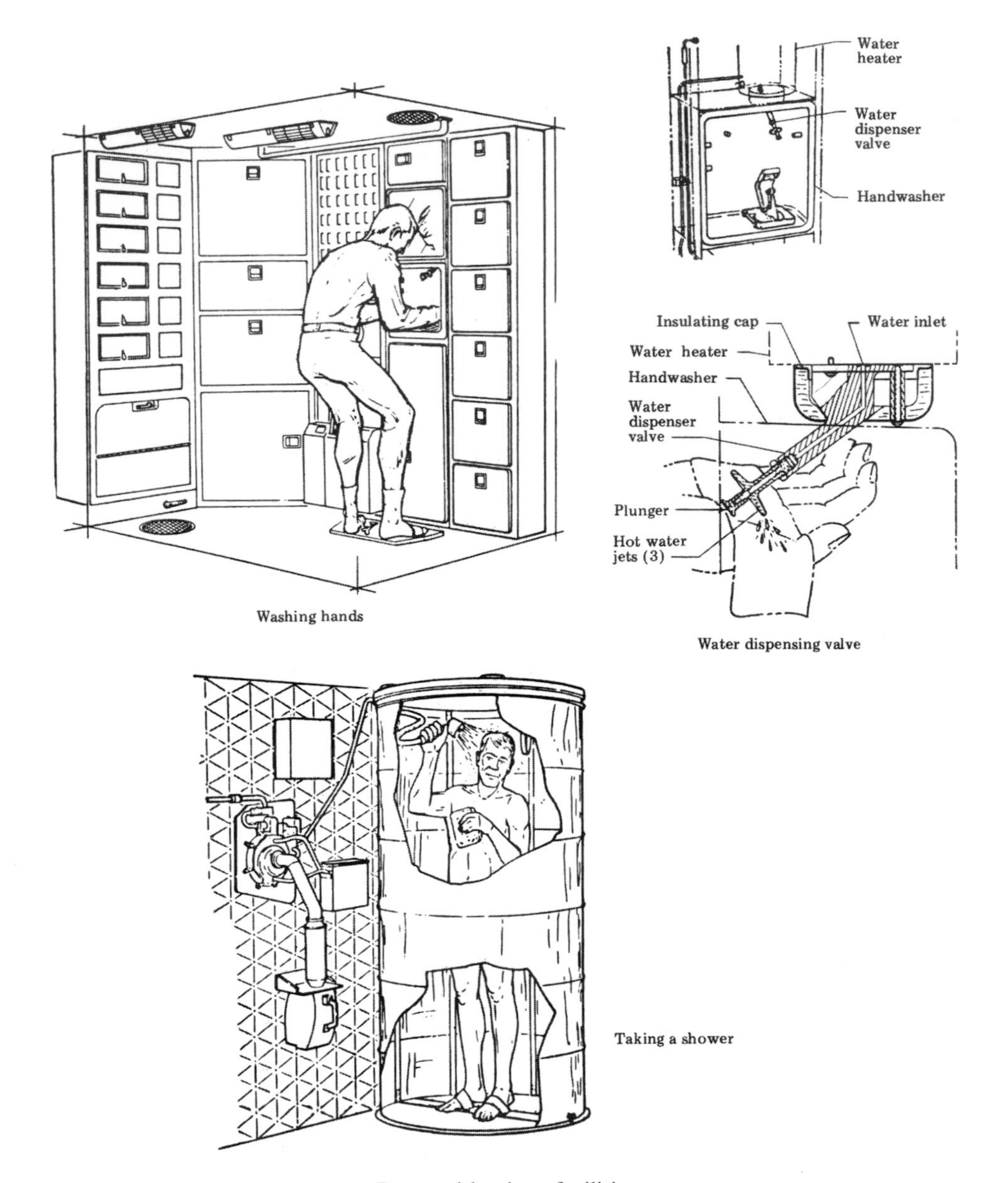

Personal hygiene facilities.

guidelines on carbohydrates, minerals, vitamins and fats. The selection of food from the varied menu was decided by the individual astronauts, although the guidelines had to be met, with the final selection governed by the nutritionists.

The crewman chef of the day removed the chosen menu from the food lockers, following the day's menu and colour identification, to the food trays, which could hold four large cans and four small cans. Three of the large holes were heated.

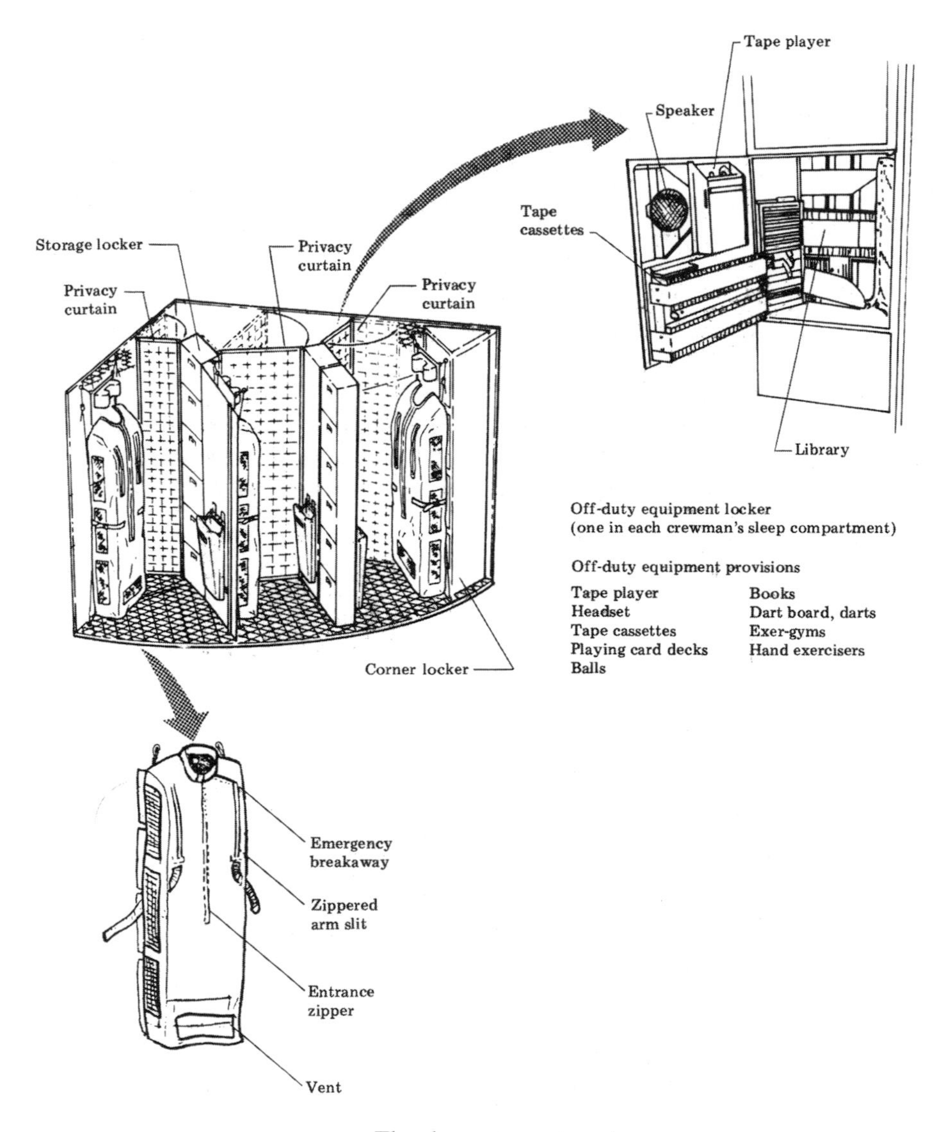

The sleep compartment.

Utensils were magnetically attached to the table, and plastic covers were used to retain the food in the open can to prevent leakage. The crew could 'sit' at the table in special thigh restraints, with their feet in portable foot-restraints.

Housekeeping Despite participating in the trip of a lifetime, these modern-day explorers of a new frontier also had to be efficient at housework and daily chores. They may not have had to mow the lawn or wash the car, but the vacuuming still had to be done, along with dusting, disinfecting, putting out the trash, collecting the mail, cleaning the windows and shower, and cleaning the food preparation area and waste management compartment.

Returning home

The final days of the mission – following the ATM EVA on MD26 – were concerned with the final experiment data runs, stowing and configuring the station for a period of unmanned flight, and preparing the CSM for the trip home and a press conference with the ground on MD27.

The final day of the mission – 22 June – began with the crew waking at 07.30 Conrad focused on checking and activating the CSM, while Kerwin set up the ATM to run in its unmanned mode for the next five weeks before the second crew arrived, Weitz helped both astronauts as required. By 00.45 on 23 June all three were in the CM to complete the close-out of the station. With the hatches closed, the tunnels vented, after some final adjustments to the trim of the station, and waiting for the cessation of the perturbations in the station's stabilisation gyros, at 03.58 the CSM slipped its mooring and drifted away from Skylab.

The CSM was commanded to fly around the station for a visual and photographic inspection before initiating a separation burn to prepare for the re-entry sequence. At 05.05 the Service Propulsion Systems (SPS) 1 shaping burn was initiated for 10 seconds, which removed 264 feet/sec from the spacecraft orbital velocity as Conrad's crew flew over the Philippine Sea. At 08.10 the SPS was fired for 7 seconds, reducing forward velocity by a further 190 feet/sec, which would initiate the descent of the spacecraft toward a Pacific Ocean splashdown.

Pete Conrad, with Weitz next to him, waves to the crew on the recovery vessel. Kerwin is just emerging from the CM's open hatch.

The SL-2 crew leaves the spacecraft on the deck of the prime recovery vessel, with members of the recovery and medical team, after 28 days in space. Weitz is helped by a steadying hand as all three astronauts carefully descend the steps.

At 08.15 the CM was separated from the SM, pitched around so that the blunt end heatshield faced the direction of flight, and began descent into the upper layers of the atmosphere. Following re-entry black-out, communication was established with an ARIA aircraft, and the spacecraft splashed down at 08.49.48 on June 23 1973, ending an historic flight of 28 days 00 hrs 49 minutes 48 seconds, and 404 orbits. Initial reports indicated that the spacecraft landed just 6.5 miles from the prime recovery vessel *USS Ticonderoga*, and that the ship was just 6.5 miles from the planned splashdown co-ordinates. 39 minutes later, the carrier moved alongside the floating spacecraft with the three astronauts still inside, winched it up onto the deck, and placed it on a transport dolly. The three crew were helped out of the spacecraft to begin the exhaustive post-flight medical and crew debriefs that would continue for several days

Of the three, Conrad – veteran of three previous flights, and the new world spaceflight record-holder at 1,179 hours 39 minutes (the first man to break the 1,000-hour level) – readjusted quite well – better than Weitz, and far better than Kerwin, who was markedly affected by his readaptation to 1 g, although by day five, even he was quickly returning to pre-flight performance levels on the LBNP and ergometer.

Despite the difficulty encountered early on in the mission, the success of Conrad and his crew provided much confidence in the success of the second and third missions.

The second unmanned period

Following the departure of the Skylab 2 crew, the station entered what was termed the 'second period of unmanned activity'. The first period had been from the launch to the arrival of the Conrad crew. Originally intended to last just 24 hours, this period became eleven days due to the problems with the solar array and micrometeoroid shield. To many of those working on the problems before the launch of SL-2, the length of time seem much longer than it really was. This period was often referred to as the 'eleven years in May'.

NASA classified the Saturn Workshop Mission Day operations from Day 1 (launch on 14 May) through Day 273 (completion of tests, 10 February 1974). From Day 274, the period of orbital storage was conducted through to the de-orbit of the station in 1979. In addition, each of the three manned missions was assigned its own mission days from launch (MD1) to recovery, and it is these that are referred to in this text. (Full reference for SL-OWS Mission Days is shown in the Table on p. 250.)

Shortly after the departure of the first crew, the workshop was depressurised over six hours to 2 psia, and, allowing for normal leakage rate, was allowed to reach as low as 1.9 psia, with the internal temperature held at 35 F to prevent condensation. The only housekeeping task was the recycling of the refrigeration primary and secondary loops after an abnormal temperature rise had been revealed during the final close-out of the SL-2 mission. After two days the system was restored to normal operations.

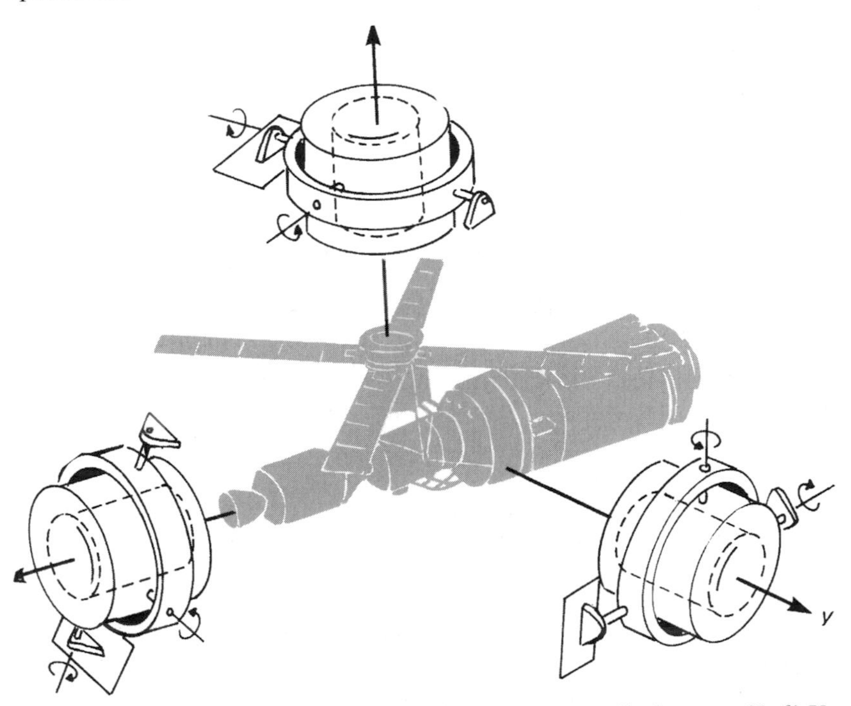

Three gyroscopes each applied torque about two perpendicular axes. (*Left*) X axis, (*top*) Z axis, (*right*) Y axis.

Even though there were no astronauts onboard Skylab, ground control could still monitor solar observations made by the ATM. This continued until 19 July, when the primary up–down rate gyro failed. Three days later a test was performed on the secondary gyro. It was successful, but to prevent serious malfunction until the next crew arrived with a replacement package of six supplementary rate gyros (called the 'six-pack') limited solar observations were conducted until the station was remanned.

SKYLAB 3: THE SECOND MANNED MISSION

In addition to the 'six-pack', the SL-3 CM also carried an improved version of the parasol thermal shield developed at JSC. The crew had trained for the deployment of this and the twin-pole shield developed by Marshall that had been taken to the station by the first crew. Also on board were experiment film, extra food for a three-day extension (from 56 to 59 days), spares for failed experiment components, and other items of hardware such as two laboratory tape recorders.

Some forty hours before the launch of SL-3, the workshop was depressurised to 0.65 psia, and then immediately repressurised up to 5 psia, and terminated just twenty hours before lift off.

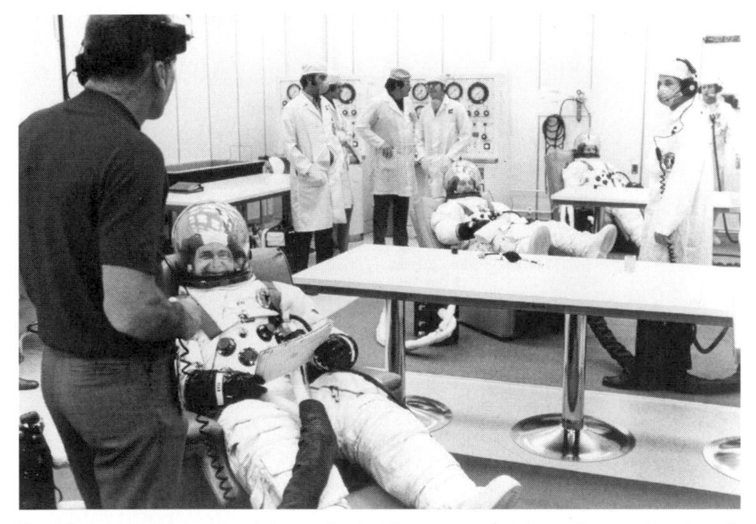

Three Skylab 3 astronaut participate in suiting up activities prior to leaving for the pad on launch day. (*Left–right*) Al Bean, Owen Garriott and Jack Lousma. Standing next to Bean is astronaut Deke Slayton, Director of Flight Crew Operations.

The launch

The launch of SL-3 had been scheduled for 17 August, but due to the degradation of the rate gyros and possible deterioration of the thermal shield deployed during SL-2, it was decided to return to the station as soon as possible. On 2 July, SL-3 was remanifested for launch on 28 July, with a recovery on 22 September after 56 days.

Skylab 3 training: (*Left–right*) Garriott, Bean and Lousma check out the flight documentation in the crew quarters of the OWS located in the Mission Simulation and Training Facility at MSC, Houston.

However, on 20 July NASA changed the duration of the mission from 56 to 59 days in order to provide more favourable conditions for recovery. SL-3 would return home on 25 September.

The launch vehicle for SL-3 had been stacked in the VAB by 28 May, as Conrad's crew were on their third mission-day in space. This was completed with the Apollo CSM by 8 June, and installed on pad LC 39B three days later. If there had been a problem with the SL-2 CSM docked to the OWS, then the SL-3 vehicle under preparation on the pad would be used for the rescue vehicle, although it would not be able to launch for another seven weeks, requiring a maximum mission extension of three weeks for the Conrad crew. If a rescue was indeed required, then Vance Brand and Don Lind would be the astronauts to fly to the OWS to bring back the Conrad – crew a role which they had trained for on all three manned missions.

In the event, all went well for Conrad, Kerwin and Weitz and as they completed their mission, and so preparations progressed, with relatively little trouble, towards the second crew's launch. On 28 July 1973 an estimated 35,000 people gathered at KSC to witness the launch of the second crew. Onlookers filled the Press Site, family viewing areas, public sites and the many causeways around the space centre area. Although this Apollo was not heading to the Moon, the excitement and drama of the first Skylab mission still caught the public's imagination enough to wish 'God speed' to the crew of Skylab 3.

As the count-down approached the appointed time for launch, the three astronauts – Alan Bean (Commander), Jack Lousma (Pilot), and Owen Garriott (Science Pilot) – were lying in the CM couches. They had completed their checks some 45 minutes before the planned lift-off, and had nothing to do but sit and wait. Only Bean had flown in space before, when he accompanied Conrad and Dick Gordon to the Moon on Apollo 12. On that mission he had completed 1.5 Earth orbits in a little over two hours, and he now prepared to spend two months orbiting the globe.

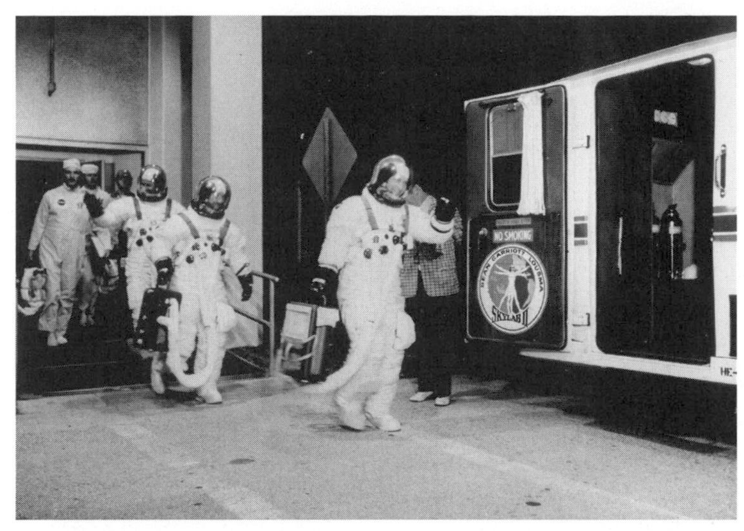

Skylab 3 astronauts leave the Operations Building at KSC and walk to the Astro-Van to take them to their launch vehicle at Pad B, Launch Complex 39. Leading is commander Al Bean, followed by Science Pilot Owen Garriott and Pilot Jack Lousma.

For Garriott and Lousma this was their first trip into space, both of them having worked on technical assignments on Apollo and the AAP since joining the astronaut programme. If the excitement and anticipation of making a flight above the atmosphere was causing extra tension, Jack Lousma certainly was not showing it. He drifted off to sleep, and was roused by Dick Truly, who would be serving as launch and rendezvous CapCom for the second time, having served in that role for Skylab 2, and was preparing to repeat the role for SL-4. Truly had also served as entry and recovery CapCom for the Conrad mission but, this role would be handled by 'Crip' Crippen for SL-3 and SL-4.

Men, mice, minnows, gnats and spiders

At 07:10:50 EDT the Saturn 1B lifted off the launch pedestal, climbed into the sky, and made an uneventful ascent towards orbit. The ascent, the separation of the launch escape tower, the first stage ignition, and the shut-down of the S-IVB second stage, all proceeded without incident, and just 9 minutes 51 seconds after leaving Florida, the spacecraft was in an orbit of 83.6 × 125 nautical miles. A few minutes later the CSM separated from the Saturn rocket stage, and the crew prepared to commence their eight-hour chase towards the OWS.

Onboard Apollo the 'crew' of SL-3 included not only three happy astronauts but also two Mummichog minnows, 50 minnow fish eggs, six pocket mice, 720 fruitfly pupae, and two Common Cross spiders named Arabella and Anita. At 783 (52 fish, 720 flies, six mice, two spiders and three humans) it was a rather large 'crew'

Return to Skylab

Computations to effect the rendezvous with the unmanned workshop were

Following a suggestion by Science Pilot Owen Garriott, two 'brackish water' minnows (Mummichog minnows) formed part of the Skylab 3 payload. Also included were fifty eggs in an experiment to reveal if fish experience any disorientation in space.

Arabella, one of two Skylab 3 Common Cross spiders *araneus diadematus*, and its web, spun in zero gravity during the mission This was Experiment ED52 – one of 25 student experiments flown on Skylab.

completed in addition to a series of burns of the large SPS engine in the back of the Service Module. These firings gradually brought the Apollo towards the station, in combination with onboard electronic equipment that provided radio contact for the VHF ranging information. The CSM transmitted a tone-modulated signal to the

OWS transponder, which received and then transmitted the signal back to the CM, where the returning signal was evaluated to determine the distance and closing rate of the two spacecraft.

The crew first saw the station about 390 miles away, identified by the four flashing lights on the ATM. Around the station there were 8 docking lights, colour-coded to determine the orientation of the station at a distance or in darkness, and there were four smaller lights on the tips of the antennae. During the occupation Skylab, these lights were turned off to conserve power.

'Here's our home in the sky,' Lousma told CapCom Truly as the CSM approached the cluster. Eight hours after launch, the TV views from the CM revealed the RCS of the Apollo repelling the parasol during a fly-around of the station. Ground controllers were concerned that the thrusters might damage the parasol or contaminate the solar array cells, and so the fly-around was terminated. Fifty-two minutes later, the crew gently nudged the Apollo docking probe into the drogue on the OWS.

Two hours after docking, with all checks completed, the hatches were opened and the three astronauts transferred into the MDA. Thirty-six days after the first crew closed the hatches and undocked, the second crew reoccupied the station. This was the first time that a space station had been reoccupied by a second crew. In 1971 the Soviet Salyut station had been crewed by the three Soyuz 11 cosmonauts, but the earlier Soyuz 10 crew had been unable to secure a hard docking and entry into the station, and a planned 'Soyuz 12' mission was abandoned after the deaths of the Soyuz 11 cosmonauts.

A sick crew and a sick spacecraft

Following the docking and transfer, the reactivation of the station was slowed due to a combination of the crew experiencing motion sickness, the reliability of the CSM, and a series of hardware and systems troubleshooting tests.

Shortly after entering orbit, Jack Lousma began to experience the symptoms of what is now called Space Adaptation Syndrome (SAS), but by taking one scopalomine-dextroamphetamine (scop/drex) anti-motion sickness capsule he was able to take an active part in the rendezvous and docking procedures, and also eat lunch. Some fourteen hours into the mission, both Bean and Garriott reported stomach awareness and their inability to move quickly around the station. Even by taking the medication and by slowing down his activities, Lousma could not prevent himself from becoming quite ill after the evening meal.

The effect was that crew activities fell behind schedule, and the following morning the breakfast remained only half eaten. Bean radioed to mission control that they were not completing the tasks assigned to them as quickly as intended, and requested that they be allowed to rest in their bunks and 'just stay still awhile', while consideration was given to moving the first planned day off from MD7 to MD3. Initially MCC agreed to the afternoon off, but even that was not fully achieved, as the crew were busy solving electrical problems in the spacecraft. The controllers had, however, decided to postpone the planned ATM EVA on FD4 by 24 hours.

The medical team on the ground suggested that the crew continue to take the

Looking 'down' from the dome, through the grid floor, towards the trash air-lock.

scop/dex and complete a series of head movements, tipping the head to one side then the other 30–40 time for periods of ten minutes. The idea was that the medication would help alleviate the feeling of nausea, while completing the physical head movements would help the crew pass the barrier of irritation of adjusting to the weightless environment. However, when the crew tried this they found that the nausea increased. While the medics were assuring the crew that this was the only option, the astronauts and some flight directorate officials were not so sure. The crew endured the discomfort. Garriott managed two sessions on FD3, and Bean just one, while Lousma just completely avoided moving his head more than was necessary.

By the fourth day, all three were feeling marginally better, but Garriott and Lousma still took medication, and with activation a day behind schedule the EVA was again postponed to MD6. Despite the awareness of the nausea, the crews' work-rate began to improve, and the activation slowly returned to schedule.

By MD5, all three were feeling much better, Garriott was able to conduct 150 head movements with no ill effects, Bean – a keen gymnast – demonstrated to TV viewers his new skills and ability to eat while upside down, while even Lousma commented that the food tasted much better.

The doctors had suggested that instead of eating a full meal, the crew should have several smaller meals so that their stomachs did not initiate the feeling of nausea. The problem of space sickness was still relatively new in 1973, and the symptoms were thought to be the result of the larger volume inside the spacecraft, over the earlier, more restricted spacecraft, allowing more rapid movements before the body had time to fully adjust to the lack of gravity.

Earlier mission experiences, and the results from SL-3, indicated that it took three or four days for the human body to fully adjust to orbital motion. In addition, the rapid change (ten minutes) from normal gravity on Earth to increased gravity forces during launch, and finally to microgravity in orbit, had the most disorientating effect on the body systems. Coversely, this was also experienced when, within an hour, microgravity was replaced by the high g-forces of re-entry and the '1 g' on Earth; plus the effects of the rolling ocean.

One of Skylab's main scientific aims, therefore, was to study this phenomenon of human adjustment to spaceflight and return to Earth. After Conrad's crew had remained relatively untroubled by the motion sickness, the SL-3 crew now presented the medics with three new sets of data to investigate an alternative experience.

Bean later tried to reason that one of the problems in feeling so ill was that they were rushing around trying to get all the activate everything as soon as possible: 'We were not eating on time, we were not getting to bed on time, and we were not exercising.' For Carr's crew – none of whom had yet flown in space, preparing for

This Skylab 3 image reveals England almost free of cloud. Cornwall, the south coast, the Isle of Wight and the English Channel are all visible, as are South Wales and the Bristol Channel, while the Midlands and East Anglia stretch out towards a cloud-covered North Sea. (The author wrote this book near the centre of this photograph.)

SL-4 – Bean suggested that meal-time and rest should be given priority over activation, and if necessary it should take a day or two extra to complete while they adjusted.

The illness of Bean's crew was a worry for the medics and for NASA officials planning a longer stay for SL-4 and more frequent flights of the Space Shuttle. Although several astronauts had reported feeling ill on earlier missions, never had an entire crew become ill. What was even more surprising was that on his first flight, Bean never felt ill; and in ground tests Lousma had been one of the astronauts who demonstrated a high resistance to motion sickness. Clearly, space motion sickness would require further study.

'Some kind of sparklers'

When Skylab 3 had been launched, work was well advanced to rollout the Saturn 1B/CSM stack for the SL-4 manned mission on 20 August. About three hours into the mission, as the Apollo established communications with the Goldstone tracking station Bean reported seeing 'some kind of sparklers' streaming past the No. 5 CM window after the first SPS engine burn. At the same time, controllers at mission control also noted a visual drop in pressure in one of the four RCS quads. Both the oxidiser and the pressurising helium on Quad B (port side) were rapidly falling off scale. The crew were told to isolate the whole quad, and to compensate for the loss by using the other three available quads.

It was clear that the Quad B was leaking fuel, and further leakage would be alleviated by isolating supplies of manoeuvring gas. While the Apollo continued to approach Skylab, malfunction procedures were conducted to determine whether the problem was mechanical, electrical or procedural. The spacecraft was quite operable with three quads but the problem still caused concern, as the RCS on Apollo had always been one of the most reliable systems in the spacecraft. However, this did not prevent the docking of the Apollo to the station, and with the crew safely onboard Skylab – although a little ill – any serious concern to the ground diminished while the problem was further examined.

Six days later, at 06.37 on 2 August a caution and warning alarm in the Skylab sent the three astronauts hurrying into the CM, while on the ground the G&N controller informed the Flight Director that temperatures in the Quad D RCS pack (opposite that of the isolated Quad B) were falling, and advised switching to secondary heaters. The problem did not seem to be too serious, and after activating the heaters the crew continued with other duties. An hour later, ground data recorded a sharp drop in both temperature and pressure in the same Quad D. It appeared that there was a second leak, which was confirmed by visual observations of the crew, who reported streams of propellant crystals emitting into space. They immediately shut down the second troublesome quad.

Data indicated 12 lbs of oxidiser had been lost at 0.0023 lb/sec – less than 10% of its available propellant – but there could be no indication of how fast the leak would increase. While the astronauts could use two or even one quad, it was not a situation which was acceptable for re-entry without further investigation. The astronauts were perfectly safe onboard the station, with a supply far in excess of their planned two-month visit, so there was no immediate danger. As JSC engineers reviewed the situation and their options, the crew feared a shortened mission.

It was initially thought that there could be a contamination of the nitrogen tetroxide that, in a connected problem, affected the two quads. If so, the worst that could be expected was the loss of both Quad A and Quad C. If the leaks persisted, other internal circuits could be soon render the SM unusable. By mid-morning the press was informed that a rescue mission might be launched – emphasising the prudence of such an early investment in the Skylab programme. Had it not been an option, then the Bean crew would have returned home as quickly as possible, before losing all the RCS quads.

Preparation of the Skylab rescue mission

Within hours of the first reports of the problem, the Cape reacted to the development with lightning speed. It was decided to move the CSM to the VAB by 10 August and, in an abbreviated and accelerated launch, mate the Apollo to the Saturn 1B and install the stack on the pad – all within 72 hours. Flight readiness tests could be achieved by 24 August, with propellant loading on the 27th and a launch on 5 September for a maximum five-day rescue mission landing on 10 September.

In Houston, the rescue crew of Vance Brand and Don Lind headed straight for the CM simulators to begin rehearsing for the rescue flight, while engineers evaluated the data on the faulty quads. For a while it seemed as though Lind would finally obtain his ride into space on Skylab. He had been hoping for a flight as a

The crew that never flew. Skylab Rescue crew-member Vance Brand (*left*) and Don Lind. They talked themselves out of a mission to rescue the three Skylab 3 astronauts, and could have flown the Skylab 5 reboost mission.

prime Science Pilot, but had been assigned as back-up Pilot for the second and third missions, to replace either Lousma or Pogue in the event of illness or injury. Much to his disappointment, both were 'depressingly healthy', and it seemed that he would not be called upon to replace them – and despite his thoughts of applying a back-up crewman's technique of straw dolls and long pins, they still remained healthy.

However, with the problem on the SL-3 SM, Lind and Brand were suddenly thrust to the forefront, with the chance of flying a rescue mission to return Bean's crew. Despite being a short mission, it would certainly capture the headlines, and would be remembered as the first space rescue mission

'Very efficient but perfectly stupid'

After just a few hours, the engineers were beginning to confirm that the two leaks were unrelated, and that contamination was improbable.

As a result, Brand and Lind, as the SL-3 back-up crew, worked simulators in an attempt to solve the problems of re-entry with only two active RCS quads, and other issues, while the rescue hardware was being prepared.

Since that day, Lind has always thought that as a back-up crew, he and Brand were very committed, but as a rescue crew they were not very smart. The two astronauts had devised contingency procedures which proved that the crew could use the SM with restricted RCS operations by relying on redundancy from the other quads, and that in any configuration of firing, the risk was quite small and acceptable. It was decided later the same day that the rescue mission would be cancelled and its crew (Brand and Lind) stood down, and the SL-3 crew could then return at the end of a nominal mission, using the procedures devised by their back-up crew (Brand and Lind) in the simulators.

Each member of the two back-up crews completed a full programme of mission training. Here, Don Lind sets up a station for the onboard data acquisition camera during a simulation in the workshop trainer at JSC.

In achieving this, Brand and Lind agreed that they had been 'very efficient but perfectly stupid, because we have literary worked ourselves out of the mission.' They knew that they had made the right decision, and there were no regrets apart from not being able to fly in space. With a rescue flight, most of the experiment results could not have been returned due to spacecraft weight and balance considerations for entry with five instead of three crewmen. There could be no guarantees that SL-4 could return all the data left behind and from their own mission, so what was brought back had to be carefully chosen.

Had they flown the rescue mission, Lind was to be the 'cargo master' who decided what would be returned and what would be left behind. There would be no advice from the ground, and no input from the Bean crew, and Lind would decide which science data would be returned. There was no question of compromising crew safety. In the event as a scientist as well as pilot Lind was pleased that he did not have to make that decision.

Later that day the crew was informed that the mission would continue as normal but that the SL-4 vehicle would be prepared for rescue if required and that the EVA had also been delayed to MD10. Meanwhile, Brand and Lind continued their back-up crew training with Lenoir for SL-4, and the pair continued CM training for a two-man rescue launch for SL-3 and SL-4, if necessary – but it was never required. The subsequent investigation indicated that loose fittings in the oxidiser lines had gone undetected in two years of ground testing.

Shortly after retiring for the night, the crew worked another master alarm in the CM. It was caused by a major short in the ATM system, which spread through the C&W system and produced ominous readings at mission control. This was the latest

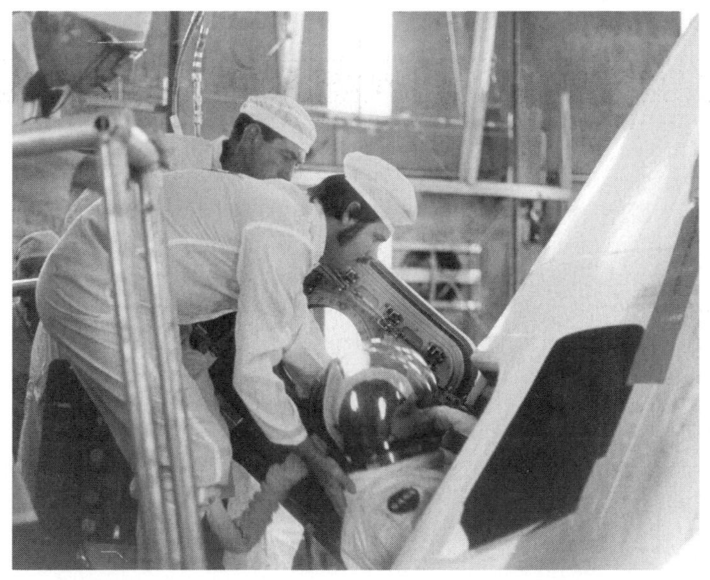

Astronaut Don Lind – Skylab back-up crewman and Skylab Rescue Pilot – is squeezed into a CM during training for his dual roles on the Skylab programme.

in a series of systems faults early in the mission, but they were unrelated to the CSM leaks. Nevertheless, they led to increase tension on the ground and in orbit.

The crew, however, remained optimistic, and Garriott reflected that he should perhaps have asked for a reduction in his life insurance payments, as it was highly unlikely that he would drown in space!

Systems faults

As with the first mission – and indeed the third mission – the second crew had their share of hardware failures to rectify, and maintenance tasks, to perform. Each of the crew had trained for both routine and unexpected maintenance tasks and with the help of the ground the repair and maintenance of numerous items of hardware and onboard systems.

During the first crew occupation, one of the ATM power conditioners – which controlled the energy received from the solar arrays – failed, and then one of the rate sensing gyros that controlled spacecraft attitude failed. Both of these problems were worked and solved during EVAs by the Conrad crew. There had been several rate gyro problems during the first three weeks of OWS orbital activities, inducing fundamental problems in drift, oscillations, indications of high temperatures, and other problems. It had been decided that the placing of an additional 'six-pack' of gyros in the MDA would alleviate at least one of the problems.

Another concern had been the Thruster Attitude Control System (TACS) that assisted in maintaining the correct attitude of the station. Launched with 80,000 ft/lb at launch, only 35,000 ft/lb remained at the launch of the second crew. Most of this had been used during the first ten days of the unmanned flight, with the problems of maintaining adequate temperature before the Conrad crew could deploy the parasol and the remaining solar array. Fortunately, no firing of TACS had been required since SL-2 had returned home, and with the installation of the 'six-pack' by the Bean crew, it was hoped that little or no further excessive use would be required, so that the contingency margin that had been allowed for would not further reduce.

Other unscheduled maintenance tasks performed by the Bean crew during their stay included replacement and electrical continuity tests of the heated water dump probe; disassembly, inspection and the replacement of laboratory tape recorders; the removal of four printed-card circuits inside a video tape recorder; repair of an ergometer pedal; installation of the gyro 'six-pack'; and tightening of the chain linkage on a mirror system for one of the experiments.

System failures included a pressure leak in the condensate dump system which, after troubleshooting by the crew early in the mission, was located, isolated and repaired. A saturated gyro-control resulted in a temporary loss of gyro attitude control after an attempted manoeuvre following two back-to-back EREP passes that used an additional 2,584 lb/sec of thruster propellant. Then the pressure in the primary laboratory coolant loop became so low on 23 August that the pump was shut off and the second loop implemented, which then too revealed a leak. This, however, was smaller, allowing the system to remain in operation; and post-flight analysis of returned carbon dioxide sensor cartridges indicated that the leak was so small that the coolant evaporated into the atmosphere of the station as soon as it leaked. A reservice kit was prepared to be used by the third crew.

Installation of the twin-pole shield

The first EVA of the second mission occurred on MD10 (6 August), and was conducted by Garriott and Lousma. All three astronauts were feeling much better, having overcome the adjustment to weightlessness that had affected them early in the mission. Before going outside, the crew pulled in the parasol closer to the hull of the OWS so that the twin-pole assembly they were to deploy would lie on top of the earlier shield. In this way, the unprotected OWS would not be exposed to the Sun.

The problem with inadequate foot restraints – which Kerwin encountered on the 7 June EVA – would be solved on this excursion by Garriott standing in the foot-locking devices near the EVA hatch. The science pilot assembled the eleven 5-foot sections of each pole, and passed them to Lousma, who was stationed on the ATM central station, secured by tethers and a portable foot-restraint.

The Skylab twin-pole thermal shield is deployed during the second manned mission, Skylab 3, in August 1973.

From his lofty position, the Pilot installed the two poles on a base-plate that he had installed earlier to a handrail, forming a 'V' shape. From here he attached the folded shade material supplied by Garriott, using the ATM transfer boom to ropes running through eyelets along the length of the two poles that were lying across the top of the parasol. Reefing lines secured the outriggers on the ATM structure, and once secure, Lousma simply hoisted the sail that stretched across the area to within four feet of the far end of the OWS, covering the parasol, now just eight inches above the skin of the workshop.

The crew had spent many hours in the water tank practising the task, but in space, plans do not always proceed as scheduled. It took longer than planned to fix the pole sections together, as the rope became twisted. It was then suggested that they separate the poles to run the ropes through first and then reconnect the segments. As

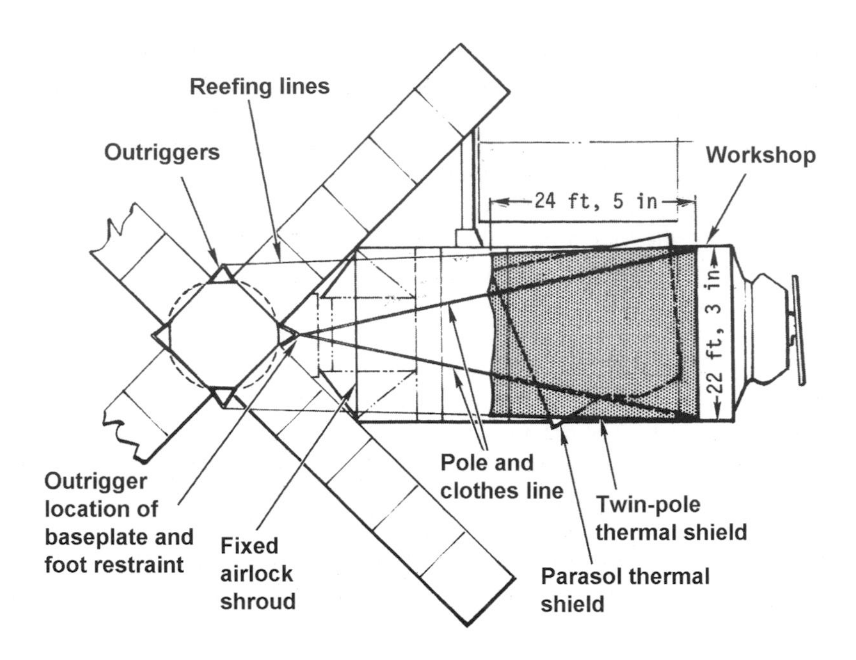

Thermal shield deployment configuration.

with the parasol, the sail did not at first unfurl flat, but solar heating would gradually expand the material to its full dimensions.

The deployment consumed four hours of the EVA, and although it was very demanding, both men seemed to enjoy the experience. The EVA was monitored by Bean from inside the pressurised MDA. With the twin-pole shade deployed, the next tasks were the replacement of ATM film cassettes, the deployment of experiment samples, and the search for visual evidence of the leaking RCS Quads on the SM. He could not see any visible scorching due to the earlier ATM electrical short, nor could Lousma find any discolouration around the CSM quads from the outgassing, which

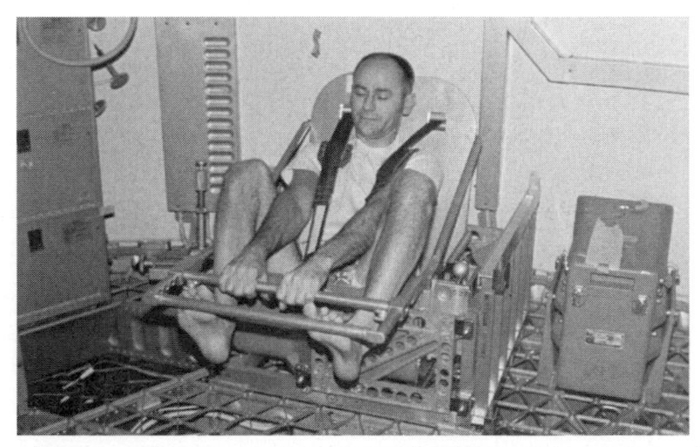

Alan Bean conducts tests to measure body mass in zero gravity.

was suspected to have initiated the ATM short. Lousma, however, commented that the sunward side of the cluster was certainly discoloured by extended solar exposure, and even the CSM, that had been in orbit for less than two weeks, was showing signs of solar-ray heating.

Planned for 3.5 hours, the EVA lasted 6 hrs 29 minutes, and resulted in an immediate reduction of internal temperatures as soon as the shade had been unfurled. Once the EVA had been completed, the crew rested. The following day they returned to the experiment programme.

In a packed programme ...

Once the space sickness had passed, and after the reality of a rescue mission subsided, the second crew soon settled down to the their science programme. They had an extensive experiment programme to complete, and soon the space environment seemed to suit the three astronauts as they settled into their new home and increased their work capacity. By MD40 their physiological changes had levelled out, but their appetite for work never diminished. They only seemed to need six or seven hours' sleep, and often forwent the option of eating meals together to each work on different experiments or to review work assigned for the following day's activities – a self-induced discipline.

They skipped main meals to take a snack late at night, and even pushed the sleep period back an hour or so, although they still seemed to consume the food. (Lousma had the best appetite but he did not like tuna or the bread, and tuna sandwiches were therefore a 'double no-no'). Even some of the leisure time was filled with extra experiment hours, with the crew taking only half days off to catch up if they fell behind the schedule.

The crew completed 39 EREP passes, and conducted 69 separate TV transmissions. The +Z airlock still retained the parasol pole, and the –Z airlock

Jack Lousma vacuums an air-conditioning screen.

was precluded from further use when part of an instrument used in the airlock was jettisoned after it would not retract after use.

Extensive ATM and EREP observations were completed, as well as several student experiments including monitoring the progress of the spiders spinning webs in zero g. Several technological experiments were conducted, including the first flights of the M509 Astronaut Manoeuvring Unit inside the OWS. Both Bean and Lousma flew the unit on the morning of 13 August, and then Bean tried a hand-held unit fed from the AMU. Later, both of them used a foot-controlled unit, which was far more difficult to control.

Sometimes the crew were ahead of their flight plan, and asked the ground for more work. Following the 6 August EVA, Bean asked if they were behind in the flight plan, and in a desire to catch up he frequently pushed for more activities: 'We're working less hard at the moment than we were prior to flight [and could] do a little bit more. We've got the ability, and time, and energy.' Lousma stated that they were determined not to go home before they had accomplished 100% or more of mission goals. By the end of the flight they had achieved an impressive 150%

Stepping outside

During their residency, the Bean crew completed two further EVAs to change out ATM film cassettes. The first, on 24 August, saw Garriott and Lousma again venturing outside the workshop in a 4 hour 30 minute EVA. During the spacewalk the two astronauts also installed a new 24-foot cable for the newly activated 'six-pack' gyros, and attached two samples of the parasol material, to a handrail to be retrieved during later EVAs.

The third EVA occurred on 22 September and lasted 2 hours 45 minutes, and this

Jack Lousma participates with Owen Garriott in the 6 August EVA to deploy the twin-pole solar shield.

Owen Garriott performs EVA at the ATM location on Skylab. He has just deployed the Particle Collection Experiment (S149) the cross-like structure immediately above his head.

time Bean accompanied Garriott outside, while Lousma remained in the MDA. In addition to the ATM film cassette exchange, the two astronauts retrieved one of the parasol samples for return to Earth after a month's exposure outside. Unfortunately, leaks prevented the AM suit cooling system water to circulate through the suits, but Garriott reported being only slightly warm during EVA, while Bean's hands were constantly warm.

During his EVAs, Lousma was impressed by the whole experience, and compared it with being on a fast train. Inside the workshop, the Earth was always an impressive sight, but the grandeuristic view was still restricted, even when looking from close to the window pane. Outside, however, all perspectives changed: 'It's like a brand new world out there.' Standing on the outside of the workshop was more like standing on the very front of a locomotive hurtling down the track, with the whole panorama presented. The difference was that there was no sound, no vibration, and no visible signs of forward movement to indicate the flying speed of 17,500 mph! During the brief times that there were no communications from the ground or from other astronauts, even the gentle hiss of the life support system, or just the silence, could be deafening.

For Lousma, looking 'down' at the Earth was even more impressive, although he found it difficult to distinguish the curved horizon, as the Earth was so big, and he could see it roll beneath him, with the blue horizon dropping away as they orbited a living, moving planet, and not a picture 'like you're really going around something. God! I'm glad that Saturn worked. It's great to be here.'

'Hello, space fans'

As the mission progressed, it was difficult to imagine that Jack Lousma could have been anywhere but in space. All signs of the space sickness had passed, and as the 'systems man' he became one of the first 'space reporters' with his narratives on many TV casts from the station revealing orbital life to his audience of ground-based 'space fans'. He narrated the broadcasts with enthusiasm and fun, including a step-by-step account of Bean's use of the washing facilities, in a way in which a sports commentate would report on a football game.

Later he narrated the TV pictures of the 'Distinguished Professor' Owen Garriott giving 'Captain Alan Bean, who is the leader of this mob' a haircut including the bandages ready for 'when we have to patch him up ... it's not going to be a professional job, but there's no waiting, and the price is right.' He then pulled the camera back to show the use of the vacuum cleaner to catch the loose hair-clippings that 'do not fall down, or fall on your shoulders'. He also explained the reason why they chose Garriott for the job: 'We figured you could always trust a barber with a moustache.' He then closed the sequence, reminding the audience that there are not many people who can say they have had a haircut at 17,500 mph !

Lousma also 'performed' on the other side of the camera, including one famous scene which is often included in documentaries about the fun of space. He was seen lifting a 'barbell' off the deck of the station and over his head, with physical strain etched on his face, and then soaring into the air, holding the weights high above his head as he silently drifted out of frame. As a Major in the USMC, Lousma also liked to promote the service as often as he could, as an added bonus for the recruitment effort!

Al Bean uses a battery-powered shaver in the crew quarters of the workshop during his 59-day mission Note the floating face towels secured to the bulkhead.

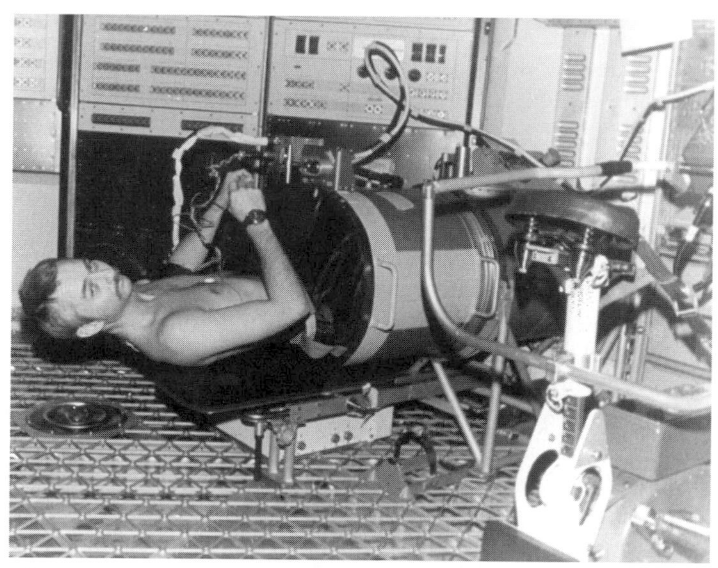

Owen Garriott in the LBNP chamber during Skylab 3. At right, in the foreground, is the bicycle ergometer. This view also clearly shows the triangular grid floor used for retaining position on the work area floor.

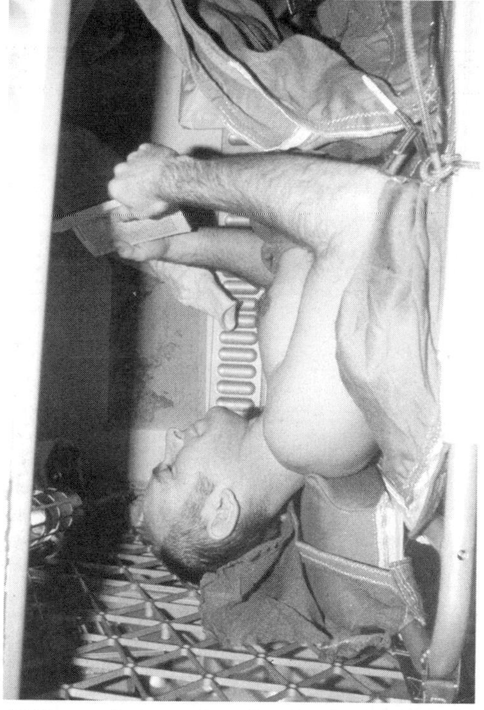

Sleeping on the wall. Al Bean relaxes and reads a book during the 59-day mission, Each of the three astronauts had their own sleeping location in the wardroom.

Garriott, on the other hand, was ever the scientist. Each week he found time to demonstrate the effects of weightlessness to the Earth-bound audience, using water drops, magnets and spinning objects to demonstrate the forces of gravity and orbital flight. Although they were not part of the main scientific research objectives of Skylab, these small, simple demonstrations probably more than anything provided the laymen on the ground with a better understanding of the environment in which the astronauts were living and working.

For Al Bean, the only diversion he needed was more work, and on MD36 he asked for a further extension of the mission duration to 65 or more than 70 days; but this was refused on the grounds that the medical data on them was required on the ground to evaluate flights beyond two months; and there also had to be careful management of food and film reserves for the third crew in training. Bean would later remark that despite having good scientific experiments they were not enough to fill their day, despite working twelve-hour days and 70-hour weeks in an attempt to produce the best return from the investment in Skylab. He was also mindful that in August the decision had been made to stand down the back-up OWS. Skylab B would not be flying and, so all that was left – probably for the remainder of the decade – was the rest of his mission, the SL-4 flight, and the shorter ASTP with the Russians.

'This is Helen here in Skylab'

Air-to-ground communication on Skylab was more frequent and more extensive than in any other previous manned flight programme. Teams of engineers, scientist, doctors, flight planners, managers and controllers all linked through the CapCom, who rotated with the flight control shifts throughout the three missions. These were astronauts who had trained on the Skylab systems and procedures and with the flight control teams, and were the vocal link to the crew in space, acting as the crew representative in Mission Control.

Al Bean compares data on the swirling teleprinter tape, sent up from Mission Control, with that in the documentation in Flight Data Files (FDF) onboard the station.

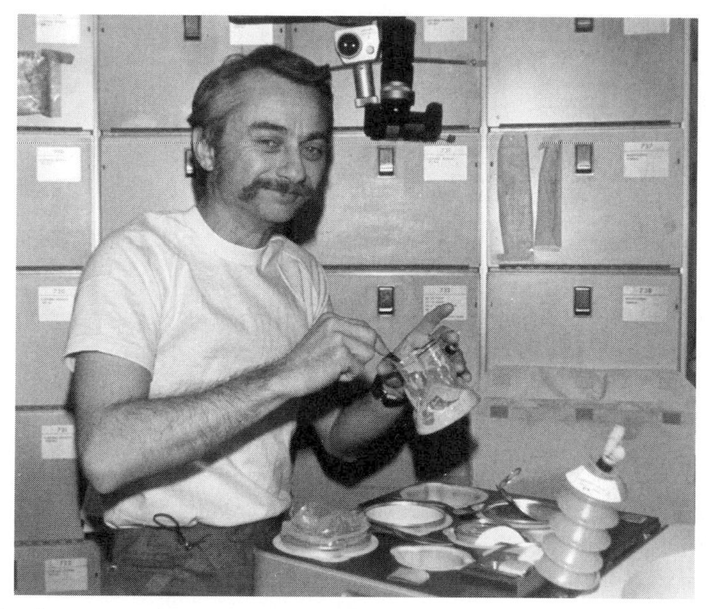

Owen Garriott reconstitutes a pre-packaged container of food on the crew quarters' wardroom table during the second mission. On the tray is a knife and fork, and the astronaut uses a spoon to mix the contents in the container.

Open communications from the spacecraft were via the 'A' channel, supplemented by input by the Public Affairs Officer seated in MCC, who expanded on what the astronauts were saying or doing for the general public. The more technical and scientific reports, and the crew's personal interpretations, comments and evaluations of procedures and hardware, were handled on the restricted 'B' Channel, and were recorded on tape for later downlink to the ground. Private discussions with family and doctors were also reserved for the Command Module B channel, and were not monitored by other controllers.

Downlinks from Skylab were always male in origin, and Flight Control was therefore surprised, during one broadcast from Skylab during the second mission, when the voice coming over the speakers was definitely female! 'Hello, Houston this is Skylab. Are you reading me down there? ... Hello Houston, are you reading Skylab?' CapCom on duty was Bob Crippen, who, after a pause, carefully replied: 'Skylab, this is Houston, I hear you all right, but I had a little difficulty recognising your voice. Who have we got on the line here?' From Skylab came the reply: 'Houston, Roger. I haven't talked with you for a while. Is that you down there Bob? This is Helen here in Skylab. The boys hadn't had a home cooked meal in so long, I thought I'd just bring one up. Over.' Crippen was really confused! The voice was Owen Garriott's wife Helen, and with a gathering audience at the CapCom console, Crip was reluctant to take the bait: 'Roger ,Skylab. I think somebody has got to be pulling my leg. Helen, is that really you? Where are you?' Skylab: 'Just a few orbits ago we were looking down on the forest fires in California. You know the smoke

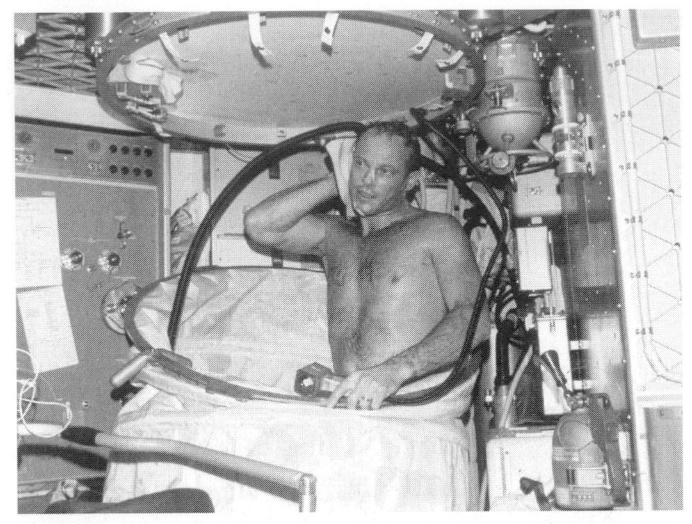

Jack Lousma takes a shower during the second mission. He uses a wash cloth to clean excess soap off himself. In his left hand is the push-button showerhead attached to a flexible hose. To draw off the water he needed to vacuum himself.

Al Bean works with the S019 Ultraviolet Stellar Astronomy experiment camera in the scientific airlock, taking UV photographs of large areas of the Milky Way to record the large number of hot stars located there. Working harder increased the astronauts' efficiency, and they requested additional tasks to occupy the resultant extra free time.

sure does cover a lot of territory. And, oh Bob, the sunrises are sure beautiful ... Oh-ho, I have to cut off now, I see the boys floating up towards the command module, and I'm not supposed to be talking to you. See you later Bob.' As Crippen replied with a shaken 'Bye-bye,' the sounds of the crew laughter replaced the feminine voice. Apparently Garriott had earlier spoken to his wife on the private channel, and had asked her to read the lines back to him, which he taped, and waited for the right moment to replay them, when Crip was at the console.

The recovery of Skylab 3

The deactivation of the station for the second time followed the procedures similar to those of the first crew, with Bean's crew adding a portable fan in the MDA to circulate air over the 'six-pack' gyros to keep them cool until the arrival of the third crew.

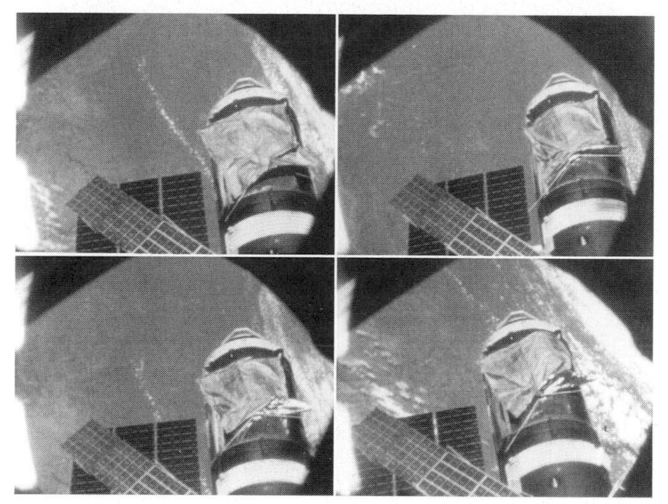

A sequence of four frames showing the 'flapping in the wind' of RCS thruster firing from a departing CSM.

Activation of the CSM took more time than normal due to the checking of the RCS. The activation and checking of the CSM actually began on MD55 – a few days before re-entry. In addition to the RCS leak, the CSM was also found to have developed a small coolant loop leak, back in August, but at only 1% per day leak rate. The mission would be concluded before the problem became serious, although the levels were constantly monitored.

The crew rose very early on MD60, 25 September, to begin the final preparations for separation and re-entry. The CSM had remained passive for 58 days while docked to the OWS, and although its systems were monitored on the ground and the spacecraft received routine inspections by the crew, there remained some concern about the performance of the faulty RCS quads.

After the final deactivation takes were completed, the hatches were sealed, and the three astronauts, with their cargo of experiment and film cassettes safely stowed away undocked from the MDA and performed the separation manoeuvre. However, no fly-around was completed because of the problems with the RCS earlier in the mission. A little after 90 minutes after leaving Skylab, the SPS fired to initiated de-orbit, and a little more than 40 minutes later, Skylab 3's CSM was floating in the Pacific Ocean, 250 miles south-west of San Diego. The mission had lasted 59 days 11

Skylab 3 crewmen onboard *USS New Orleans*, following the end of their 59-day mission. (*Left– right*) Lousma, Garriott and Bean are seated on top of the platform of a forklift dolly. Recovery support personnel are wearing facemasks to prevent exposing the crew to infection.

hours 9 minutes, and surpassed that of the Conrad crew, setting a new world endurance record for the mission and an individual time in space record for Bean of over 1,671 hours on two missions.

Forty-three minutes later the astronauts were hoisted onto the deck of the prime recovery ship, *USS New Orleans*, and 48 hours after recovery they were back in Houston to receive welcoming celebrations and to continue the postflight debriefing and medical evaluations. The mission, crew and hardware, despite early problems, had performed well in light of over-achieved pre-mission expectations. With two flights completed and only one flight due, expectations were high that the last mission would be even better than the first two. For Carr, Gibson and Pogue, it would be a hard act to follow, but they were determined that their flight would be the most memorable and the most rewarding.

The third unmanned period

On 25 September – the day that the Bean crew came home – the OWS was depressurised to 2 psia to lower the dew-point, and then raised to 5 psia with nitrogen to aid in the cooling of the 'six-pack' in the MDA. By 24 October, normal leakage had lowered this to 4.05 psia. Controllers initiated a repressurisation to 4.5 psia, and then allowed it to decay to 3.75 psia by 14 November.

As with the period between the first and second visits, ATM experiments were performed during the third unmanned period until 14 November, when the primary experiment pointing control failed. The use of a secondary pointing controller allowed normal operations, but these were curtailed to protect the system until the third crew arrived. Only one solar experiment did not use the controller, and it continued to gather data.

On 14 November the pressure inside the workshop was allowed to decrease to 0.7 psia to purge the abnormal mixture of nitrogen and oxygen that had be used for rate gyro cooling, and was then repressurised with the normal mixture of these two gases to 5 psia, pending the arrival of the third and final crew.

SKYLAB 4: THE THIRD MANNED MISSION

A new comet

For more than five years, the duration of the third visit to the Saturn workshop had been set at 56 days, and that plan continued into the spring of 1973. When the Bean crew was launched to Skylab, at last the Carr crew was 'number one on the runway', and was able to take full advantage of all the simulators to complete final preparations for a two-month visit that would begin in October and bring them home before Christmas.

However, on 7 March, Czech astronomer Dr Lubos Kohoutek, working in his Hamburg observatory, discovered a new comet. Skylab became an obvious platform to conduct manned observations from orbit, and the NASA management reviewed

Comet Kohoutek, photographed from Catalina Observatory on 11 January 1974.

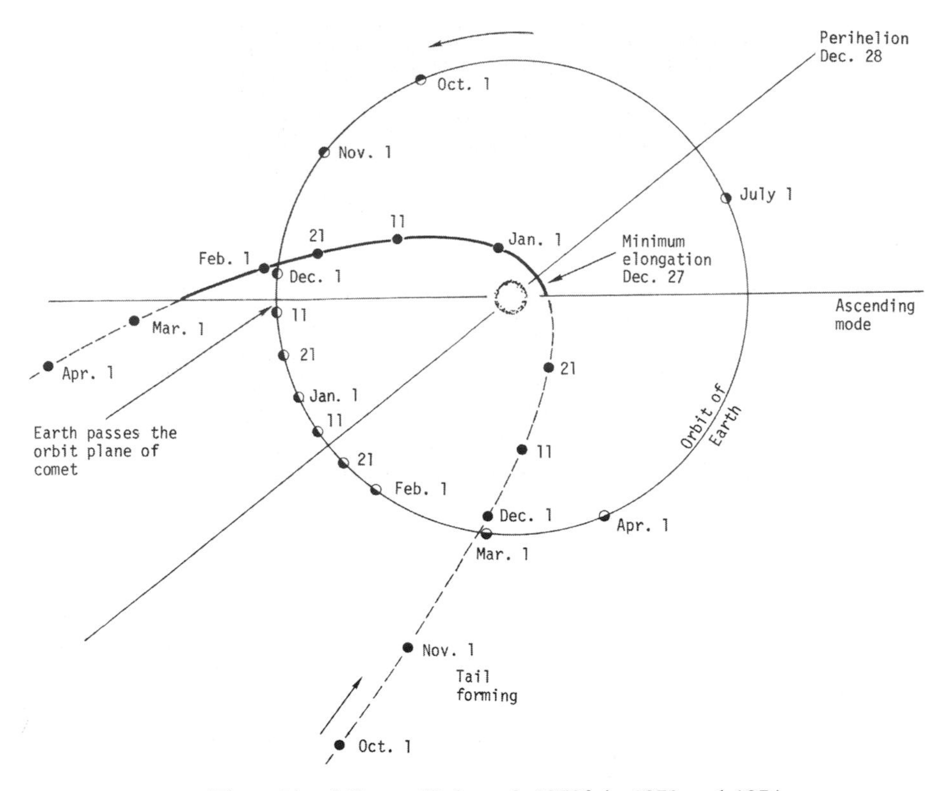

The orbit of Comet Kohoutek 1973f, in 1973 and 1974.

the potential of using the station for observations by the third crew. Subsequent calculations indicated that the comet would pass closest to the Sun (perihelion) on or around 28 December, at 13.4 million miles and travelling at 250,000 mph.

Under the original plan, Carr's crew would be home well before the comet reached perihelion, and therefore in April it was decided to delay the launch of the third crew until 9 November (subsequently slipped to 11 November) and effect a recovery early in January to take full advantage of the location of the comet near the Sun and during the first few days as it passed from behind the Sun. However, this would require a new programme of training to achieve complex manoeuvres of the station.

Mission extension

The achievements of the Skylab 3 mission had led to high expectations from the third visit, with the Skylab manager Ken Kleinknecht hinting that NASA believed that in space man was far more capable of doing more on a mission than at first thought. As a result, NASA management kept the option open for the third crew to extend their mission up to 70 days, and in order to quantify this the flight planners had to devise a suitable crew activity plan that justified the added expense of keeping Carr and his crew in space.

The crew discusses the personal recreation devices that would be onboard Skylab, including softballs, Velcro-tipped darts, books, and audio cassettes.

Consequently, JSC devised a new flight plan that included 28 man-hours of experiment work each day (with one day off every seven days), and added twelve new observing programmes to the ATM schedule and another fourteen EREP passes to the twenty for which the crew was already training. There was also quite a heavy maintenance programme for the crew, including the changing of the coolant in the refrigeration system, and troubleshooting the failed Earth resources microwave antenna. But at least they did not have to install a new solar shield, as the combined parasol and twin-pole assemblies were holding up sufficiently for their mission. What was not known as the November launch date approached, however, was that the gyro system that recorded a brief slow-down in one of the units was an indication of a potentially much larger problem that faced the third crew early in their mission.

Cracks on the Saturn

The launch vehicle for SL-4 had arrived at LC 39B on 14 August – somewhat earlier than originally planned because of the need to support the SL-3 rescue mission requirements. When the need for the rescue flight was averted, the vehicle remained on the pad in a state of readiness, should it be required to rescue the Bean crew. As the vehicle was not called up on to do this, however, it stayed on the pad in preparation for the third and final mission.

Launch preparations continued to progress well until 6 November, just five days before launch, when a pre-launch management meeting at the Cape was interrupted with the news that an inspection team conducting routine structural integrity checks on the launch vehicle at the pad had discovered hairline cracks in the aft attachments of the eight stabilising fins at the base of the first stage of the Saturn 1B. Closer inspection revealed cracks on all eight attachments – two cracks each of the seven fins, and one crack on the eighth fin, the longest crack measuring 1.5 inches.

The cracks were probably caused by stress corrosion due to the age of the stage,

A fin on the Saturn 1B is removed to examine the cracks on the vehicle.

which had been out of the post-manufacturer check-out programme for more than seven years. With the possibility that the fins could be ripped off as the Saturn passed through Max Q, the management team decided to delay the launch to 15 November while the fins were replaced. It took 35 hours to rig the support platforms to remove the first 485-lb fin, but confidence remained that the start of the count-down could be met in the early hours of 13 November.

On the morning of 12 November, a further problem was discovered – more cracks, this time in seven of the eight S-1B/S-IVB interstage reaction beams. Stress analysis indicated that 'beefing up' the beams with heavy aluminium plates would be sufficient, and would remain within safety limits. But the launch had to be delayed a further 24 hours to 16 November

Meanwhile, the removal of the fins continued during twelve-hour shifts, with new fins being flown in from NASA's Michoud, Louisiana, facility. By 07.04, 13 November, workers had replaced all eight fins, and had fitted reinforcing blocks around the mounting blocks on each fitting to ensure an adequate load path if new cracks appeared in the bolt fittings after final inspection.

Stress analysis tests on both the new fins and the old fins, before fitting and after removal, provided new sets of data to confirm that the new fins were within safety limits for launch. Director of Launch Operations Walter Kapryan added that he was told that 'age makes a difference' even in metal!

A full spacecraft

The launch vehicle for the Saturn 1B/Apollo combination was the heaviest of the series at 1,310,021 lbs, and reflected the additional items that the Apollo was carrying to orbit for the mission, which was scheduled to return in mid-January. Mission management had allowed for a tentative extension to 70 days (ten weeks), or a maximum 84 days (twelve weeks) if conditions were suitable during the mission.

Taking into account the reserves of supplies and consumables left onboard the station after the first two missions, the CM was loaded with extra film and magnetic tape for the variety of cameras and sensors; a kit containing further inflight demonstration equipment; additional items for the reserving or replacement of equipment that had failed, a selection of hardware that was of a design improved since the original had been launched; contingency items for changing the flight plan, and for a higher-than-planned usage rate of certain of items, replacement of lost items from the first two missions; items for better crew comfort; and improvements for the communication, TV and photographic equipment.

SL-4 also carried an additional 159 lbs of extra food, of which 59 lbs was of high-density foodstuffs that would supplement the current food stocks. In addition, a new experiment was installed in the cramped CM – S-201, the Far Ultraviolet Electronographic Camera, and two film magazines for observations of Comet Kohoutek. This was the flight hardware back-up to the unit carried on Apollo 16 to the lunar surface in April 1972.

Two vials containing 500 gypsy moth eggs supplied by the US Department of Agriculture were also carefully packed in the CM, as well as the Airlock Module primary coolant loop servicing kit. To make room for all of this additional equipment and supplies, the vibration padding used for packing equipment inside the CM had been removed, and was replaced with extra clothing!

With all of this packed into the CM, there seemed to be little room to include three pressure-suited astronauts. In fact, on the day of launch, *four* astronauts managed to enter to the spacecraft, although the fourth found it a little difficult to get out again!

The beginning of the end

The three astronauts for SL-4 were the first all-rookie crew since Gemini 8 in March 1966. Commander Gerald (Jerry) Carr and Pilot William (Bill) Pogue were both from the Class of 1966 astronauts (Group 5), and were reassigned from the cancelled Apollo 19 crew to Skylab in 1970. Science Pilot was Edward (Ed) Gibson, who had been working on AAP/Skylab almost since joining NASA in 1965.

Carr was a former LM specialist, and was only the fourth American rookie 'crew' Commander since 1965. (Apart from the six solo Mercury astronauts, the other three were McDivitt (Gemini 4), Borman (Gemini 7) and Armstrong (Gemini 8).) Pogue had worked on AAP before Apollo lunar assignments, and was a CSM specialist with additional training in the EREP experiments. Both Carr and Pogue had devised the Skylab crew-training programme for the CB group when it was formed in 1970. Gibson, the solar scientist, had spent much time working on ATM development, and had found time to write a textbook on solar science (*The Quiet Sun*, NASA SP-303

The spacecraft now departing from Pad 39B... is the Skylab 4 Saturn 1B carrying the final crew to the station.

1973) before the flight! What this crew lacked in spaceflight experience, they certainly made up for in preparation.

After suiting up on launch day, 16 November, the crew completed the three-mile trip to the pad, and took the elevator to the swing arm of the launch tower that would take them to the CM. Unlike previous visits when there was hustle and bustle around the vehicle and pad, this time it was deserted except for the launch team, who were present in the White Room surroundings in the CM. As the astronauts approached the CM, they saw a picture of the latest 'Playmate' taped to the wall, with the added inscription, 'Keep This In Mind!'

Assisting the crew into their couches was support crew-member Henry 'Hank' Hartsfield, who had positioned himself in the lower equipment bay of the CM. Carr took his place – assisted by Hartsfield from inside and launch support team-members from outside – in the left couch. Pogue then took the right-hand position, and finally Gibson took his place in the centre seat.

With the three men safely strapped in, it was time for Hartsfield to leave the CM. The support crewman normally exited the CM under the centre couch (Gibson) and up between the headrest and lip of the hatch, but on this mission there was so much additional stowage that the only was out for Hartsfield was to crawl over the top of Gibson, who offered several glib comments as he crawled over him, trying not to hit the switches on the control panel as he went.

With the hatch in place and the count moving towards zero, all three astronauts looked at each other and grinned like Cheshire cats. At 09:09:23 EST, 16 November 1973, the Saturn left the pad at the Cape to begin the third and last Skylab mission – to the exclamation 'Here we go!' from all three crew. The crew commented on a rough ride as the Saturn ascended, pinning them in their seats as the g-forces steadily grew. At staging they were flung forward against the seat harnesses as the first stage shut down, and were then slammed back to the seats as the second stage kicked in to continue the ascent. Many descriptions of this event have been offered by those who have experienced it. Carr's impressions were captured in his occasional personal diary which he kept during the mission: '... staging ... HOLY COW!'

It took 9 minutes and 47 seconds to reach orbit, and approximately fifteen minutes after that the CSM separated from the S-IVB. By the first pass over the United States, Carr commented that the windows were smudged by the crew looking out at the Earth for the first time. The astronauts realised that there would be plenty of time to admire the view in their long flight, as at the moment it was all 'rush, rush, rush' to prepare to dock with Skylab. It was not until Carr recognised the familiar boot shape of Italy that he realised his position over the globe, and he was amazed to find that it did indeed look like a boot!

Skylab was sighted on the fifth orbit, seven hours after launch: 'She looks pretty as a picture' Carr said, as he closed in for docking noticing the TACs 'puffing back ' at them as the station was stabilised. On the first attempt, Carr approached gently, and although the crew thought they had soft-docked, the side-to-side movement revealed that they had not done so. Carr therefore backed away and tried again, only to bounce off the drogue. At his third attempt, he Carr 'goosed it', and hit the workshop hard – but he heard the familiar shotgun blast of the latch capturing the station. At last they were on the front porch of their home for the next two to three months.

The plan was for the crew to rest before entering the station, as it had been a long day – and undoubtedly there were more lengthy days ahead.

An error in judgement

Following the experiences of the second crew in adjusting to the reduced gravity early in the mission, the NASA management became concerned that, as a precaution against the sickness, the third crew should take medication shortly after entering orbit. Although this might alleviate the onset of the condition, there were known side-effects that could be just as unpleasant as a short bout of space sickness. After discussion with the crew it was agreed that, due to the importance of the renedezvous, Carr would delay his medication until docking had been achieved, and for the first few days each would follow a course of medication. After that they would take capsules only as required, and try to minimise quick head movements. According to Gibson, although the doctors believed that this would help to alleviate the problem the crew never really believed that the medical team properly understood the problem – but they would follow the instructions.

Before they had left the pad, Carr had requested to begin the activation that evening, instead of spending a night in the CM, but with the experience of SL-3, mission planners were not convinced that there was an added advantage to this, and so the crew would spend the night in the CSM and would transfer to the OSW the next day.

As the crew began to stow away gear in preparation for the night in the CM, all three took their medication and settled down for a meal. Out of communication with mission control for about 40 minutes, Pogue was overcome with nausea, despite the medication. A small amount of the stewed tomatoes that went down at mealtime, came back up into the vomit bag.

During a brief communication link with the ground, the crew were reminded of the adverse affects of beginning the activation too early. Pogue's illness was not mentioned, and, though trying to remain still as much as he could, he was not feeling at all well. While he tried to rest, Carr and Gibson prepared their second meal and discussed their predicament. Officially they should have reported the nausea and the vomiting, and although Pogue was definitely ill he was not so bad that it might affect the mission, which was their overriding fear.

In discussing the matter between the two of them, Carr and Gibson reasoned that their colleague would begin to feel better in a few hours – well before entering the workshop. All food taken or left was to be reported each day, and in a desire to prevent an over-reaction on the ground, Carr reported only that Pogue had not felt hungry, and had left some of his food – delaying a true account of events until he had considered how best to tell the medics and not raise an alarm.

Pogue, meanwhile, had been helped into the CM docking tunnel, where, it was thought, the cabin air might be fresher and make him feel more comfortable. Both Carr and Gibson were amazed that of the three of them, it had been Pogue that felt the worst. In the astronaut office he was known as 'Iron Belly' with 'cement in his inner ear' due to his considerable resistance to motion sickness during training. As a former member of the USAF 'Thunderbirds' aerobatics demonstration team, many of his colleagues felt much worse when they flew as a passenger in the T-38 with Pogue. In trips across the country he frequently demonstrated his well-practised aerobatics skills, resulting in some colleagues feeling 'just plain scared to death' every time Pogue took the controls!

Carr asked to postpone the evening status report while they ate, and with agreement from Houston, Carr had two hours to decide what to do next. Forgetting that the onboard CM tape-recorder was running, the crew discussed the matter. Carr decided to tell the ground in the conference that Pogue was nauseated, but not that he had vomited. There was the question of the vomit bag that had to be accounted for, but Carr reasoned that it could simply be thrown down the trash airlock. Gibson agreed, adding that probably the management on the ground would be happier if they disposed of the bag without reporting the true events. If the medics knew that the crew was vomiting, their influence over the rest of the flight could be quite serious. That decision was certainly debatable, but the crew reasoned that they were acting in the best interests of the mission; and as the evening status report was read down to the ground, Carr reported that Pogue had 'not eaten the strawberries for lunch, and has not eaten meal C.'

The next day, after a night's sleep, Pogue felt better, but was still restricting his movements. As the crew prepared to enter the station, on the ground the onboard CM tapes, which had been running, and recording onboard communications – were was routinely dumped and transcribed. Had Carr remembered the tape recorder he would have reported the incident, but decided to keep quite about it, and while they discussed the best course of action, the tape of the earlier comments about throwing away the evidence was being revealed as the tapes were transcribed.

In a subsequent medical conference later the second day, the tape content was discussed, and generated inevitable results. The medics were not happy that the data were not only being kept from them, but that it could also have been hidden. The fall-out went straight back to the Astronaut Chief Al Shepard, who later that same day opened a channel on the uplink, and gave Carr and his crew a mild but very public and official reprimand: 'I just want to tell you that on the matter of your status reports, we think you made a fairly serious error in judgement here in the reporting of your condition.' Carr replied: 'OK Al. I agree with you. It was a dumb decision.'

There was some further discussion on the incident, but there the matter rested. Journalists tried to draw NASA about whether this represented a serious breakdown on the frank and open communications between the crew and the controllers, but it was not thought to be so. Indeed, analysts of the air-to-ground transcripts revealed how open the crew were in using the B channels in debriefings for the rest of the mission.

Fully aware that their comments were on VOX on channel A, the crew remained guarded. Even the channel B tapes were later made public, and the crew was fully aware that any critical comments about the hardware, procedures or activities could be misunderstood, and be used in incorrect context – which, in fact, did happen (as discussed in the next chapter).

Reactivating Skylab

For the first few days on the station, Carr and Gibson took over some of Pogue's work schedules whilst he handled the light tasks of his colleagues. Although the nausea was beginning to pass, the effects of the incident were to put an edge on the first few weeks onboard as the crew chased around trying to activate the station and begin work, almost at the pace at which the previous crew ended their mission.

Carr rebuked himself for letting the vomiting issue go beyond good sense. In his personal diary he reflected on how the ground had the crew 'dead to rights' on the audio tapes, and that what happened was a 'slap on the wrist' by Shepard. A lack of rapport with the ground, and an attempt to over-work to improve matters, did not help. He reflected that it was indeed a dumb decision, and that he should have known better. All it had achieved was the discredit of his crew – which hurt.

On the ground, the overall feeling was that it was satisfying to have a crew once more on the station, even though it was the last mission planned; and as soon as the activation was completed then work would commence. The uninformed understanding that setting up the OWS was only a little more complicated than unpacking after a vacation, was certainly disproved by all three crews as they chased their tails trying to complete all their tasks on time and to the flight plan. Carr noted that the first few days had been so busy that he had not had time to look out the window. New tasks were added to the activities, and Houston persistently interrupted them to monitor their progress, all of which gave the impression of 'a fire drill ever since we got here'.

Revealing the evidence of the length of time that they had spent in space, Pogue (*left*) and Carr sport healthy beards during their three-month mission.

Jerry Carr feeds three bags into the trash airlock, while Pogue prepares to hold down the lid as the bags are ejected into the oxygen tank. By the end of the third mission, disposing of the trash was becoming a difficult task.

A new Skylab 'crew' greets the Skylab 4 crew as they enter the workshop for the first time – Commander 'see no evil', crewman 'hear no evil', and crewman 'speak no evil'.

Skylab stowaways

On entering the station, one of the first things that the crew found was a note on the teleprinter from MCC, left there by Bean's crew at the end of the second visit: 'Jerry, Ed and Bill, welcome aboard the space station Skylab. Hope you enjoy your stay. We're looking forward to several months of interesting and productive work. Signed, Flight Control.'

As they continued to activate the station, the first three stowaways were found onboard the station: three flight suits stuffed with spare clothing, left by the Bean crew. One was on the bicycle ergometer, another was in the LBNP – and the third was sitting on the toilet in the waste management compartment. The crew relocated the results of their colleagues humour to the upper dome of the station, as a response to the events of the previous days: 'Hear no evil, see no evil, speak no evil!'

As well as activating the station, the crew were assigned maintenance tasks and additional medical experiments to perform in the first few days, which pushed them even more. Pogue was beginning to feel much better by the fourth day, and successful recharged the primary coolant loop, allowing the first of the EVAs to proceed as planned – but not before Carr gave the ground his opinion that the number of tasks in the flight plan had to be reduced if they were ever to catch up and stow away everything.

'The great outdoors'

By MD7 (22 November – Thanksgiving Day) Pogue was feeling much better, and with Gibson, was suited up for the first of four excursions outside, leaving Carr in the MDA. Gibson would complete three of the four EVAs, and during the mission was really impressed with the grandeur of the view outside, as had been Lousma before him: 'Boy, if this isn't the great outdoors.'

The two astronauts replaced film cassettes, deployed the Coronagraph Contamination Experiments on the ATM truss, and used a camera that had

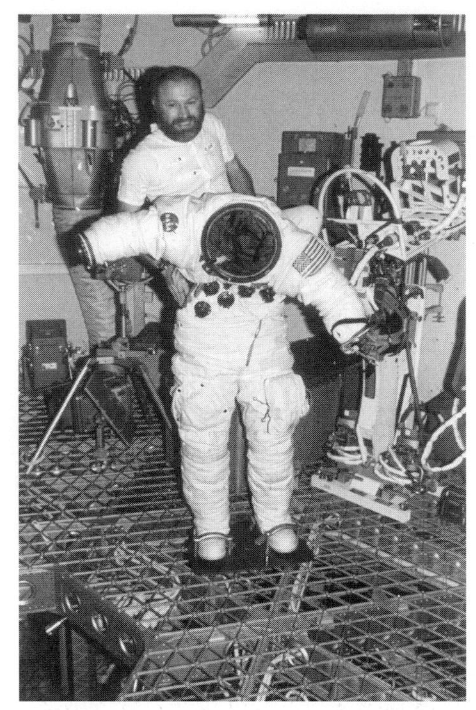

Preparing for work, Jerry Carr floats into his pressure garment for an EVA outside the station.

originally been intended for use from the airlock, now blocked by the parasol. They attempted taking photographs of the Earth's atmosphere, but it failed after only five of forty photographs had been taken. As they tried to repair the Microwave Radiometer/Scattermeter/Altimeter antenna in an area not designed for EVA operations, they found it very difficult to maintain position and prevent their umbilical from tangling. The top of the ATM structure reminded Pogue of being in the crow's nest of a ship.

After 6 hours and 33 minutes they had completed their task, and headed back inside. They had completed almost 100% of their tasks, and had hardly raised a sweat. The EVA confirmed that, since Gemini, a great deal had been learned in performing spacewalks in Earth-orbit. However, although the EVA was a success, it tired the crew, and on squaring away they left some of the tasks to the next day.

Carr did not wish to admit that his crew were unable to keep up with the flight plan, and the ground also later admitted that they had made an error in assigning too much to the crew, and in not realising how much time would be required to complete even the simplest tasks, which took much longer to accomplish in space.

All in a spin

Early in the mission, the first gyro motor current showed a rapid rise which indicated failure, and so it was turned off. This action enabled the switch to a two-gyro

Jerry Carr puts up a sign in the crew compartments of the wardroom during the third mission. Note the floating towels and wash cloths to the left.

Bruce McCandless, back-up Skylab 2 pilot and co-investigator for Experiment M509, models what the well-dressed astronauts should wear during EVA. Here he tests the balance and control of the manoeuvring unit back-pack test model in 1971. The flight unit was tested inside the workshop two years later, and in 1984 McCandless became the first to fly an untethered EVA, with the Space Shuttle MMU during his first spaceflight.

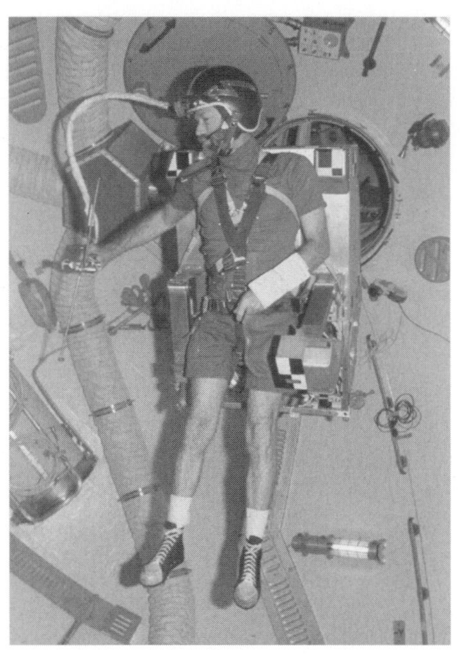

Jerry Carr flies the M509 Astronaut Manoeuvring Equipment experiment on the thirtieth day of the third mission (15 December 1973). He is holding the manoeuvring unit in his right-hand during a test flight, and is actually on the 'roof' of the dome of the forward OWS. A cuff check-list is on his left forearm. Tests of this device led to the development of the MMU, which was flown successfully from the Space Shuttle by six astronauts during three missions in 1984.

Sporting a crash helmet and goggles, Jerry Carr test-flies the third mission's M509 experiment and foot-controlled unit combination.

operation for orientation, instead of three. Normal operation was with two gyros, backed up by the third, and so losing one was not the main concern; but with so much attitude fuel being used early in the mission, with only a third of the original amount remaining, and with significant EREP passes to be completed, as well as the planned complex manoeuvring to photograph comet Kohoutek, it might be a close call to complete the mission as planned, although the crew remained confident that this would be the case.

Twelve days later, gyro 2 recorded irregular operation, and threatened early termination of the mission. The troublesome gyro ran normally for a while, and then recorded lower speed and higher temperatures before returning to normal operations; but the abnormal operations were lenghtening in duration. Extensive ground-testing and analysis was carried out to control the operation of the gyro heaters and monitor the operation of the gyro for the remainder of the mission.

As part of the normal mission operations, the back-up rescue vehicle for SL-4 was rolled to the pad on MD18 (3 December); but it would not be needed, and by the end of the third mission the gyro was still operating in the same irregular mode.

Down to work

With the crew adjusting to life on orbit, and the ground monitoring the gyros, the astronauts began serious work with the ATM, EREP and the rest of the science investigations.

The third crew completed 45 EREP passes, in addition to most of the student experiments, and seventeen science demonstrations were completed, together with 70 TV transmissions showing life on board, their work, and views outside, including the Earth and the comet, all of which attracted the attention of the media

The crew was beginning to gain in proficiency and workload, and soon began to make their own mark on Skylab. Carr estimated that the first 28 days were the most difficult, with heavy work-loads and long days. They often responded to questionnaires on habitation and activities in direct answers, indicating that there was no way to train for unexpected activities that occurred, and by being rushed by the ground was not the way to overcome the problem. They also found it unhelpful to include housekeeping activities in the time-line, and found that by setting a set of primary daily activities to be achieved, with a shopping list of housekeeping chores that needed to be done, or delayed to the next day, the performance rate increased. They also found that eating after exercise was not advisable, and by discussing this with the ground they significantly increased their output for the next two months – so much so, that when analysed, there seemed to be no difference from that of the second crew.

The medical experiments continued, and included a new device designed by Bill Thornton – after analysing data from the first two missions – to condition the crew's leg muscles to help maintain their condition for the return to Earth. Bill Thornton strived to obtain the best value per $ from spaceflight hardware, and this was certainly the case with his design for a treadmill. A sheet of shiny Teflon was attached to the floor, and wearing socks on the feet, the astronaut, restrained by bungee cords, could slide the feet over to condition the muscles. Loads

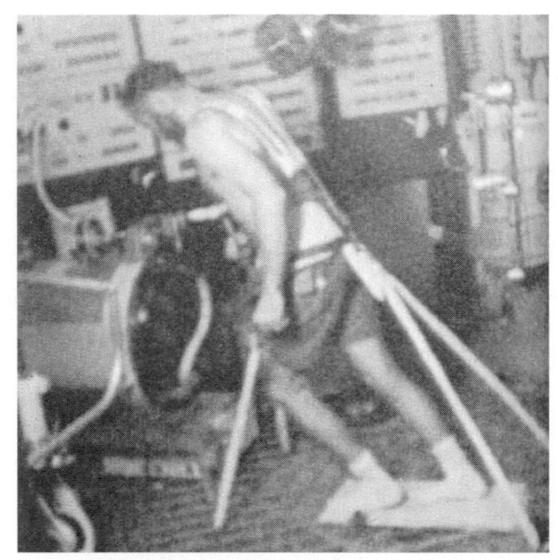

Jerry Carr keeps in shape on the treadmill – 'Thornton's Revenge'.

Jerry Carr opens one of the scientific airlocks.

could also be added to the 'run' and the third crew quickly named this device 'Thornton's Revenge'. The effect of this device in conditioning the crews' lower limbs was dramatic, and led Thornton to design a more sophisticated treadmill for Space Shuttle and space station crews. Thornton was particularly impressed with the data from Carr, and asked if he could arrange for his legs to be donated to Thornton for medical science!

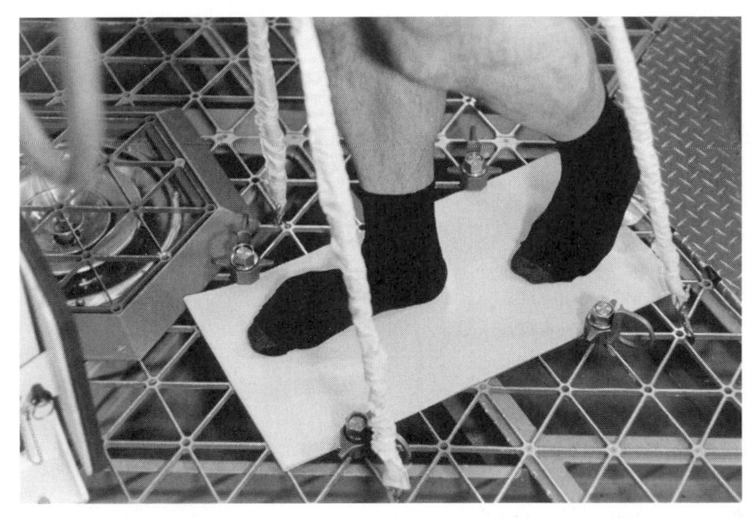

'Thornton's Revenge' is demonstrated by its inventor Dr Bill Thornton in the workshop mock-up at JSC prior to Skylab 4. This Teflon-coated device, bolted to the floor, allows the crewman, wearing adjustable bungee cords attached to the floor, to apply load to the back or leg muscles as he runs in a treadmill-like exercise.

Bean and his crew had taken about 30 days to reach a peak of 16-hour work-days; but the flight controllers apparently forgot this, and considerably increased the workload on Carr's crew from almost the very beginning. The third crew began voicing their concerns and their frustration in attempting to learn to move around the huge workshop and at the same time keep up with the flight plan, and they consequently became very demoralised. By the end of the second week they felt that they were being driven beyond their limits, and the following week they clearly stated that they did not expect to work 16-hour days for 85 days on the ground, and should not be expected to do so in space.

The controllers heard their complaints, but always expected them to break through the barrier. However, by the sixth week, the crew could take no more, and they told the ground that they wanted a rest. Using the B channel, Carr requested a frank appraisal of the status of the mission, and reminded the ground that during the first six weeks, the time off amounted to a couple of hours and a shower. The TV/ press conference on 2 January was not ideal for public relations, as it concentrated mainly on Pogue's illness, the mistakes, and the demands for time-off which made the crew appear to be 'screw-ups and slackers' and the management to be not fully supportive. After a call from Deke Slayton, during which these concerns were discussed, the second six weeks seemed to progress much better. The ground eased the demands on the crew, the exercise and housekeeping chores were removed from the working priorities, and the crew was then able to revert to the normal schedule.

One of the mission tasks was to record comments on aspects of the habitability of Skylab, using the tape recorders, to critically analyse the spacecraft environment; but unfortunately – and as normal – only the negative comments were reported by the media, and together with the above reports and earlier events they were misquoted and placed out of context. These 'subjective evaluations' included a wide range of observations and comments: 'Getting to where you want to go and then anchoring yourself, getting what tools you need and learning to cope with your new environment takes a great deal of time ... the best part [well-designed for living and working in zero g] is the MDA ... that's because we've made good use of the wall ... one thing that is lacking throughout the spacecraft is provision for temporary stowage or restraint of equipment ... another unanticipated problem was that at least two dozen brand new tasks which have never [been trained for] or had just been talked through ... I think the flight planners had a lot of this stuff jammed down their throats by the flight activities people ... it's nothing personal, it's just sort of caught in the system ... I think were getting an awful lot of good hand-held photography and visual observations by virtue of the fact the wardroom window is in the wardroom ... the dome and wardroom handrails are good when we use them ... the triangular shoes' cleats/grid are good to excellent ... the EREP foot platform is very good, except it's very limited ... the shower is good to excellent ... the towel holders are excellent ... general utility and wet wipes are great, man, we use a lot of them ... food cans are adequate ... the food tray I don't like, the magnets aren't strong enough ... I've had some problems with the trash airlock, but it's functional ...'. The positive comments were often overlooked: 'Busy, busy day today. We had our first EREP and carried it off really well ... had another EREP today and did a good job again ... today was a real barn-burner ... We did a little of everything, solar viewing, exercise and medicine in the first half of the day and EREP for the rest ... Deke called as today and said we're doing an outstanding job ... Looks like we're on top of the schedule now, we're getting everything done on time or earlier.'

Stepping out for Christmas

On 18 December the three Skylab astronauts were joined in orbit by two Soviet cosmonauts onboard Soyuz 13. It was the first time that astronauts and cosmonauts had been in space simultaneously, and although no joint activites were planned, the two nations were working towards a joint docking mission in the summer of 1975. Onboard Soyuz 13 were the Orion 2 telescope and life science experiments, but this was a short flight of only a week. By Christmas Eve the pair were back on Earth. Since the first Salyut in 1971, the Soviets had experienced difficulty in their space programme with the orbital break-up of Salyut 2 in April 1973, just prior to the launch of Skylab and Cosmos 557 (Salyut) in May. It was some months before a new Salyut was readied, and by then, Skylab had established new endurance records that the Soviets were unable to exceed for the next four years.

No-one had been in space for Christmas for five years – since the historic Apollo 8 mission, in which the crew became the first humans to spend Christmas off the planet – and around the Moon. By Christmas 1973, Carr, Gibson and Pogue were also to spend the festive season in orbit, but also to work; but it was a fine present – an EVA for Carr and Pogue to replace film cassettes on the ATM experiments and to photograph the comet from outside the station.

During the EVA, which lasted 7 hours 1 minute, the astronauts accomplished

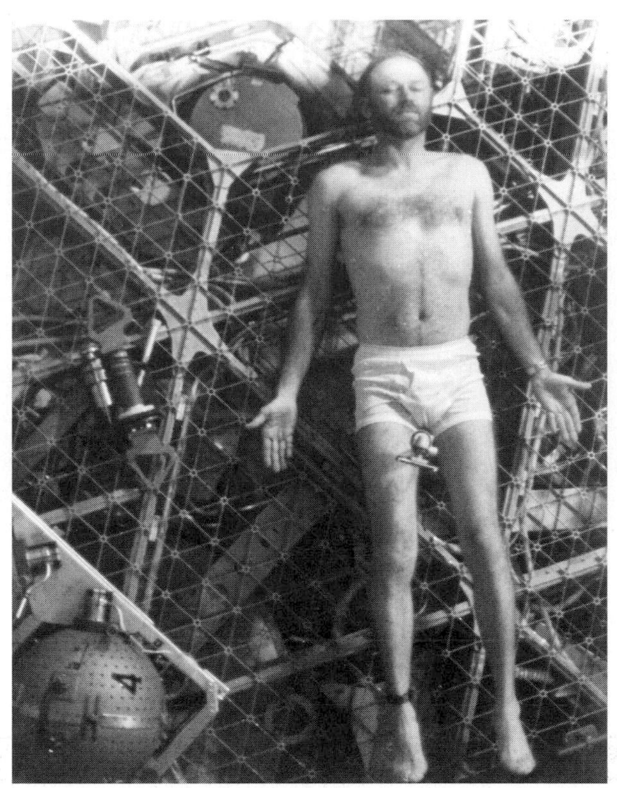

Jerry Carr participates in one of the biomedical experiments that recorded the physiological changes of the body in zero g. The experiment required a sequence of photographs of the human body from various angles, to reveal changes in leg volume and limb positioning. Note the strap around his right ankle, which prevented him from floating away.

partial film replacement, and took forty photographs of the comet, as well as repairing the telescope filter wheel, which required detailed work, and was challenging due to the restrictions of the pressure suit gloves and the use of a dental mirror, a thin pen-light to illuminate the work area, and a screwdriver to place the wheel. As they worked, a slow leak of yellowed cooling water from the chromate corrosion inhibitor flaked off.

Four days later, Carr was back outside – this time with Gibson – for the third EVA, of 3 hours 39 minutes, to repeat the cometary observations; and he also retrieved a piece of the AM micrometeoroid cover for return to Earth. He again to encountered ice during an EVA, but this time on his suit, due to a small coolant water leak.

New year and new records

The crew had decorated their orbital home for the season with a tree – not supplied from Earth (although family presents were stored in the CM), but made from empty food containers!

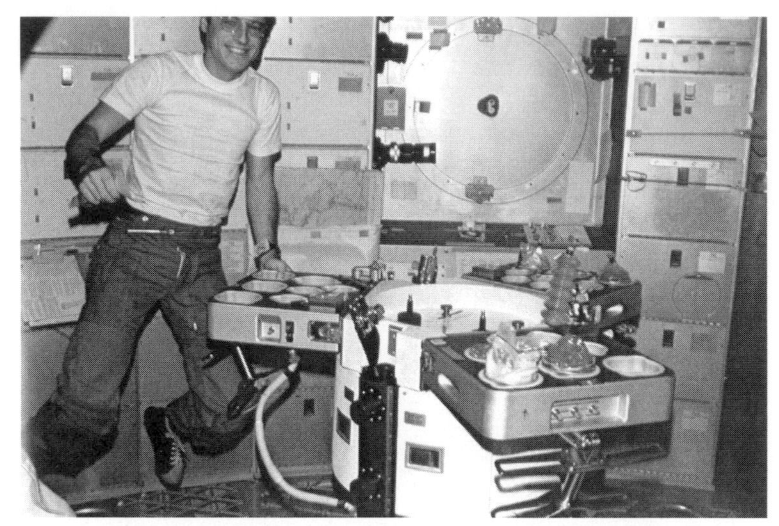

Ed Gibson prepares to eat in the wardroom.

The following week they became the first crew to celebrate the new year in space – seventeen times in one day, as the orbited the Earth. On 4 January they surpassed Pete Conrad's accumulated time in space, and headed for Al Bean's record.

On the ground, mission planners were evaluating the situation concerning Skylab's gyros and consumables during a 56-day mission review on 10 January. The following day the crew received the good news that they had been cleared to surpass Bean's record, the 70-day target, and set a new endurance record of 84 days – the full twelve weeks planned at the beginning of the mission. Just 48 hours later they surpassed the SL-3 duration of 59 days 11 hours, and finally, on MD70 (24 January), surpassed Bean's accumulated record of 69 days 15 hours. They were now world endurance record holders on their first mission, and they still had two weeks left! On 23 January, Pogue celebrated his 44th birthday.

The last EVA

On 3 February 1974, Carr and Gibson opened the EVA hatch – for the last time on their mission and on Skylab – for an EVA of 5 hour 19 minutes, to collect all samples and film cassettes from the ATM instruments. Up on the ATM, Gibson had secured himself in the gold-painted 'golden slipper' foot restraints. During quiet moments on the EVA, when he looked down at Earth he had the impression that gravity might pull him down; but if he leaned backwards and stared into the cosmos, the vista before him was wonderful.

As Carr and Gibson returned to the airlock to close the hatch, they also closed the door on the final EVA of the pioneering era of American space missions. There were no EVAs planned for ASTP in 1975. The next time an American would stretch his legs outside of a spacecraft would be on a Space Shuttle flight – and that was eventually nine years late. In less than a decade, from June 1965 to February 1974, American astronauts had learned to work effectively in Earth orbit, had left their footprints on the Moon, and had walked in space. The first American EVA, by Ed White, was through a Gemini hatch, and it seemed fitting that the last of the era should be through a similar hatch.

Returning home

If activating the station on arrival was a challenge, then packing up to return home was a major event that also took several days. For Carr and his crew, this was the last planned visit, although they provided for the possibility of a revisit some time in the future.

On 31 January, the updates to the deactivation and re-entry checklist – all 15 feet of it, and all to be entered into the FDF by hand – were teleprinted to the crew. Despite this being a spaceflight, it was not an era of lap-top computers, e-mail and digital processing, and pencil or pen had to suffice! The event caused good-natured comments for a couple of days, and Carr remarked: 'I understand you're going to teleprinter up the Old Testament tonight.'

The closing down of a few last experiments was combined with the packing up of the CM, which proved as difficult as unpacking it. Attempting to fit everything in the lockers was difficult in zero g, but, as Carr reported, using the suitcase method worked as well in zero g as it does on Earth: 'It fits if you force it!'

Just prior to leaving Skylab on 8 February, the crew fired the RCS on the Service Module to raise the OWS orbit, increasing the lifetime and allowing it to remain in orbit for the next eight years. Skylab was now expected to be in orbit until at least 1981 or 1982.

As they undocked, they performed a fly-around inspection before departing for the trip home. 'It's been a good home, I hate to think we're the last guys to use it,'

Skylab Orbital Workshop in orbit, taken from the third crew CSM.

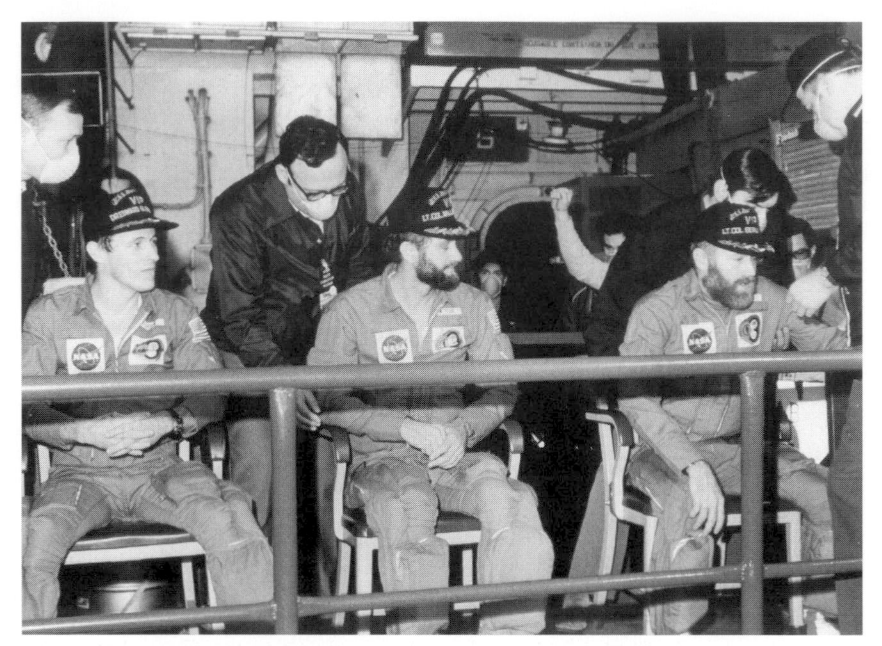

Looking tired, the last Skylab crew sit on the recovery ship at the end of their mission.

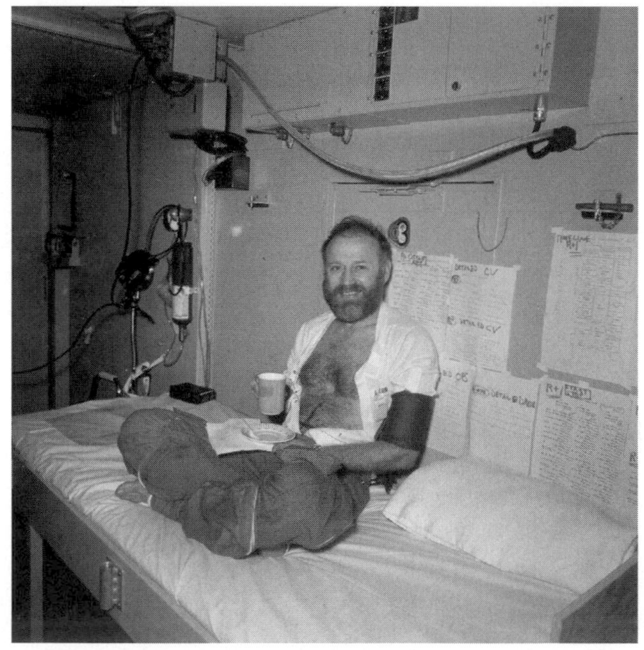

Jerry Carr sits in the recovery ship with a drink. On the wall are the time-lines for operations for the first day or so after recovery.

The 'primary support crew' is the astronaut's family. In this post-splashdown photograph the Carr family celebrate the success of the mission and the safe return of dad. (*Left–right*) John Carr, 11; Joshua, 9; Jennifer, 18; Jamee, 15; Rev Dean Woodruff, family pastor; Mrs Jo Ann Carr; Jeff, 15; and astronaut Rusty Schweickart. Out of frame is Jessica, 9. The family pet, Tags, seems unimpressed by the achievement. (From the Jerry Carr collection.)

Gibson radioed. CapCom Bob Crippen voiced the opinion of the ground: 'Say goodbye for us. She's been a good bird.'

Just under five hours after undocking, and after 84 days 1 hour 15 minutes 32 seconds, Carr, Gibson and Pogue were back on Earth, upside down in the Pacific Ocean, before the self-righting bag turned the capsule for pick-up by the recovery ship.

Despite adversity at the beginning of the mission, it had ended on a highly successful and rewarding note. Skylab had lived up to and surpassed all that was expected, and there was hope for a future return visit. Carr recalled: 'I'd like to think of our mission as one where three guys went up ... and tried to live a normal life in space to show that it could be done.'

Tests and orbital storage

Shortly after the last crew undocked and departed for home at the end of mission, engineering tests began on the OWS, and continued for the next 32 hours. Internal pressure was reduced to 0.5 psia, and was allowed to decay.

Attempts to power up gyro 1 failed, indicating seized bearing wheels. Controllers ran a series of electrical tests, and each battery in the AM was discharged down to 30 V before a test was run on the coolant loop system. After beginning the power-down sequence, it took about two months to completely deplete the oxygen/nitrogen environment, after which the station was orientated to a gravity gradient position (with the MDA facing out to space) and the TCS and CMG were shut down. The

Welcoming the last crew home from their mission, stepping off an Air Force transport plane are Carr, Gibson and Pogue. Carr's wife, JoAnn, waits to greet her husband. (From the Jerry Carr collection.)

The scene in MCC, Houston, following splashdown in the Pacific Ocean. In the foreground are (*left to right*) Flight Director Neil Hutchinson; Flight Director Don Puddy, and CapCom astronaut Bob Crippen.

last commands to turn off the telemetry were sent to the station, and the controllers then departed the Skylab mission control room for the last time.

Zero gravity transformed Skylab's crewmen into skilled acrobats.

Skylab CM locations (2001)

SL-2 National Museum of Naval Aviation, Pensecola, Florida.
SL-3 Glenn Research Center, Cleveland, Ohio.
SL-4 National Air and Space Museum, Washington DC.

Skylab workshop mission day reference

1973 May 14 – 1974 February 10 – 1979 July 11			
Day	Date	DOY	Mission period
	1973		
1	5–14	134	
First unmanned period begins			
Skylab 1 (OWS) launched			
2	5–15	135	
3	5–16	136	
4	5–17	137	
5	5–18	138	
6	5–19	139	
7	5–20	140	
8	5–21	141	
9	5–22	142	
10	5–23	143	
11	5–24	144	
First unmanned period ends			
12	5–25	145	
First manned period, MD1			
Skylab 2 launched; docked to OWS			
SUEVA, 37 min (Weitz)			
13	5–26	146	2
14	5–27	147	3
15	5–28	148	4
16	5–29	149	5
17	5–30	150	6
18	5–31	151	7
19	6–1	152	8
20	6–2	153	9
21	6–3	154	10
22	6–4	155	11
23	6–5	156	12
24	6–6	157	13
25	6–7	158	14
EVA, 3 hrs 30 min (Conrad/Kerwin)			
26	6–8	159	15
27	6–9	160	16
28	6–10	161	17
29	6–11	162	18
30	6–12	163	19
31	6–13	164	20
32	6–14	165	21
33	6–15	166	22
34	6–16	167	23
35	6–17	168	24
36	6–18	169	25
37	6–19	170	26
EVA, 1 hr 44 min (Conrad/Weitz)			
38	6–20	171	27
39	6–21	172	28
40	6–22	173	29
First manned period ends			
Skylab 2 recovered (28 days 00 hrs 49 min)			
41	6–23	174	
Second unmanned period begins			
42	6–24	175	
43	6–25	176	
44	6–26	177	
45	6–27	178	
46	6–28	179	
47	6–29	180	
48	6–30	181	
49	7–1	182	
50	7–2	183	
51	7–3	184	
52	7–4	185	
53	7–5	186	
54	7–6	187	
55	7–7	188	
56	7–8	189	
57	7–9	190	
58	7–10	191	
59	7–11	192	
60	7–12	193	
61	7–13	194	
62	7–14	195	
63	7–15	196	
64	7–16	197	
65	7–17	198	
66	7–18	199	
67	7–19	200	
68	7–20	201	
69	7–21	202	
70	7–22	203	
71	7–23	204	
72	7–24	205	
73	7–25	206	
74	7–26	207	

Day	Date	DOY	Mission period	Day	Date	DOY	Mission period
		1973 May 14 – 1974 February 10 – 1979 July 11					
75	7–27	208		114	9–4	247	39
Second unmanned period ends				115	9–5	248	40
76	7–28	209		116	9–6	249	41
Second manned period begins, MD1				117	9–7	250	42
Skylab 3 launched; docked to OWS				118	9–8	251	43
77	7–29	210	2	119	9–9	252	44
78	7–30	211	3	120	9–10	253	45
79	7–31	212	4	121	9–11	254	46
80	8–1	213	5	122	9–12	255	47
81	8–2	214	6	123	9–13	256	48
82	8–3	215	7	124	9–14	257	49
83	8–4	216	8	125	9–15	258	50
84	8–5	217	9	126	9–16	259	51
85	8–6	218	10	127	9–17	260	52
EVA, 6 hrs 29 min (Garriott/Lousma)				128	9–18	261	53
86	8–7	219	11	129	9–19	262	54
87	8–8	220	12	130	9–20	263	55
88	8–9	221	13	131	9–21	264	56
89	8–10	222	14	132	9–22	265	57
90	8–11	223	15	EVA, 2 hrs 45 min (Bean/Garriott)			
91	8–12	224	16	133	9–23	266	58
92	8–13	225	17	134	9–24	267	59
93	8–14	226	18	135	9–25	268	60
94	8–15	227	19	Second manned period ends			
95	8–16	228	20	Skylab 3 recovered (59 days 11 hrs 09 min)			
96	8–17	229	21	136	9–26	269	
97	8–18	230	22	Third unmanned period begins			
98	8–19	231	23	137	9–27	270	
99	8–20	232	24	138	9–28	271	
100	8–21	233	25	139	9–29	272	
101	8–22	234	26	140	9–30	273	
102	8–23	235	27	141	10–1	274	
103	8–24	236	28	142	10–2	275	
EVA, 4 hrs 30 min (Garriott/Lousma)				143	10–3	276	
104	8–25	237	29	144	10–4	277	
105	8–26	238	30	145	10–5	278	
106	8–27	239	31	146	10–6	279	
107	8–28	240	32	147	10–7	280	
108	8–29	241	33	148	10–8	281	
109	8–30	242	34	149	10–9	282	
110	8–31	243	35	150	10–10	283	
111	9–1	244	36	151	10–11	284	
112	9–2	245	37	152	10–12	285	
113	9–3	246	38	153	10–13	286	

1973 May 14 – 1974 February 10 – 1979 July 11

Day	Date	DOY	Mission period
154	10–14	287	
155	10–15	288	
156	10–16	289	
157	10–17	290	
158	10–18	291	
159	10–19	292	
160	10–20	293	
161	10–21	294	
162	10–22	295	
163	10–23	296	
164	10–24	297	
165	10–25	298	
166	10–26	299	
167	10–27	300	
168	10–28	301	
169	10–29	302	
170	10–30	303	
171	10–31	304	
172	11–1	305	
173	11–2	306	
174	11–3	307	
175	11–4	308	
176	11–5	309	
177	11–6	310	
178	11–7	311	
179	11–8	312	
180	11–9	313	
181	11–10	314	
182	11–11	315	
183	11–12	316	
184	11–13	317	
185	11–14	318	
186	11–15	319	
Third unmanned period ends			
187	11–16	320	
Third manned period begins, MD1			
Skylab 4 launched; docked to OWS			
188	11–17	321	2
189	11–18	322	3
190	11–19	323	4
191	11–20	324	5
192	11–21	325	6
193	11–22	326	7
EVA, 6 hrs 33 min (Pogue/Gibson)			
194	11–23	327	8
195	11–24	328	9
196	11–25	329	10
197	11–26	330	11
198	11–27	331	12
199	11–28	332	13
200	11–29	333	14
201	11–30	334	15
202	12–1	335	16
203	12–2	336	17
204	12–3	337	18
205	12–4	338	19
206	12–5	339	20
207	12–6	340	21
208	12–7	341	22
209	12–8	342	23
210	12–9	343	24
211	12–10	344	25
212	12–11	345	26
213	12–12	346	27
214	12–13	347	28
215	12–14	348	29
216	12–15	349	30
217	12–16	350	31
218	12–17	351	32
219	12–18	352	33
220	12–19	353	34
221	12–20	354	35
222	12–21	355	36
223	12–22	356	37
224	12–23	357	38
225	12–24	358	39
226	12–25	359	40
EVA, 7 hrs 0 min (Carr/Pogue)			
227	12–26	360	41
228	12–27	361	42
229	12–28	362	43
230	12–29	363	44
EVA, 3 hr 28 min (Carr/Gibson)			
231	12–30	364	45
232	12–31	365	46

1973 May 14 – 1974 February 10 – 1979 July 11

Day	Date	DOY	Mission period
	1974		
233	1–1	1	47
234	1–2	2	48
235	1–3	3	49
236	1–4	4	50
237	1–5	5	51
238	1–6	6	52
239	1–7	7	53
240	1–8	8	54
241	1–9	9	55
242	1–10	10	56
243	1–11	11	57
244	1–12	12	58
245	1–13	13	59
246	1–14	14	60
247	1–15	15	61
248	1–16	16	62
249	1–17	17	63
250	1–18	18	64
251	1–19	19	65
252	1–20	20	66
253	1–21	21	67
254	1–22	22	68
255	1–23	23	69
256	1–24	24	70
257	1–25	25	71
258	1–26	26	72
259	1–27	27	73
260	1–28	28	74
261	1–29	29	75
262	1–30	30	76
263	1–31	31	77
264	2–1	32	78
265	2–2	33	79
266	2–3	34	80
EVA, 5 hr 19 min (Carr/Gibson)			
267	2–4	35	81
268	2–5	36	82
269	2–6	37	83
270	2–7	38	84
271	2–8	39	85
Third manned period ends			
Skylab 4 recovered (84 days 01 hr 16 min)			
272	2–9	40	
EOM Engineering tests			
273	2–10	41	
EOM Engineering tests			
274	2–11	42	
Orbital storage until July 1979			
	1979		
2249	7–11	192	
SL-1 (OWS) re-entry and burn-up			

SL-1 (OWS) orbital duration: 6 years 1 month 27 days

Research fields

With the completion of the three manned missions, the operational phase of Skylab was over. Proposals to add an extra mission to the station, schedule a reboost mission to prolong its orbital life, and launch a second OWS, were all unsuccessful (as discussed in Chapter 6). Flight operations were over, and there then began the huge task of analysing the information gathered during the nine months of activity. As a direct result of the efforts of the astronauts, and the amount of experiment results, some final reports would take several years to complete, and in many cases the analysis and use of the Skylab material still continues.

One of the priorities for each of the crews, after returning to Houston, was to generate the series of post-crew debriefs that supplemented the scientific data from their missions. Each of the major research fields is discussed in this chapter, with a summary of the experiments and operations, and a review of lessons learned from the Skylab missions.

Skylab was also to provide a wealth of data on human adaptation to spaceflight, including biomedical studies, habitability provisions, the ability to operate the experiments, inflight maintenance, and EVA. Because of the duration of the missions, a significant level of experience in flight control was obtained that was very different from all the previously flown American missions, including the use of a large manned vehicle in orbit over a period of several months.

SCIENCE

Solar physics

By far the largest and most visible experiment hardware on Skylab was the ATM, which operationally consisted of the cylinder containing the major scientific instruments: the attitude, pointing and control system, for experiment pointing; the four solar array wings, to provide the power supply; an octagonal rack assembly that surrounded the canister and also provided attachments for solar and thermal shields; the solar arrays, outriggers and several subsystem components; and the control and display (C&D) console inside the MDA, which was used by the crew for control and monitoring during experiment operations.

The major predecessor to the ATM was the unmanned Orbiting Solar Observatory (OSO), built by Ball Aerospace for the NASA Goddard Spaceflight Center, and launched by Delta rocket into a 33°-inclination Earth orbit. Each carried a range of instruments to study the Sun for up to one year, although several exceeded this, and allowed studies of the Sun throughout its 11-year cycle. Although extremely successful, the setback with the series was the limitation of size due to power and telemetry capacity, and the ability of the launch vehicle to lift the payload. In general each OSO weighed 600 lbs (except OSO-7, which weighed 1,250 lbs), with only 300 lbs allocated to the solar instruments, each of which weighed less than 55 lbs and did not measure much more than three feet long. In comparison, the ATM weighed 24,656 lbs, with a principle experiment housing measuring about ten feet long and 4.5 feet across to carry the full-scale 10-foot, 2,000-lb telescope, and with its own solar arrays could generate 2,000 W of experiment power compared to 20 W on OSO-7. An advanced OSO-8 was built by Hughes, and was launched after Skylab in 1975.

Orbiting Solar Observatories, 1962–1975

OSO	Launched	Notes
OSO-1	1962 Mar 7	Non-scanning; X-rays and gamma rays, interplanetary dust; observed more than 140 solar flares over two years, and recorded 77 days of 'near perfect' solar observations
OSO-2	1965 Feb 3	UV spectrometer, white-light coronagraph, ray spectrheliograph; operated for ten months, and raised total data hours to more than 6,000
OSO-C	1965 Aug 25	Lost in launch accident
OSO-3	1967 Mar 8	X-ray spectrometer, UV spectrometer
OSO-4	1967 Oct 18	UV spectrometer, spectroheliograph, ray spectrometer
OSO-5	1969 Jan 22	UV spectrometer, spectroheliograph, ray spectrometer; carried British, Italian and American experiments
OSO-6	1969 Aug 9	UV spectrometer, spectroheliograph, ray spectrometer; carried British, Italian and American experiments
OSO-7	1971 Sep 29	X-ray and extreme UV spectrometer, visible and UV polychromator; decayed July 1974
OSO-8	1975 Jun 21	UV spectrometer, visible and UV polychromater; weighed 2,345 lbs; carried American and French instruments; operated until September 1978

The other advantage of the ATM over OSO was the capacity to store a significant amount of data on photographic film, due to the ability of the astronauts to replace, retrieve and return the film cassettes from the ATM structure during EVA. There was a selection of specially designed cartridges that could hold 16,000 frames or advance a series of frames to create a filmstrip. The cassettes weighed 88 lbs on Earth, but in space – nothing at all. However, due to the bulk of the canisters the exchange was assisted with either a telescoping boom or 'clothes-line' device that greatly assisted movement of the cartridges to and from the ATM Sun-end.

Highly sensitive instruments, mounted on the spar in the solar observatory canister, studied the Sun in great detail.

Film canisters were stored in Skylab at launch, and extra canisters were included in the third crew CM, which resulted in thirty of them holding more than 150,000 exposures, being returned to Earth.

The primary experiments located in the ATM were:

S052 White-Light Coronagraph Designed to take photographs twice daily or, in the case of high limb activity up to one every 13 seconds, in visible light, of the solar corona to a maximum of six solar radii (3 million miles), using 35-mm film for approximately 8,000 data frames loaded in one camera at launch and three replacement cameras in the OWS for exchange during EVA. The duration of Skylab allowed for observations of the corona for several rotations of the Sun (each rotation taking approximately four weeks), and despite the designed 56-day orbital operating life, the instrument was still operating at the end of the third mission. During the duration of the Skylab mission, it captured 35,918 frames.

S054 X-Ray Spectrographic Telescope Took sequential photographs of X-ray events such as flares and active regions of the Sun, in order to determine the corona temperatures, and densities of energy particles. Two concentric highly polished metal alloy mirrors intercepted the X-radiation, and filters selected the wavelengths to be photographed. The image projected onto the console allowed the astronaut to view a 'live' image of the Sun in order to complement observations made with the other instruments. The first flare was recorded on 15 June (SL-2 MD22), and again the instrument exceeded its orbital design life of 42 days.

S055 UV Scanning Polychromator Spectroheliometer Designed to record temporal changes in the UV radiation that is emitted by several solar regions. Data was recorded electronically, and no film was required. The total operating time for this instrument was 2,292 hours, which exceeded its orbital design life of 56 days.

S056 X-Ray Event Analyser/X-Ray Telescope Designed to obtain solar radiation data in the X-ray region of the solar spectrum. This was to provide data on active phenomena (solar flares) within the solar atmosphere, the data from which was to increase the knowledge of how the magnetic field of the Sun influences the development of flares, and relationships between the formation of flares and sunspots. Each film cassette held 7,200 data frames, and was exchanged by EVA. More than 27,000 X-ray photographs were recorded, and more than 1,100 hours of spectral data were displayed and transmitted to the ground. It was the first instrument to detect a flare, on 15 June 1973.

S082 XUV Spectrograph/Spectroheliograph A two-part experiment designed to photograph the Sun sequentially over a long period in selected wavelengths. Instrument 'A' was the spectroheliograph that took photographs of the solar disk in white light (EUV) regions. The astronaut selected the wavelength range and exposure time of the 'B' instrument, the spectrograph which took high-resolution wavelength data in the middle and near UV ranges of the spectra of the limb or solar disk. The spectroheliograms were to reveal specific emission features, greatly enhanced over the white light photographs, giving the appearance of a 'blotchy Sun'. The field of view was restricted to 1,000 miles wide anywhere on the solar disk. The 'A' instrument was designed for a 56-day life requirement, but was still operating at the end of the third mission, and obtained 1,024 high-resolution photographs; while the 'B' instrument also exceeded the 56-day design life and recorded 6,411 photographs.

Hα Telescopes Used for primary boresighting of the ATM package. It contained two telescopes designed to be sensitive to red Hα light on the Sun. One of these also featured a beam-splitter for simultaneous photographic and TV pictures, while the second operated in TV only. These telescopes included a filter to allow precise observations at desired wavelengths, and a zoom capability for close-up viewing. These two telescopes took TV and photographic images of the solar disk, revealing flare activity as enlarged Hα sources. As the astronaut traversed the solar disc, these telescopes served as his 'eyes', and when a flare was observed, the level of Hα emission was correlated with emission levels from other energy regions. Each videocon accumulated 1,402 hours of operation, and the camera obtained more than 68,000 high-resolution photographs.

S020 X-Ray/UV Solar Photography Part of the solar physics studies, but not located in the ATM. Designed to be installed in the solar SAL, it could not be operated on the first or second mission due to the deployment of the solar parasol from the solar ATM. The third crew took the instrument outside on an EVA. It was designed to obtain detailed spectra of X-ray and UV radiation from both normal and explosive areas in the solar atmosphere.

In preparing the ATM instruments to fly, solar scientists had to decide on exactly what they wanted to observe to take full advantage of the largest manned observatory ever placed in orbit. Each of the instruments was aligned before the station was launched, and was provided with the facility of checking the realignment from onboard to refine the observing capabilities as the mission proceeded. In addition, a programme of joint operating procedures was developed, allowing examination of each major solar feature, and including contingency planning for unexpected events – such as solar flares – during the course of the Skylab mission.

Not only was it important to draw up a precise programme of observations and provide the appropriate instruments with flexibility; it was also necessary to decide on the priority of observations of the various features on the solar disc. After months of evolution, simulation and discussion involving solar scientists, NASA and the astronauts devised an observation book for each instrument, and a copy of each was placed on the station for use by the first crew. Their success in operating the ATM, along with experience in mission training, planning and adapting results from the first mission, allowed the second crew to take a second edition of the ATM observation book to Skylab, and in turn contributed to the third edition, flown with the last crew. These books, NASA stated, were 'the best books on solar observation that had ever been written'.

Each day during the manned occupation of Skylab, a meeting was held between the senior representative of each experiment and members of an advisory panel, to discuss the next day's planned observations, in order to review the real-time activities of the crew, the status of instruments, and activities on the Sun, and to plan for predicted events advised by solar forecast staff of the National Oceanic and Atmospheric Administration (NOAA) from its world-wide network of solar observatories. In all, 300 solar scientists and almost all the solar observatories and several satellites supported ATM activities.

These meetings resulted in a solar observation plan that was sent to Skylab teleprinter – usually during the night – to be read by the crew the next morning. In the early years of ATM development, the instruments were to be upgraded versions of automated instruments flown on a satellite, leaving the astronauts to serve in pointing the telescopes and changing film. However, as the ATM evolved, so did the role of the crew-member – to one of an observer and telescope controller able to change the observation plan to respond to opportunities and events almost as they occurred, thus broadening the scope of data collection.

The training of the crew had to reflect this, and all three astronauts – not just the Science Pilot – were trained to operate the ATM package, recognise significant solar events and align and operate the instruments to best record them. So enthusiastic were the astronauts that their involvement went far beyond what was expected of them, and they often scheduled ATM observation periods on their days off, or even during scheduled sleep periods.

The astronauts' control console was located in the MDA, and featured the switches, indicators and controls to direct power to the instruments, to open and close each of the telescope doors, alter the speed of the instrument, and record data, which was rapid and intense during explosive activity on the solar disc, and much slower and relaxed during the quiet Sun and more routine observations. The console also indicated which instrument was ready, how many frames were left in the camera film, and the instrument's alignment in relationship to the other telescopes.

A TV screen allowed the astronauts to safely view what the instruments saw at different layers of the atmospheric regions of the Sun. In the ATM TV system there were five channels of solar viewing available:

- Channel 1 offered a white-light view of the full disc, revealing sunspots.
- Channels 2 and 3 offered two views of the chromosphere in Hα light via the cameras on the Hα telescopes, using power-zoom for either high or low magnification to enable closer examination or survey of the entire solar disc.
- Channel 4 provided a view of the transition region and lower corona over the whole disc in extreme UV, as seen by the UV spectrograph.
- Channel 5 provided a white-light view of the outer corona.

Skylab astronauts were the first to use live video of the Sun. By flipping from channel to channel they were rewarded with the most comprehensive view of our nearest star ever experienced by man, and by using the instruments they captured this data for later analysis on Earth.

It had been known long before Skylab flew that the Sun follows two periods of activity: a low and high activity, with a peak (solar maximum) at which there are more violent and intense features, and a trough (solar minimum) called the 'quiet Sun' during periods of inactivity. The solar cycle lasts about eleven terrestrial years, and prior to Skylab the maximum was in 1968–1969. When Skylab flew, therefore, the solar cycle was in 'decline' towards a quiet Sun period and the minimum of 1976–1977. The Skylab observations were made during a time when the Sun was becoming 'quieter', although by looking at some of the pictures returned by the astronauts, it could hardly be thought that this was the case.

The package of instruments allowed high-resolution data in the 2,000–7,000 Å spectral range. In addition, the ATM was employed to observe Comet Kohoutek, Mercury, lunar libration, and Earth's atmosphere.

With the Skylab data, the solar astronomy textbooks were almost completely rewritten, as the series of flights had provided an unprecedented amount of new knowledge about the Sun. This generated a flood of research papers and experiment results, findings and a new understanding of how the Sun's atmosphere works and how different layers interact with each other. This was impossible before flying the ATM, and provided scientists with new data in X-ray and UV pictures of flares that led them to completely reappraise the older data. Skylab missions also revealed in detail the coronal transits of solar matter that stream outwards into the solar system and beyond the planets, which had gone largely unnoticed before Skylab.

With sunspots, flares, the corona and the photosphere, Skylab results offered new and stunning data to reward the years of effort by the experimenters and planners of the ATM programme. But in addition to the instruments, it was the dedication of the nine astronauts that truly made Skylab an outstanding success for solar physics, with their unprecedented view of our nearest star, which certainly revealed itself to be intensely active.

(Further reading: Skylab solar research and some of the early results appear in the NASA publication *A New Sun: Solar Results from Skylab*, SP-402 1979.)

Earth observations

In addition to the ATM experiments, the Earth Resources Experiment Package (EREP) represented a significant percentage of Skylab's scientific investigations. The package of experiments carried on Skylab consisted of six remote sensing systems that supplemented the already existing programme of satellite and aircraft remote sensing, but in a far more precise and thorough way, offering a chance to evaluate test sites off the planned ground track and to verify and adapt to the acquired data. In addition, the ability of the astronauts to 'eyeball' the Earth below them again offered unique opportunities to concentrate on new targets of opportunity and on phenomena, in addition to the programmed series of data passes.

The EREP instruments could be operated individually or as a group, depending on the requirements, weather and vehicle capability. The data were also used in conjunction with satellite observations and with aircraft and ground-based studies to provide a tiered level of data capture, providing multi-point sets of data for analysing the results with greater accuracy.

For EREP operations, Skylab was launched with sixteen rolls of 5-inch film (450 frames per roll) 78 rolls of 70-mm film (400 frames per roll) nine rolls of 16-mm film for viewfinder tracking systems (V/TS) (5,600 frames per roll) and 25 reels of 28-track magnetic tape (7,200 feet per reel). The crews also carried additional film to the station on the CM, as required.

Most of the EREP hardware was mounted on the exterior of the MDA, and was operated by the astronauts from an EREP control and display panel, with the data recorded on film and magnetic tape, supplemented by housekeeping data records and the astronauts' voice annotations.

The instruments on the EREP were:

S190 Multispectral Photographic Facility Recorded regions of the Earth by photography in a wavelengths ranging from near infrared to the visible. A two-part experiment – the Multispectral Photographic Camera (S190A) – was a six-channel 70-mm camera that simultaneously viewed the same area in different wavelengths, with a forward motion compensation to allow for the motion of the station. The field of view was 21°.1 across flats, which provide an 88 nautical mile square surface coverage from 235 nautical miles orbit. The Earth Terrain Camera (S190B) was a 5-inch single-lens camera with an 18-inch focal-length lens offering a 14°.2 field of view across the flats, and provided a 59 nautical mile square coverage using high-resolution film. It was operated from the SAL window, with selectable shutter speed and compensation for spacecraft forward velocity.

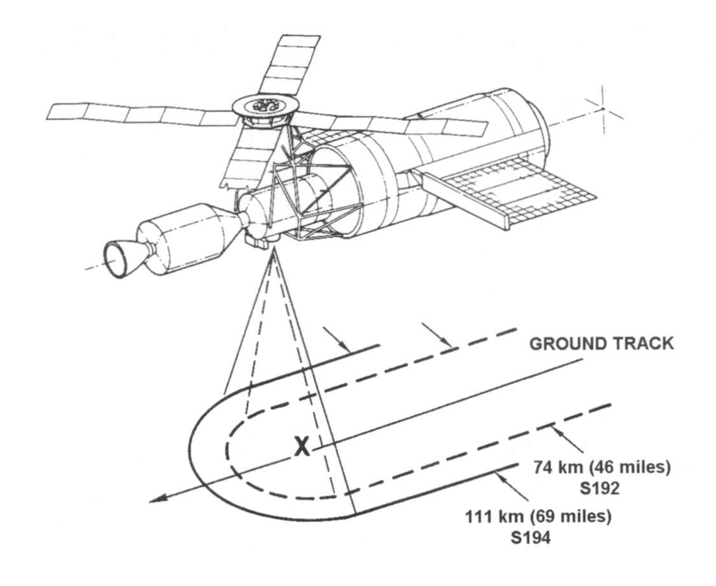

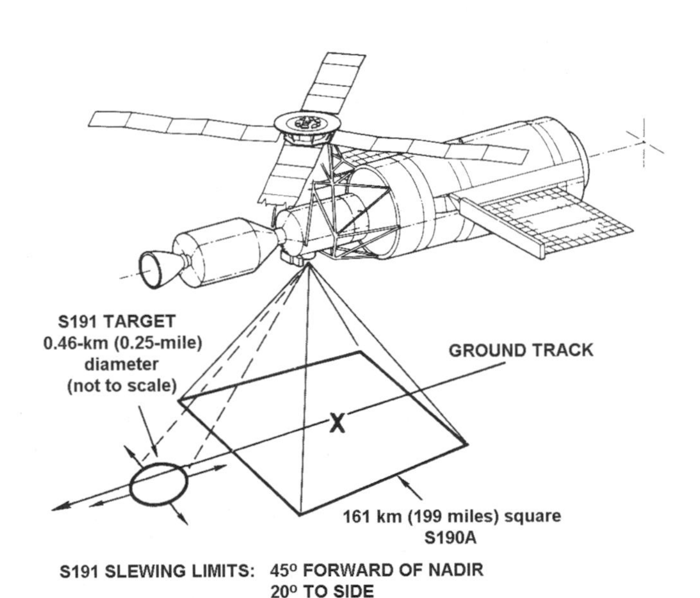

Ground coverage of the EREP instrument package.

S191 Infrared Spectrometer Used to assess the applications and usefulness of remote Earth sensing from orbital altitudes in the visible to infrared and far infrared spectral regions. A supplementary goal of this experiment was to evaluate the value of real-time identification of ground sites by the crew-member. Using two spectral bands of 0.4–2.4 μm in the short wavelength band, to 6.2–15.5 μm in the long

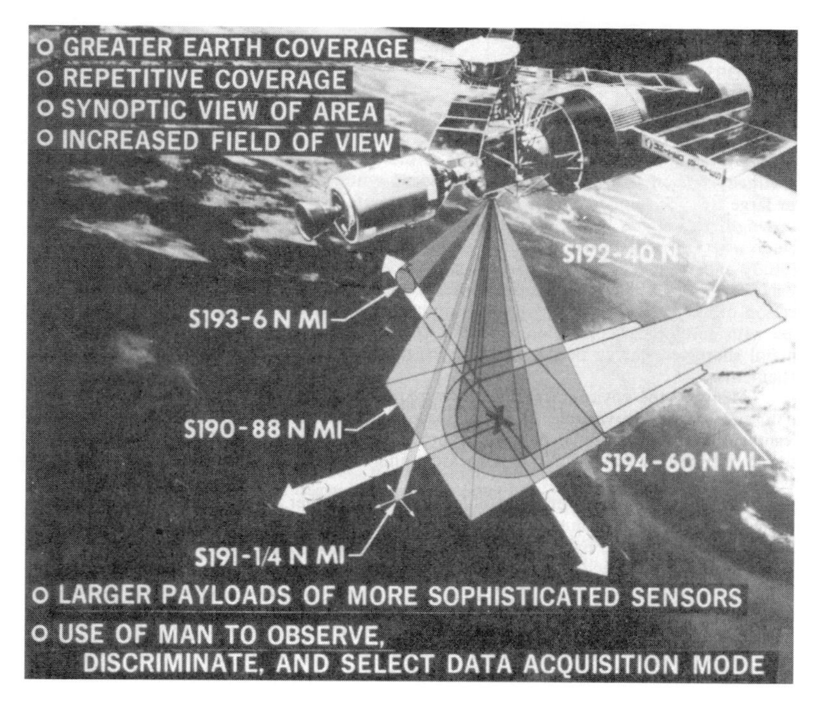

The advantages of Skylab EREP.

wavelength band. This allowed viewing of objects down to ¼ nautical mile, with a 16-mm camera used to photograph the sites while data was captured on magnetic tape in conjunction with other EREP sensors.

S192 Multispectral Scanner Based on the development of multispectral techniques evolved from the aircraft resource data collection programme, and adapted for use in space. On Skylab, spectral signature identification and mapping of agriculture, forestry, geology, hydrology and oceanography would use a range of thirteen spectral bands from 0.4 to 12.5 μm in the visible, near infrared and thermal infrared ranges. The system used a circular scan motion of the sensor with a 22.6 nautical mile radius, and swept a track 37 nautical miles in front of the orbiting station, which provided a 260-foot resolution from 235 nautical miles altitude. The data provided by the 12-inch reflecting telescope and rotating mirror provided high-spatial-resolution line-scan images from reflected radiation emitted from preselected ground sites both in America and other regions of the world. The EREP tape recorder was used, but with the tape speed increased to 60 inches per second to compensate for the high data collection rate of the scanner.

S193 Microwave Radiometer/Spectrometer/Altimeter Used for near-simultaneous measurement of radar differential back-scattering cross-sections and passive microwave thermal emission of the land and ocean, in a programme which provided

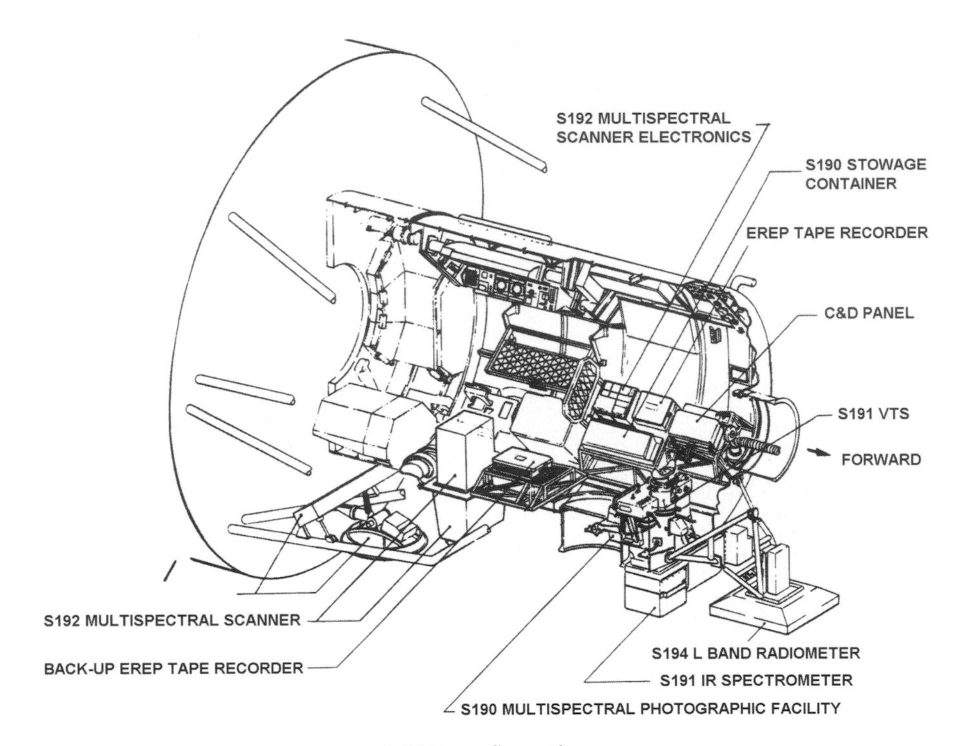

EREP configuration.

operational engineering data for the design of future radar altimeters. It investigated varying ocean surfaces, erosion, sea and lake ice, snow coverage, seasonal vegetation changes, flooding, rainfall, and soil types. In addition to areas over the ocean and land masses, for which ground data were known, the instrument was used for targets of opportunity such as hurricanes and storms. The radiometer, scatterometer and altimeter all operated at the same frequency of 13.9 GHz, with the scatterometer measuring the back-scattering coefficient of the ocean and terrain in an incident angle that ranged from 0° to 52°. While the radiometer was a passive sensor that measured the brightness of temperature from a function of incidence angle from the surface, ground overage extended to 48° forward and to either side of the ground track, and recorded on magnetic tape on one digitised channel at 5.33 kilobits per second for the radiometer/spectrometer, and 10 kilobits per second for the altimeter.

S194 L-Band Microwave Radiometer Supplemented the S193, and measured the brightness of the surface along the ground track, providing data on ocean surface features, meteorological winds and terrain surface features, recorded on magnetic tape at 200 bits per second. The passive experiment used a microwave sensor fixed planar array antenna, and recorded data in the L-band range with digital output which provided an absolute antenna temperature accuracy of 1 K, with the system sampling known sources via a built-in calibration system. Half-power beam provided a width of 15°, with a first null beam width at 97% of power at 37°. This allowed circular footprints of 60 nautical miles at half power from the operational altitude of 235 nautical miles.

Supplementing the astronauts' operation of the EREP package on Skylab was a series of underflights of remote-sensing aircraft to gather additional data to correlate with the data obtained on the Skylab. This, of course, had to be co-ordinated with the real-time flight plan, due to the changing nature of the flight programme, and whereever possible, simultaneous coverage of Skylab and aircraft data-gathering passes were part of the programme. Only US ground sites were underflown in more than 200 aircraft operations (equating to about 87,000 data miles), sponsored by NASA and the University of Michigan. Flights at medium to high altitudes was completed by several WB57F (U2) aircraft, and from low to medium altitude by a fleet of C-130B, P3A, OV1-C and C-47 aircraft and a H-47-G helicopter. The use of this fleet of aircraft allowed comparative data (ground truth) to be obtained to calibrate the effect of the atmosphere on the Skylab data.

The remote sensing programme from orbit began with the early Tiros weather satellites in the early 1960s, was later included in the Gemini series of manned missions in 1965–1966, when it was used to investigate the opportunity of obtaining photographic and experiment data on the Earth from manned spacecraft on missions of up to fourteen days. This research was continued on the two ten-day Earth-orbital missions of Apollo 7 (October 1968) and Apollo 9 (March 1969). Following these pioneering programmes came the Earth Resources Technology Satellite (later renamed Landsat), which provided repetitive multispectral scanner data in visible as well as infrared bands. The first of a series of Landsat satellites was launched in 1972, the year before Skylab, continuing the series of complementary programmes of data collection.

Earth observation experiments had been part of the AAP plans since the late 1960s, but when AAP-1A – carrying the Lunar Mapping and Surveying System (LMSS) that would be tested in a programme of Earth observations – was terminated in late 1967, it seemed that all Earth observation experiments had been lost from AAP. At the same time, observations from spacecraft, weather forecasting and the first views of Earth from the Moon from Apollo, beginning in 1968, excited the public interest in learning about the environment in which they were living. In promoting space applications in mineral prospecting, weather forecasting, agriculture and forestry, observations were counterbalanced by early studies in air and water pollution, erosion, deforestation, urban development expansion, and the depletion of fishery stocks. The awareness of the fragility of our planet began expanding at a time when extended observations from space were becoming more available. Some of the first to notice the public mood in this new 'science' of environmental research were politicians who, by touring their states, soon became aware of the growing interest in, and awareness of, 'saving the planet'.

As a result, several agencies began reviewing the use of space-borne observations for intensive Earth observations – now termed Earth Resources Observations – that evolved into the Landsat programme. NASA was at the forefront of this new area of research, and was fully aware of the benefits of promoting such research to the public, who were pressing their congressmen to adapt an Earth-friendly programme, and who of course authorised NASA's budget. Earth resources experiments were of interest to the DoD which, having abandoned its own MOL programme and adapted unmanned observation spacecraft, was interested in the military and strategic information which the EREP package could offer the armed forces in topography, mapping, and updating intelligence estimates. By 1969 the Earth Resources programme on the OWS was revived, and by the end of the year a package of experiments was suggested for inclusion on what was soon to be called Skylab.

Skylab's ground track provided repetitive coverage every five days, so that duplicate data could be obtained, over a period of months, on changes of season, missed opportunities, mission requirements and weather patterns. In December 1970 the Skylab EREP package expanded when NASA announced that data would be made available to qualified investigators for Earth resources investigation. In response to this announcement of opportunity, a total of 164 tasks was chosen from 148 Principle Investigators that included nineteen countries in addition to the United States. The fields included agriculture, range and forestry; geological applications; continental water resources; oceanographic and atmospheric investigations; coastal zones, shoals and bays; development of remote sensing techniques; regional planning and development; and cartography, with target sites using EREP data, Landsat aircraft and ground measurement data in a co-ordinated programme.

As with the other scientific research programmes, the investigators collaborated in a Mission Requirements Document that allowed flight planning for each EREP pass in real-time corollation with Mission Control and the National Weather Service, to establish flight constraints and weather conditions for each pass or series of passes, and adjust the flight plan accordingly, to compensate. Uplinked data included a detailed time schedule for the crew to revise the onboard schedule and time-line.

To conduct an EREP pass, the crew orientated Skylab to point the sensors towards Earth. In 171 days of manned operations, a total of 110 EREP passes were completed (thirteen on SL-2, 48 on SL-3 and 49 on SL-4). The ground track allowed for coverage across the United States and the continents of South America, Africa and Australia, In addition, southern Europe and Asia as well as large areas of the Atlantic and Pacific Oceans were covered.

Each pass lasted between 15 and 25 minutes, and covered 3,500–6,000 nautical miles. In the MDA, two astronauts would man the EREP equipment, and one crew station handled the operation of the S190, a battery of six cameras. The second station operated the radar instruments, and both astronauts prepared the instruments by setting switches, lenses, film cassettes and focus as per the planning documents (PAD) for the required pass. When there was cloud cover, the pass was scrubbed, and was rescheduled, if possible, for five days later, as the station flew over the same areas again. If there was clarity, the pass went ahead, with the crew flipping switches as required, operating optical instruments by day, or recording geothermal data during night passes. When the six cameras of S190 were not in use or were being reloaded with film and were out of the way, the large 'picture window' of optical glass allowed a spectacular view of the ground passing directly below. While the two astronauts worked on the EREP instruments in the MDA, the third astronaut would be in the OWS, working the S190B camera from the anti-solar airlock, pointed to the Earth and tracking the ozone layers.

To select an approaching target, the crew used the viewfinders to target the instruments about 45° ahead, and set the automatic tracking device to zero on the target as the station approached. They also found that use of the zoom lens actually placed the instruments out of focus, which annoyed them, as they had to reset the focus each time they used the zoom prior to data collection, 'tweaking' the settings as the target was automatically tracked to the point directly above the target zone where the data was collected. When the observations were completed, the instruments were set for the next target.

Skylab 2 obtained data on descending passes (north to south) only over the US, the Gulf of Mexico, the Caribbean Sea, and the northern regions of the United States during the summer months, and the crew were able to track Hurricane Ava about 55 nautical miles south-west of Acapulco, Mexico, in conjunction with USAF reconnaissance flight using EREP data to provide comparative data.

During Skylab 3, which performed descending and ascending passes (south to north) in the summer and autumn of the northern hemisphere, the crew obtained data over the US and 28 other countries in Central and South America, Europe, western Africa, Asia and Australia. Single data passes were also executed over Japan and the adjacent ocean, and over Israel, Ethiopia, Malaysia, Australia and New Zealand. The crew was also to observe the active volcano Mount Etna, in Sicily, drought regions of Mali and adjacent countries, and the tropical storm Christine in the Atlantic Ocean, north-east of Venezuela.

Skylab 4, flying in the winter months of the northern hemisphere, was hampered in its EREP data collection by unfavourable lighting conditions, but the crew was still able to obtain comparative data on regions flown over by SL-3 some months earlier, and some unique passes for thermal mapping were completed over selected areas of California, over-flight of the largest tropical cyclone in the North Atlantic for a decade, and measurements of the configuration of the Earth for S192 that resulted in a 360° altimeter data pass from 39° west to 61° west, and duplicated SL-3 topographical mapping over Paraguay.

(Further reading: EREP and Earth observation activities, experiments and preliminary results are included in the NASA publications *Skylab Explores the Earth*, NASA SP-380 (1977), and *Skylab EREP Investigations Summary*, NASA SP-399 (1978).

Astrophysics

There were nine astrophysical experiments onboard Skylab, of which three (S063, S073 and S149) investigated the interplanetary medium and the outer atmosphere of Earth, while the other six (S009, S019, S150, S228, S230 and S183) studied objects external to the solar system.

S009 Nuclear Emulsion Recorded cosmic-ray flux outside the Earth's atmosphere, and the relative abundance of high-energy primary heavy nuclei. The instrument resembled a 'book' of two adjacent hinged stacks of nuclear emulsion strips, numbered and located inside the MDA, separated from space by a thin wall of the hull. High-energy particles entered the wall and passed through the emulsion. Each crew returned the strips to Earth, where they were peeled apart for analysis. The experiment was not operated by the second crew, but the third crew installed a replacement 'book'.

S019 UV Stellar Astronomy Principle Investigator was scientist-astronaut Karl Henize, who spent many hours working on the experiment with a team of fellow astronomers. It was designed to take UV photographs of large areas of the Milky Way, where young, hot stars are abundant. The experiment was planned to photograph fifty 5° × 4° star-fields two or three times, with exposures of 30, 90 and 270 seconds. The film magazine held 164 frames of special UV-sensitive film, and allowed the astronaut to take a fourth exposure if OWS stability permitted. The experiment included a 6-inch reflecting telescope and moveable mirror located in an SAL and operated manually. It was operated by all three crews, and resulted in 1,600 individual photographs covering 188 star-fields.

S063 UV Airglow Horizon Photography A two-part experiment, one part of which carried out ozone photography, while the other completed twilight glow photography. The ozone instrument took two simultaneous series of photographs with a 35-mm camera with and without UV filters, to determine the amount of thermal absorption of the atmosphere. The twilight airglow experiment photographed the glow in the upper atmosphere caused by chemical reactions in ozone, oxygen and other gases. Part of this study was a series of photographs of the southern auroral zone taken by Owen Garriott during the second mission. About 2½ days later, after a large solar flare on 7 September, particles from the flare interacted with the magnetosphere to produce the spectacular aurorae that enabled the first photographs from space to be captured by Garriott.

S073 Gegenschein/Zodiacal Light An experiment to measure the brightness and polarisation of the visible background of the sky from above the atmosphere. It continued a programme of similar experiments from the ground, rockets and satellites, hand-held photographs from Gemini missions, and an Apollo lunar orbit experiment from the dark side of the Moon. The astronauts used a T027 photometer and a standard 16-mm data acquisition camera (DAC) operated on the night-pass of the orbit. Skylab 2 operated the experiment for a total of 14 hours 37 minutes on six different days, while the second crew managed to secure a further six hours of data despite a fault that required ejection of the photometer and camera during the first day of operation. The second crew then re-rigged the hardware of the coronagraphic measuring experiment (T025) to a 35-mm camera, and took a further seven exposures of the Gegenschein. The third crew also used this method to take 96 exposures that included the Zodiacal Light, the Gegenschein, lunar libration, galaxies, the ecliptic pole, and Comet Kohoutek.

S149 Particle Collection Studied the nature of interplanetary dust. Specially prepared surfaces (gold-covered smooth plates 6 inches square), and layers of film were installed in reusable cassettes that were exposed through the antisolar airlock between the first and second manned missions.

S150 Galactic X-Ray Mapping Conducted a survey of faint X-ray sources in the 0.2–12 KeV energy range, supplementing satellites such as SAS-A, launched in December 1970, which investigated sources in the 1–10 KeV range. This experiment was not located on Skylab, but on the instrument unit of the Saturn 1B that took the second crew to the station in July 1973. It was planned to operate for 265 minutes (almost three orbits), but after separation of the CSM twelve minutes into the mission, it was activated, and operated for only 103 minutes (one orbit plus eleven minutes). This was due to a partial hardware failure when the experiment was exposed to direct sunlight longer than planned, but not before returning data on attitude and orientation to the Sun. The exposure was explained as a coincidence of a less than 10% risk at that particular launch time! The increased radiation resulted in a failure in the plastic window of the experiment, which caused a decay of internal experiment pressure below 12 psia, and resulted in a high-voltage shut-down of the experiment.

S183 UV Panorama Measured the ultraviolet brightness of a large number of stars, by photographing a number of star-fields in three spectral bands, with fine spatial and photometric resolution not previously available. The quality of the photographs taken during the first manned period was not as good as expected, due to the high temperatures inside the OWS after launch and before the thermal parasol was deployed. Four different cameras used by the second crew produced out-of-focus photographs, indicating that the problem was on the spectrograph side of the mounting interface, and the third crew therefore took new carousels of film to use during their stay. The results included photographs of Comet Kohoutek. Thirteen exposed plates were returned from the first mission, and 43 from the third, and in addition, the first (fifteen frames) second (twelve frames) and third (35 frames) missions returned magazines containing exposed 16-mm film. It was the complexity of this experiment that made it susceptible to carousel jamming, film plate protrusions and other problems which hindered operation.

S228 Trans-Uranic Cosmic Rays Provided detailed knowledge of the abundance of nuclei with an atomic number greater than 26, in the cosmic radiation, using plastic (Lexan) detectors (two-harness assemblies, each containing eighteen detector modules, each module containing 32 sheets of 7 × 9 × 0.010-inch thick Lexan), mounted in the OWS, and exposed to the cosmic radiation passing through the hull to impact the plastic sheets of the experiment. By examining the chemical etching of the cosmic-ray tracks in the plastic, the atomic number and energy of each particle could be determined. The hardware was stowed onboard the OWS at launch, and was deployed by the first crew on 29 May, (SL-2, MD5). After 116 days, one module was returned to Earth by the SL-3 crew, and a second module was returned on the SL-4 CM after an exposure of 251 days. A third module was deployed during EVA on the seventh mission day of SL-4 retrieved on MD79, and then returned at the end of the mission. A fourth detector was installed in the MDA at the end of the SL-4 mission, and was planned to be retrieved during a revisit; but this never occurred, and the unit was destroyed during the re-entry of the workshop in 1979.

Within NASA, astronaut Don Lind was, like Bill Thornton, known as one of the astronauts who obtained the best lb per $ for his space experiments. Prior to Skylab, Lind had worked with Dr Johannes Greiss, of the University of Bern, Switzerland, on the solar wind composition experiment flown on the Apollo lunar landing missions to trap rare ions on a sheet of aluminium foil, exposed during the EVAs. Lind and Greiss thought that they could deploy similar aluminium collection devices outside Skylab, and determined that a multi-layered 'cuff' of aluminium, aluminium oxide and platinum could be wrapped around an ATM deployment truss, measuring 14 × 19 inches, to achieve this at low cost and with minimum technical hurdles.

Some opposition arose when it was proposed to fly the experiment, which, it was agreed, was relatively cheap compared with the other instruments, but very close to the launch of the OWS, where the first collectors were to be installed. Each responsible field centre told Lind that it was a good experiment, and should indeed fly, but then refused approval to fly the experiment due to its being so close to the launch. Lind countered this argument by stating that it was not a complicated experiment, and needed only to be strapped to the truss with Velcro. It was a strong argument, which he won – and the experiment flew. The first collectors were exposed at shroud jettison, 15 minutes after launch, and on 6 August were retrieved by the second crew, who exposed the inner collector sheets on the same EVA. They were due for return to Earth on 25 September, but the RCS contamination from the fly-around and docking of the first two crews entailed the dispatch of a second collector with the third crew. It was deployed on 22 November, and retrieved on 3 February. From these samples it was possible to determine that particles of helium atoms with energy values greater than 3 KeV are transported by the solar wind and are accelerated by the magnetosphere. For the first time the experiment also provided data on isotopic fractionation in the upper atmosphere.

Cometary Physics Experiments were conducted on Comet Kohoutek during the third mission. To view the comet it was decided to use the SAL and various windows, using hand-held cameras, the ATM telescopes, and by taking instruments outside during EVA. When the comet was in the vicinity of the Sun, the ATM instruments were pointed at the disc for solar studies. The SAL that was 180° from the Sun-side was used, because the Sun-facing airlock was still filled by the thermal parasol hardware, and was blocked by the shields. Turning the whole station would result in an undesirable orientation, so instead, the vehicle was rolled to 90° and then the articulated mirror system in the UV stellar astronomy camera in the airlock was used to obtain the further 90° observation angle. The comet observation programme entailed the use of existing experiment hardware (S019, S063, S073, S183, and T025) through the SAL or by EVA. The ATM instruments included S052, S054, S056,

S082A and S082B; and in addition, the third crew took two new experiments to Skylab: *S201 Far UV Electronographic Camera*, used in the SAL and during EVA; and *S233 Kohoutek Photometric Photography*, using a 35-mm Nikon camera of 55-mm focal length, with the astronauts taking pictures twice a day, whenever possible, through three different windows.

Airlock observations began on 25 November, and ended on 1 February, ATM operations were planned around solar observations, and were centred on 14 December to 10 January but actually occurred between 19 December and 6 January, while the EVA observations occurred on 25 December and 29 December.

(Further reading: Experiments in this field are included in *Skylab's Astronomy and Space Sciences*, NASA SP–404 (1979).)

Materials science and space manufacturing

This series of experiments was designed to take advantage of the microgravity and vacuum that allowed a range of experiments to be performed during orbital flight, and which were difficult or impossible to reproduce on Earth. The melting and mixing of components without contamination from containers, the control of buoyancy and convection in liquids, and the use of electrostatic and magnetic forces, were a huge step in the evaluation of using these processes in the production of new materials, or in processes that had applications on Earth.

M512 Materials Processing Facility Used as a common interface for a range of metallic and non-metallic materials experiments. The facility was located in the MDA, and consisted of a vacuum work chamber, associated mechanical and electronic controls, an electron beam subsystem, and a control and display panel. The hinged 16¼-inch vacuum work chamber connected to the electron beam, and operated at 20 kV, from the control panel.

The operation procedure was a multi-step evaluation and experimentation process. Initially, the facility itself was analysed as a workable facility, and later the electron beam was evaluated for heat and welding on two experiments. The exothermic brazing process was then evaluated, with one experiment containing four samples. Next, eleven separate experiments used a common furnace that was mounted inside the work chamber, inside of which 54 samples, involving material phase changes, were processed. The final set of experiments featured flammability tests, in which about forty samples were ignited in the work chamber, and either allowed to burn freely, or be quenched by evacuating the chamber or by spraying water directly on the sample.

M551 Metals Melting; M552 Exothermic Brazing; M553 Sphere Forming Performed in the Material Processing Facility. Operation of the facility occurred during all three manned missions with the metals melting, exothermic brazing and sphere forming experiments performed during the first visit. The multipurpose electric furnace series of experiments was performed during the second and third visits and the flammability experiments only on the third mission. Overall the facility worked very well, and significant progress was made in materials science that provided baseline data for further study in later programmes.

M518 Multipurpose Electric Furnace Used for experiments on solidification, crystal growth and other processes which involved material phase changes. The main component was a furnace that was designed to interface with M512, a programmed electronic temperature controller and experiment cartridges that contained the sample materials. Three samples could be processed simultaneously. There were three different temperature zones available: a constant-temperature hot zone (1,000° C), a gradient zone (ranging from 20° to 200° C per half inch), and a cool zone in which heat conducted along a sample was diverted by radiation to a conducting path out of the system.

Each sample of material was enclosed in a cartridge, three of which could be mounted in the facility. There were eleven different processes performed in the furnace, which featured its own electrical and instrumentation and communication interfaces with the MDA.

During the second and third manned missions, the furnace was operated with eleven processes (33 carriages) performed by the second crew, and seven processes (21 cartridges) performed by the third crew. The experiments performed in this facility were:

M557 Immiscible Alloy Composition
M558 Radioactive Tracer Diffusion
M559 Microsegregation in Germanium
M560 Growth of Spherical Crystals
M561 Whisker-Reinforced Composites
M562 Indium Antimonide Crystals
M563 Mixed III–V Crystal Growth
M564 Halide Eutectics
M565 Silver Grids Melted in Space
M566 Aluminium–Copper Eutectics

All samples were returned to Earth for analysis, and even the preliminary results indicated that samples from many of these processes were far superior to any obtained on Earth as that time.

M479 Zero-Gravity Flammability Experiments in the ignition of various materials in the atmosphere, conducted to observe the flash over propagation to adjacent materials, rate of surface and bulk flame propagation under zero convection, and the evaluation of extinguishing methods by vacuum, water spray and self-extinguishing. Six substances were tested as sample materials: aluminised Mylar film, polyurethane foam, nylon sheet, neoprene-coated nylon fabric, bleached cellulose paper, and Teflon fabric. In all, 37 samples were used, supported by a metallic frame and ignited by an electrically heated filament.

(Further reading: samples, operation and preliminary results appear in NASA Technical Memorandum NASA TM X-64814 MSFC Skylab Mission Report – Saturn Workshop, Skylab Program Office, October 1974.)

Engineering and technology experiments

This series of experiments was aimed at providing data for use in the future development of space systems, and for further experiments during follow-on missions. These were broadly grouped to understand man's role in space, how he performs, what tools would be best suited to complete assigned tasks, and the nature of the human influence on the space environment. The experiments were:

M487 Habitability and Crew Quarters, *M516 Crew Activities/Maintenance*, *T013 Crew Vehicle Disturbance*, are covered in a separate section under habitability. *M509 Astronaut Manoeuvring Equipment*, *T020 Foot Controlled Manoeuvring Unit*, which were related to the development of EVA equipment, are covered under the EVA section. *T002 Manual Navigation Sighting* was designed to evaluate the effects of prolonged spacecraft habitation on the ability to take space navigation sightings and measurements using hand held instruments. It had already been determined on Gemini missions that man could carry out such an exercise, and the advancement of instrumentation helped to determine position in space without a computer, which combined with operational experience on Apollo lunar missions, extended this work outside Earth orbit. For Skylab, the additional objective was to evaluate whether this could be done after a longer period in space, in excess of fourteen days. The experiment was comprised of two hand-held instruments: a sextant for measuring the angles between two stars, and single stars and the edge of the Moon; and a stadimeter – an optical device used for measuring the apparent curve of the Earth's horizon to determine spacecraft attitude. Readings were recorded in a logbook, and were supplemented by crew comments on tape.

In addition, six experiments investigated the natural and induced spacecraft environment:

D008 Radiation in Spacecraft A test of an advanced radiation instrument, and techniques to determine the effects of radiation on a man, with a view to providing data for protection on longer-duration spaceflights. The experiment flew in the CM for the SL-2 mission.

D024 Thermal Control Coatings This experiment, in connection with M415, provided data on exposing selected experimental thermal coatings (extensively ground-tested in simulated space environments) to the space environment, correlate the effects with those found on the flown samples, and understand the mechanism of the degradation of the coatings caused by space radiation. The experiment packages consisted of four panels, two of which contained 36 thermal coating samples, each an inch in diameter. The other pair contained strips of polymeric plastic approximately 5 mm thick. The panel plates were 6½ inches square and ¼-inch thick, and were attached by snap fasteners to the AM truss assembly. Exposed to the vacuum after payload jettison, they were retrieved by the SL-2 crew on the 19 June EVA, and a second set was retrieved by the SL-3 crew on the 22 September EVA.

M415 Thermal Control Coatings Associated with the previous experiment to determine the degradation of pre-launch, launch and spaceflight on thermal absorption and emission characteristics of a range of coatings commonly used for thermal control. This hardware featured two panels of twelve thermal sensors arranged in four rows of three. Temperature sensors recorded and transmitted data, but the coatings were not retrievable.

T003 Inflight Aerosol Analysis This experiment measured the size, concentration and composition of particles within the internal atmosphere of the OWS, while in the *T025 Coronagraph Contamination Monitoring* experiment, the astronauts had to visually and photographically observe and record the amount of light particles from thruster firing and waste water dumps, in an evaluation of how such contaminants affect optical experiments.

T027 ATM Contamination Measurements A two-part experiment. The sample array was to obtain controlled data on the degradation of contaminants on Skylab windows and mirrors in the quality of their optical properties, while the photometer system measured the brightness and polarisation of scattered sunlight from solar illumination of the contaminate cloud that surround Skylab.

(Further reading on these experiments is included above hardware, operations and early findings in the NASA Technical Memorandum NASA TM X-64814 MFSC Skylab Mission report – Saturn Workshop, Skylab Program Office, October 1974.)

Student experiments

The solar physics and astronomy experiments, the Earth resources package, the manufacturing and material processing experiments, and the engineering and technology experiments, were logical fields of research that had a long history in the space programme. As Skylab evolved, it was the Earth resources experiments that above all were promoted as offering the most useful benefits 'for all mankind', and they caught the public imagination.

Student Investigations A second package of investigations that captured the public attention during the Skylab missions. It had evolved from an idea by Ken Timmons, of Martin Marietta, who early in 1971 thought that it would be advantageous to the programme if small experiment opportunities were offered to high-school students, designed to broaden the range of smaller experiments flown, raise public awareness, and generate interest in space science for members of the community who were about to make the next step in their careers. If they had the chance to fly their own experiments on Skylab, then perhaps that career might move towards the space programme as an investment in the future. If anything, a high school involvement in technology would be beneficial to the American educational system, and if successful, could be applied to other programmes.

Following discussions with education officials, the idea was passed to Marshall, where it was adopted with enthusiasm. The resulting contract between NASA and the National Science Teachers Association set out to manage and organise a nationwide search for proposals from students, in the form of a competition. 100,000 announcements were issued in October 1971, and by the 4 February 1972 deadline for receipt of proposals, some 55,000 teachers had requested entry material, of which 3,409 proposals were received. This involved 4,000 Grade 9–12 students from all States, all of which were evaluated and judged by NSTA before announcing 300 finalists from twelve regional screening committees by 1 March 1972. NASA engineers then judged each experiment on its technical feasibility, before the final selection of the winners, on grounds of scientific merit, on 15 March. A further 22 were listed as 'special mention' entries.

The experiments were judged on the limitations of power, weight, volume and the time that it would take crewmen to operate the experiment. Some of the mountain of paperwork was diverted from the normal NASA experiment route, and each student investigator became the Principle Investigator for his or her experiment, with the guidance and assistance of a Science Advisor from MSFC and consulting Science Advisors from MSC in Houston.

During the second week of May 1972, a preliminary design review was held at Huntsville, in order to evaluate which proposals could be categorised as those needing separate elements of hardware, those that could use existing Skylab hardware, and those that could obtain the same results by using already developed hardware and procedures. From this review, eleven required hardware to be developed for the experiment, eight used existing hardware, and six could not be performed or flown because of operation requirements and time constraints. Of these, four were provided with data from existing Skylab experiments, and the remaining two were associated with NASA researchers with a similar research interest or objective.

All experiments had been determined early in 1973, and completed a flight acceptance review at Marshall on 23–24 January, before shipment to the Cape for payload integration on 26 January. (A listing of all student experiments is presented in Appendix 3.) They were judged to include good research objectives in areas of astronomy, biology, and physics. Although some students judged, NASA's expectation had been rather low. The programme also revealed shortcomings in the students' basic understanding of scientific principles, which led to an evaluation of college teaching practises.

Although not ground-breaking science, the programme was a valuable exercise for both NASA and the education community in co-operative programmes that continued with the Space Shuttle programme. Many of the experiments gained the public attention during the Skylab programme, the most famous of them being the experiment featuring two small spiders:

ED52 Web Formation Proposed by Judith S. Miles, of Lexington High School, who proposed a determination of the effects of reduced gravity on the web-building process of a common cross spider (*arenus diadematus*), by comparing webs built in Skylab and those on the ground. Two spiders – Anita and Arabella – were chosen as a 'flight crew'. Their enclosure was provided with transport vials for the spiders, food (water and flies), an automatic motion picture actuator, a 35-mm still camera, and a 16-mm motion picture camera, with film for both.

Flown on Skylab 3, they were stowed in the Command Module on 25 July, and were transferred to the forward dome area by 31 July. When the first spider was shaken from the vial, it flicked its eight legs abnormally, and bounced across the enclosure before attaching itself to one side. It seemed that, like astronauts, spiders needed a little time to adjust to spaceflight! The first web was reported on 6 August. The second spider was deployed on 26 August and had completed its first web by 29 August – but was found dead on 16 September. By then, the first spider had also died. However, despite several early failures, both spiders adjusted to the conditions, and spun recognisable webs.

(For further information on the student projects, see Skylab *Classroom in Space*, NASA SP-401 (1977).)

Science demonstrations

This group was developed to demonstrate a range of scientific principles suitable for educational purposes. A group of thirteen was flown on the second mission, and when the crew requested more activities to fill in their time, so Marshall developed two more experiments for them from onboard hardware. For the third flight, a further fifteen demonstrations (one a repeat of SD9) were prepared. In most cases, data were returned by TV, film camera, audio, and, in a few cases, returned samples The demonstrations were:

Skylab 3

SD1 Gravity gradient effects
SD2 Magnetic torque
SD4 Momentum effects
SD5 Energy loss and angular momentum
SD6 Bead chain
SD7 Wave transmission reflection
SD8 Wilberforce pendulum
SD9 Water drop
SD10 Fish otolith
SD11 Electrostatic effects
SD12 Magnetic effects
SD13 Magnetic electrostatic effects
SD14 Airplane
SD15 Diffusion in liquids
SD16 Ice melting

Skylab 4

SD17 Ice formation
SD18 Effervescence
SD19 Immiscible liquids
SD20 Liquid floating zone
SD21 Deposition of silver crystals
SD22 Liquid films
SD23 Lens formation
SD24 Acoustic positioning
SD28 Gyroscope
SD29 Cloud formations
SD30 Orbital mechanics
SD33 Rochelle salt growth
SD34 Neutron environment
SD35 Charge particle mobility

(Further reading: details of these experiments and their operation during the missions can be found in the NASA Technical Memorandum NASA TM X-64814 MFSC Skylab Mission report – Saturn Workshop, Skylab Program Office, October 1974.)

Life sciences
One of the major reasons for placing Skylab in orbit was the study of human adaptation to reduced gravity during prolonged periods. In the twelve years since the first astronauts had ventured into space, considerable biomedical data had been obtained on the selection and training of flight crew personal. From the 27 American manned missions, varying in duration from 15 minutes to 14 days, an equally extensive database of post-flight medical studies had been compiled.

During the same period, the Soviets had obtained comparative data from 18 manned missions of between 108 minutes and 23 days, one of which included the first flight of a female crew-member. Tragically, four cosmonauts had lost their lives during two spaceflights in 1967 (Soyuz 1) and 1971 (Soyuz 11), but these were due to hardware malfunctions, and were not biomedical in origin, although the difficulties that both missions presented for the crews were a factor that was investigated.

For Skylab and its three manned missions, the Americans were offered their first opportunity to study longer-duration spaceflight for themselves, in nineteen experiments that supported 28 investigations in the life sciences field, divided into four elements:

- The habitability of the stay in space, in duration, medical monitoring, and studies of crew performance in scientific and technical assignments,
- A range of medical experiments that were to investigate physiological effects and their development in relationship to earlier flights.
- Biology experiments to study fundamental processes affected by spaceflight.
- Biotechnology experiments which would have future application in even longer flights by man, the technology to improve biotechnological instrumentation, and the conditioning, both before during and after spaceflight, to counteract adverse effects.

Prior to Skylab, astronauts had experienced a loss of body weight, a small and inconsistent loss of bone calcium and muscle mass, and a general reduction in orthostatic tolerance when they returned to Earth. After only a few days, the effects reversed themselves, and the data from flights of up to fourteen days revealed no constant relationship with flight duration, and the fear was that even longer flights could continue this effect to a point where man's effectiveness in space would be reduced, and re-adaptation to Earth's gravity become dangerous. The experiments flown on Skylab were therefore aimed at determining a criteria that gradually increased the duration of spaceflight past the fourteen days to three months, to provide additional information that would be useful in projecting longer flights of 6–12 months or beyond, prior to actually attempting them.

The main life sciences investigations and experiments on Skylab included:

M071 Mineral Balance Required a recording of daily body weight, an accurate record of all food and water intake, volume measurement of 24-hour urine output, samples of urine, and the determinination of the mass, process and storage of all faeces and vomit, with samples collected in flight for post-flight analysis, and samples of blood in pre-flight, in-flight and post-flight conditions. This allowed a better understanding of the effects of spaceflight on bones and muscles by measuring the daily gain or losse of biochemical constituents.

M073 Bio-Assay of Body Fluids Similar data recording (except of faeces and vomit) for analysing the blood and urine to determine the effects of spaceflight on the endocrine-metabolic functions.

M074 Specimen Mass Measurement Designed to weigh 1.75 oz to 2.25 lbs in a null gravity environment by using the inertial property of mass instead of gravity, determining mass. The device used a spring-mounted tray that oscillated, with the amount of mass in the tray determining the period of oscillation. Measured electro-optically, this was electronically converted to a direct mass read-out on the instrument, which could provide accurate mass measurements of faeces, vomit, and food residue.

M078 Bone Mineral Measurement Conducted pre- and post-flight, using gamma-ray measurements of the heel bone and right radius of the forearm that provided a comparison of bone density before and after flight, to determine the degree of change in bone minerals.

M092 Lower Body Negative Pressure Recorded the duration of cardiovascular adaptation during the flight. It used a cylinder into which the astronaut floated feet-first up to his stomach, and an air seal diaphragm was then placed around his waist, and the pressure lowered inside the cylinder to simulate a person standing in 1 g. Measurements were then taken of increases in leg volume as the blood pooled, while a blood-pressure cuff worn on the upper arm recorded blood-pressure signatures as the unit was depressurised.

M093 Vectrocardiogram Used to detect flight-inducted changes in the heart function, to compare with pre- and post-flight measurements. This instrument was also used in support of the M092 and M171 experiments. An eight-electrode input harness was worn to record data at regular intervals throughout the flight, at rest and at work, as well as before, during and after exercise.

M111 Cytogenic Studies of the Blood An experiment to determine chromosome aberration frequencies in pheriphal blood leukocytes recorded from blood samples. Other related experiments on these samples included radiation studies, and the determination of genetic consequences of prolonged spaceflight. Comparative samples were taken one month before launch, and three weeks after recovery. Blood samples also formed the basis of the *M112 Man's Immunity – In Vitro Aspects* experiment, which recorded changes in humoral and cellular immunity reflected in concentrations of plasma and blood cell proteins. Samples taken pre-flight at 21, 7, and 1 days were taken for indication of normal metabolism, and a control 'crew' of three men similar to the flight crew were also sampled. In flight, samples were taken four times from each astronaut on SL-2, and eight times each on SL-3 and SL-4, and again 7 and 21 days after recovery. The purpose of *M113 Blood Volume and Red Cell Life Span* was to study the effect of spaceflight on plasma volume and the populations of red blood cells. It paid particular attention to changes in the red cell mass, destruction rate life span, and production rates. *M114 Red Blood Cell Metabolism* looked for metabolic and/or membrane changes in red blood cells as a result of spaceflight; and again, samples taken pre-flight, in-flight and post-flight were to contribute. *M115 Special Haematological Effects* also took pre-flight, in-flight and post-flight samples to investigate critical physiochemical blood parameters that maintained a stable state of equilibrium between blood elements, and to determine how spaceflight affected these parameters.

M131 Human Vestibular Function Designed as a three-fold experiment that tested crew susceptibility to motion sickness, to understand the functions of human gravity receptors in prolonged absence of a gravity reference, and to test for changes in the sensitivity of the semicircular canals using a Rotating Litter Chair (RLC) that could also tilt the crewman, who was tested in his perception of rotation; in motion sickness symptoms while conducting out-of-plane (head-rocking) motions and while being rotated; and visual clue determination of the ordination with the spacecraft. As with many of these medical experiments, data were collected pre-flight, in-flight and post-flight, for complete comparison.

M133 Sleep Monitoring Aimed to evaluate the quality and quantity of sleep (of each of the science pilots) by analysing EEG and EOG activity periodically during the mission sleep periods, by wearing a self-contained cap fitted with sensors that took real-time readings, and was connected to a magnetic tape recorded the findings. The *M151 Time and Motion Study* evaluated the astronaut's adaptation to extended-duration spaceflight in comparison with identical activities performed during training or on other missions. *M171 Metabolic Activity* was designed to evaluate whether metabolic effectiveness was altered by exposure to the space environment. As the astronauts peddled the bicycle ergometer, measurements were taken of oxygen intake and carbon dioxide output, while work-rate and RPM were calibrated, and respiration and heart rate were measured, along with other parameters. The ergometer was also used for evaluation of future in-flight exercisers. A larger version of the M074 Mass Measurement device was used to determine body mass of crew-members to support experiments for which such measurements were required. *M172 Body Mass Measurement* was calibrated pre-flights, and took measurements of the astronauts, known body masses three time each during each of the missions. Three smaller experiments were also related to the biomedical investigations on Skylab. The *S015 Effect of Zero G on Single Human Cells S015* was designed to determine the effects of spaceflight on living human cells in a tissue culture, while the *S071 Circadian Rhythm – Pocket Mice* was used to determine mammals' daily physiological rhythms, which are altered in spaceflight. Six pocket mice were stored in a completely dark cage with a temperature of 60° F, 60% humidity, and a sea level equivalent pressure. *S072 Circadian Rhythm – Vinegar Gnat* was flown to determine whether the daily emerging cycle of the pupae of the vinegar gnat (*drosophila*) was altered during spaceflight.

In support of the medical experiments was an Experiment Support System that provided practical support for experiments M092, M093, M131 and M171. This central control unit regulated power, spacecraft power, controls, displays, data management, event timers, pressurised gaseous sources, and calibration commands.

Biomedical results

During 27–29 August 1974, NASA conducted a symposium of Skylab Life Sciences research and preliminary results at JSC. Richard S. Johnson, Director of Life Sciences, opened the meeting by stating that he was 'happy to report that no major medical findings will be presented which might curtail man's dreams of more extensive space exploration; rather, we have found that man can adapt to the new and wondrous environment of space'.

In addition to the medical experiments, several major subsystems were part of the life science investigations, including the food system, water management, personal hygiene, the in-flight medical support system, and ground medical support pre-flight, in-flight and post-flight.

(Detailed biomedical results are beyond the scope of this current volume, and further references are listed in the Bibliography. Further reading includes *Biomedical Results of Skylab*, NASA SP –377 (1977).)

It was generally found that motion sickness still remained an unresolved problem, with four of the nine astronauts suffering from the effects. It appeared that the space sickness was very individual, and almost unpredictable, and even the drugs supplied to alleviate the symptoms never completely prevented them.

Results of urine analysis revealed that there were higher levels of calcium in the crews' samples, along with loss of structural material in weight-bearing bones. This was indicated during the Gemini programme and by Soviet studies, and despite additional exercise on Skylab 4, this loss continued, and had been reflected in several bed-rest experiments. This indicated that although the amounts of bone mineral lost on Skylab was not dangerous, any further loss on longer missions might be irreversible on the leg bones. The recorded high concentrations of calcium in urine samples could also lead to the hazardous condition of kidney stones.

After Skylab 2, astronaut Bill Thornton realised that the exercise programme was inadequate, and although the ergometer was excellent for cardiovascular exercising (Conrad varied its use by hand-pedalling the device), it was not enough to put loads on the leg muscles. He realised that a different type of device, allowing walking or running, might be the answer, and devised a treadmill for Skylab 4.

When weight considerations for launch were overriding anything to be added to the CM, Thornton revised his plans and evolved a 'simulated treadmill', that weighed just 3½ lbs, consisting of a Teflon-coated aluminium walking surface attached to the grid flood, and four rubber bungee cords and a harness that provided the equivalent of 175 lbs. By angling, the bungee was simulating hill-runs, which added load to the muscles as they worked at pumping the legs during the exercise.

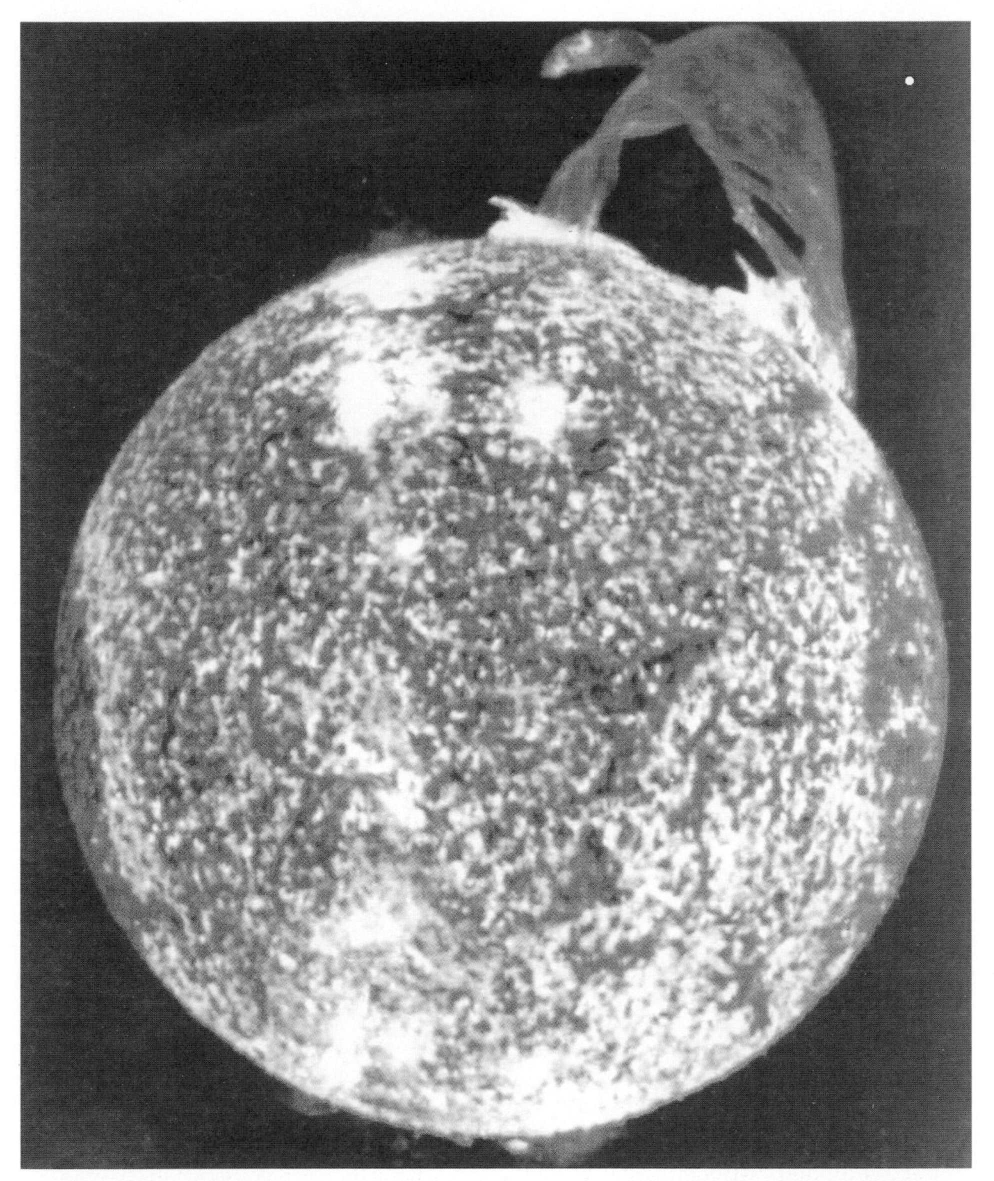

This Skylab 4 photograph, taken by the ATM on 19 December 1973, reveals one of the most spectacular flares ever recorded, extending more than 365,000 miles across the solar surface. Such photographs taken by the three crews completely rewrote the solar astronomy textbooks as they revealed the violent activity of the Sun. The diameter of the Sun is approximately 865,000 miles – 109 times the diameter of the Earth (indicated by the white disc at top right).

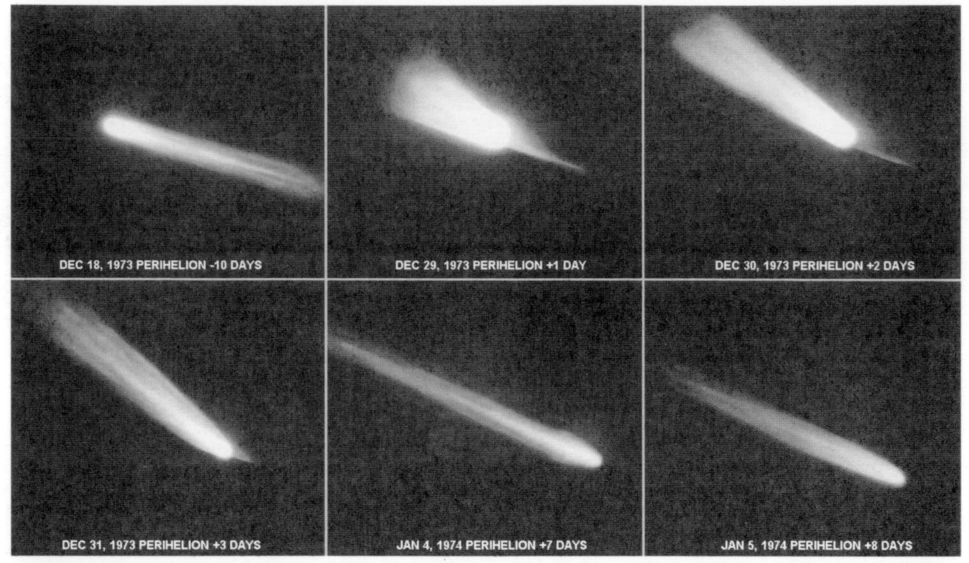

An artist's illustrations of comet Kohoutek 1973f, based on Skylab 4 astronaut observations made between 18 December 1973 and 5 January 1974.

This view of the greater metropolitan area of Chicago – obtained during the SL-3 mission in September 1973 – demonstrates the resolution of the S-190B EREP camera The photograph reveals the differences between commercial, industrial and residential areas for urban planning; transportation network for projected growth and development; agricultural land for crop studies and forestry management; airports for studies of the ecosystem balance; air and water plumes for pollution studies, recreational potential of current and planned leisure activity areas; topographical studies for accurate mapping and monitoring of erosion.

Early post-flight results – including the standing and walking of the SL-4 astronauts on the day after recovery – indicated that such a device was beneficial in maintaining the muscle loads on the lower limbs during spaceflight. The larger and more sophisticated 'Thornton's Revenge' went on to to fly on the Shuttle. He concluded that the human leg muscle in space is no different from that on Earth, and that as long as the astronaut is well nourished and exercised with reasonable load-levels, the muscle will retain its primary function.

Cardiovascular data indicated that despite exercising, the astronauts did not maintain their pre-flight levels, and took a week or more to return to pre-flight levels after the mission, although the onboard data also indicated that the body tolerance for exercise had not decreased, while the LBNP chamber took a while to adjust to in space, and was in some cases quite uncomfortable. Although the 28-day flight showed minimal adaptation, the second and third crews slowly increased their tolerances between 30 to 50 days of flight and a quicker re-adaptation after the mission.

Overall, Skylab found that several fundamental concerning man's ability to live and work in space for three months were solved, but others indicated that there remained many more questions than there were answers. Unfortunately, the American follow-up to Skylab was cancelled, and the Space Shuttle could initially support flights of up to only ten days, and eventually 17 days. It was not until 1995 that the Americans could once again obtain useful data on prolonged spaceflight first-hand – but on a Russian space station.

Some of the suggestions for future astronaut selection reflected the interpretation of the results and future research directions where 'professional subjects' could be flown during laboratory test programmes. It was also suggested that non-system responsible astronauts could be flown, and be left to deteriorate, so that the full testing of compensation and preventive measures could be fully evaluated! Some even suggested flying amputees who had lost their legs, as it seemed that the legs incurred the most problems. Indeed, during the 1966 astronaut selection, Lieutenant Frank E. Ellis, USN, who had lost both legs in a jet crash in July 1962, was one of the applicants. Ellis reasoned that despite his handicap, his flying skills were unimpaired, and being able to run and jump was totally irrelevant in space. He was not a finalist, but he did receive a nomination by NASA for special work in the space programme.

EVA operations

Almost 42 hours were spent on EVA during the three missions, and this included the work involved in trying to release the solar array and deploy the thermal shields during the first and second missions. Most of the remaining EVA activities centred on ATM film retrieval and deployment, retrieval of samples from the outside of the AM and ATM truss, and repair of experiments and hardware. Apart from the first stand up EVA from the CM hatch, two astronauts performed all other EVA activities, with the third in the MDA, monitoring the activities and photographing the event.

These were the first Earth-orbital EVAs since 1969, which followed the historic

Gemini forays outside the spacecraft during 1965–1966. Since 1969, considerable confidence and experience had been gained from the Apollo missions to the Moon, but activity in Earth orbit had not been attempted since Apollo 9, four years before Skylab flew, and with only 5 missions (Gemini 4, 9, 10, 11, 12 and Apollo 9) conducting EVA in Earth orbit prior to the Skylab 2 mission. The success of the Skylab EVAs pointed to future programmes in which EVA would be required to support space station operations, as reflected in the Soviet Salyut and Mir programmes.

During the solar array deployment EVA the major difficulty encountered was maintaining body position in areas not designed for EVA operations. It was therefore recommenced that hand-rails be provided across future spacecraft to help planned or contingency EVA. At the ATM work-station, a rotating control panel allowed for rotation of the experiment canister to facilitate removal and installation of film cassettes, while the crewman was held in a foot-retraint with a protective screen keeping his legs and feet away from the ATM gimbal rings and canister hardware during rotation. There were no problems associated with ATM film retrieval on any of the Skylab EVAs, when the astronauts found that the hand-rails, foot restraints, illumination and work-stations were suitable for the task for which they had been designed. In addition to the EVA operations outside Skylab, the crews also evaluated two new items of hardware inside, which were related to later EVA operations.

M509 Astronaut Manoeuvring Equipment A combination of two jet-powered Astronaut Manoeuvring Units – AMU – a back mounted hand controlled unit called the automatically stabilised manoeuvring unit (ASMU), and a hand-held manoeuvring unit (HHMU). The ASMU had a set of fourteen thrusters, controlled by two hand-controllers that were mounted on arms extending from the unit. The left hand controlled forward, backward, up and down and sideways movement, while the right hand controlled rotation in any direction. The crews could fly the unit either in shirtsleeve or pressure suits, with one astronaut serving as a ‘pilot’ while an observer photographed the event. The ASMU included a rechargeable/replaceable high-pressure nitrogen propellant tank and battery. It featured control moment gyro and reaction jet stabilisation, with a third option of firing the jets directly from the hand-controllers. For the HHMU, the ASMU provided propellant and instrumentation facilities.

On 13 August, Al Bean flew the first M509 unit, while Lousma photographed the exercise before flying it himself. Their comments indicated that the unit was quite good, although Bean would have preferred a faster translation and less precise attitude control. On the next mission, Carr and Pogue tried the unit, which they confirmed was a good design. A total of fourteen hours was spent flying the unit inside the upper dome of the OWS, although the hand-held unit was found to be quite difficult to control accurately.

Owen Garriott – who had never trained on the ground unit but flew it successfully inside the OWS – completed a demonstration of how easy the unit was to fly. In using the unit, the crews demonstrated a range of potential uses for on future programmes. They flew point-to-point and station-keeping, and simulated inspections of objects and close proximity flying to the walls of the station, as well as match-spin rates of a small object, grapple techniques by hand and slowing rotation by using the experimental back-pack. This, of course, was a successor to the AMU developed for Gemini, and a predecessor of the highly successful MMU on the Space Shuttle.

T020 Foot-Controlled Manoeuvring Unit A second device, not requiring the astronaut's hands to control it. In this experiment, a foot-controlled propulsion device, supplied by a high pressure nitrogen supply in a detachable propellant tank, was the mode of control. This was not as successful, as it produced unwanted rotation when used. The second crew flew it on three mission days – once with a pressure garment – and the third crew twice evaluated the unit.

HABITABILITY

Living in Skylab

Perhaps one of the most important research areas for Skylab, next to the biomedical studies, was an evaluation of the living and working conditions in the workshop in order to provided 'data that could be useful in the design of future manned spacecraft'. The three crews were asked to evaluate daily activities such as sleeping, eating, and the ease or difficulty of moving around the station and in using equipment, experiments or procedures.

M487 Habitability/Crew Quarters Included evaluations not only on architecture and environmental elements and communications but also on mobility aids, food and water, clothing and personal items, housekeeping and the working and off-duty period of the mission. In addition, *M516 Crew Activities/Maintenance* in connection with M151 and M487, evaluated crew performance during 'normal' tasks. Pre-flight training and performance established a baseline for evaluation of in-flight tasks, and then during post-flight the crew provided 'subjective and technical comments' during debriefing. *T013 Crew/Vehicle Disturbance* measured the effect of the crew while moving around the station, the dynamics and torque forces that were produced by astronaut body motions, and how this affected the control of the vehicle, the ability to perform tasks and the data recording of instruments and experiments as the crew moved around the station. To study body motion, a limb motion sensing system – a suit which followed a skeletal form, with a linear potentiometer at the joints – recorded continuous movements of the limb positions during a task. When combined with onboard film, these provided an indication of body posture and movement to complete a task, to retain position, or at rest in a relaxed state.

These investigations were also supplemented with film and audio reports – sometimes real-time, but often into the B communication tapes for later debriefings on return. These could be quite open in their critical comments, and were later released to the public domain as part of NASA open policy.

The privacy issue

The tapes were the origins of some of the adverse reports that circulated in the media, implying that the astronauts were complaining of the work-load, the systems, the scheduling of work during rest days, and so on. But that was the point of the taped reports. They were meant to be critical of the systems, being evaluated as part of the habitability programme. It is true that the crews voiced strong objections to over-scheduling, and combined with personal adjustments to the environment and the personal desire to achieve over 100% from each day, all of this contributed to the pressure to achieve mission success. However, misinterpretation of these habitability reports, along with the other events that reached the headlines, led to their being presented as dissent by the astronauts – especially the final crew.

The medical reporting issue by the third crew early in their mission was picked up by the media, but Conrad had also encountered his own communication controversy during the first mission. Prior to Skylab, all communications between the astronauts and the ground were open, with private communications reserved for special medical situations or operational emergencies. With the extended duration of Skylab, and the high medical activities, NASA planned to have the astronauts present a private medical report each day, with the flight surgeon reporting anything significant. The tapes would, however, remain within NASA for medical evaluation, and not be released or even transcribed for the media. In addition, private communications would be provided where 'a real operational need existed', and be summarised for the press by PAO if needed. Each week the crew could talk to their families, unmonitored from the CM.

The reason for the change was emphasised by the general 'doctor–patient' rule that was not generally discussed openly. Medical concerns between the astronauts and their doctors were to remain private, but summarised in the medical reports, although specifics on individuals were withheld. Public Affairs disagreed, and stated that the information on the men in the vehicle was more interest to the public than was the vehicle itself. Private family conversations were fine, but medical information should be made more open. The astronauts agreed that the public should know what NASA was doing, but were not so keen on letting the world know how fit or ill they felt, knowing that admitting such things could shorten the mission or make them even more answerable to the ground.

A few weeks before Skylab launched, James Fletcher initiated a compromise that would see routine medical reports and discussions over a private line summarised by flight surgeons, and in the event of an emergency a privacy operational communication would be operated. Therefore, on 28 May, when Conrad requested a private conversation the next day, it was believed they had a problem pending. Several officials were telephoned at home, and left their beds to discuss the request. When Conrad came on the loop the next day, he apologised for the difficulties he had experienced in riding the ergometer a few days earlier, and expressed surprise that the crew was doing so well, considering all the other problems had that they had to contend with.

When the media heard the summary of the 'private communication', they wondered if any emergency had existed at all. Public Affairs officials were not happy, and went as far as stating that the situation did not warrant a private communication. Some thought that Conrad should have been reprimanded though this was not carried out. What was decided was that in future the CapCom would ask the crew if an emergency had actually occurred that required a private communication when requested.

That same day a second problem occurred that later had repercussions on the third crew. Conrad had indicated, at the press conference earlier in the year, that the 'B channel dump data' would be released to the media; although this was an error, because NASA was not planning to release them. The events of 29 May produced a rift between Public Affairs and the Office of Manned Space Flight in releasing the B channel tapes that the media was expecting. Reluctantly, this was agreed, but the decision was then altered to censor from the transcripts all the medical experiment data, in case the wrong conclusions were drawn by the media and the public. John Donnelly the Assistant Administrator for Public Affairs, challenged this decision, citing a possible request to release the censored data under the Freedom of Information Act, which might have had serious and detrimental effect on the public image of NASA.

The request from CapCom as to the nature of the request for a private communication link was rescinded, as NASA did not want to place the astronauts in the difficult situation of not wanting to use the private communication loop when there was a real need. The decision remained to restrict the distribution of medical data, and information would not be released, which also protected the astronauts' privacy.

From then on, the astronauts were very reluctant to use the private operational channel for anything in case they initiated a new controversy. After the mission, Conrad commented that Houston had kept him in the dark regarding mission plans, and that he learned of the EVA plans to free the array during a birthday greeting from his wife, which indicated that communications from MCC were also occasionally guarded.

The private channel was not, in reality, private, and this was one of the concerns of Jerry Carr on the third mission when Bill Pogue became ill. It was not so serious as to threaten the mission, and he knew he had to report the vomiting, but he was aware that this might be misinterpreted or be seen as an 'emergency', and he waited to review the situation before reporting it. Unfortunately, the CM tape dump pre-empted his report, and the subsequent reprimand from Shepard and release of the tape contents did nothing to help the situation.

It was not until the early Space Shuttle missions that the medical conditions of each crew-member would be classed 'private', and not be released to the media other than in general reports outside of 'real emergencies'.

The 'lessons learned' documents

Shortly after the final crew returned home, a series of five documents were generated by NASA centres under the category of 'lessons learned from Skylab'. These were issued from the Engineering Directorate of the Skylab Program Office in Washington, JSC, KSC and the Skylab and Saturn Offices at Marshall. They indicated that 'authors of the lessons have been encouraged to be candid', but also indicated that recommendations were not the only, or even the best approach to questions raised, but offered sometimes a personal interpretation or an opinion which was not necessarily NASA policy. They did, however, offer topics of further investigation and discussion. Details of these documents are listed in the Bibliography, but interesting points relating to habitation include:

Communications Both the flight crew and flight control teams were reluctant to discuss the flight planning and workload schedules on the open circuit, due to misquoting or misunderstanding by the press. It was evaluated that a free exchange of thinking was frequently needed to prevent over-scheduling, but Skylab rules prevented private communication periods for this purpose. It was also evaluated that direct communications with PIs or their representative were beneficial for the crew in directing their efforts to achieve the most from each experiment.

It was also found that occasionally crewmen would omit a task or perform one incorrectly, due either to not receiving a message on the teleprinter, or to one that was received in error, prompting the suggestion for verifying onboard receipt of teleprinter messages. In addition, the sheer volume of printer messages needed to be handled more efficiently, as sometimes a ten-foot piece of paper greeted the astronauts in the morning, and they could take several days to deal with the issues or track the single sheets of paper. So many changes to the flight plan and to daily activities, and lack of communication with ground stations, meant that the teleprinter became one of the most important elements onboard the station to keep up to date with what was forthcoming on the flight plan.

Clothing The crew found that the outer clothing could be worn for one or two weeks without needing to change, which extended the use of the garments. However, underwear and socks needed to be changed daily, and they found that the toes of the soft-shoes were worn out due to rubbing against the grid floor. In addition, the pocket was inadequate carry everything from nail clippers to maintenance tools, and the duplication of aircraft flight-suit pockets was not necessarily beneficial to spaceflight requirements. It was suggested that some sort of headgear might be useful as a protective device.

Crew motion disturbance On Skylab, crew disturbance was almost undetectable on the ATM instruments, apart from when the crew initiated large disturbances and were able to conduct other activities without serious impact to ATM operations.

Exercise It is estimated that 1–1.5 hours of deliberate daily exercise was required by each crewman to maintain reasonably good physical condition during flight. Scheduling this exercise was important, as time was required to wash and freshen up, without going straight into experiments or a meal period. One element of exercise that was also discovered was that perspiration tended to pool on the body in some quantity until it broke away; and it tended to float around the compartment, which could be most inconvenient to other crew-members!

Food From the Skylab experience it was evaluated that it would be more suitable if the menu was standardised. Preparations involved a significant amount of documentation and time in selecting the menus that met nutrient and experiment requirements and crew preferences. It would also be more beneficial if the crew ate in the same place at the same time. The 'pantry' items were also of benefit, and food storage was more useful when all items were stored in the same place for preparation. It was also found that separating the operational medical requirements from 'normal' food systems reduced complexity and costs. The crew found that there was far too much handling of the food from stowage to eating, and that access to items within the pantry tray was difficult. They also commented on the shortcomings of the food can disposal system, which was very small and was difficult to keep clean, and that the utensil provisions were hard to reach and their compartments difficult to clean. Sometimes the cans simply floated away.

The food table was generally adequate, with all except the foot restraints working very well, although the third crew found it easier to remove them and use the grid floor foot restraints. One problem identified during training was that the metering dispenser for hot or cold water contained trapped or dissolved gas, which made reconstitution difficult, as the food pack was then larger than normal. Heating the food in the food tray was occasionally unsatisfactory.

Variety in the menu was important, and although there were 8,000 items available, if there were any changes to requirements or preferences, the result was a considerable amount of documentation and record-keeping in relationship to the medical requirements. Despite choosing his own menus, Lousma stated during the second mission: 'Beef hash for breakfast! I've asked my self every six days, whenever it turns up on the menu, "How come I picked beef hash for breakfast?"'

Hand-washer The crew found that as the hand-washer was not enclosed it was used mostly to dampen a cloth to use in sponge fashion to wash the body or face. Once soap contacted the rinsed rag, it became useless for further rinsing. Future design could be more user-friendly, to allow hand insertion, and actually 'working' directly with the water.

Habitably environments Known as the Skylab 'comfort box', the habitual environment inside the OWS was 'acceptable'. Although the temperature was comfortable, the humidity was found to be a little low, causing chapped lips, dry skin and nasal discomfort. The acoustic environment was pleasant, and odours were virtually non-existent. The astronauts commented that individual thermal control for sleep and the waste management compartments would have been desirable, especially after washing or showering. Portable fans would also have been of help when, after exercise, heat was not dispersed by convection.

Housekeeping The crew found that the vacuum cleaner worked satisfactory but needed more power. They found that the filter screen required a cleaning every three days, instead of once a week, as recommended. The machine was also used to collect cut hair, and was found to be very effective in preventing loose hair from floating around the crew compartment. The crew used the various wipes around the station almost as planned, but added their own variations. Clothes and towels that were not too dirty were stored in a 'rag bag' for large clean-up tasks, and the wipes were found to be still useful after initial cleaning, and were used to wipe down surfaces after a meal. A suggestion for the future was for a less tedious method – perhaps a single use wipe that did not have to be cleaned off after use, or an aerosol biocide for large cleaning areas. The astronauts also suggested a more aromatic aroma be used instead of the more clinical antiseptic wipes provided.

Inventory control The single most difficult area in updating the final flight data file for each mission was in keeping up with constant changes to the manifest onboard the station. Late changes in documentation or incorrect logging of location resulted in frequent wasted time in searching for an item in a logged area, only to find that it was not there any more and its new location had not been recorded. This required the ground to manually search transcripts and question the crew about the last usage or sighting, which took time, and occasionally resulted in uncertainty and more searching. It was suggested that a more efficient system of tracking items, usage and stowage be devised that was the same on the ground as that used in space. There was also a need for simple restraints for loose equipment on Skylab. The crews found small items really difficult to retrieve at short notice, such as the flashlights that were packed in cloth bags, inside an overbag, located within a compatrrtment that was inside a stowage locker

Loose items on air vents Several times the crew lost small items, and reported that they were later found stuck to the air return vents. It was suggested that perhaps in future spacecraft, such migration could be useful in positioning vents to collect items strategically around the spacecraft; and instead of repeated vacuuming, it might be more suitable to replace the filters.

Lighting Another area of complaint from the crews was the quality and availability of lighting in the OWS. The crew had to supplement lighting with flashlights which were effective, but inefficient and time-consuming. It was also difficult to hold the flashlight in the teeth while working with the hands. In the waste management compartment, overhead lighting prevented adequate facial illumination for grooming and hygienic chores.

Mobility The location of the pilot station on the wardroom table was placed in such a location that in order to adequately exit the area, he had to translate over the table, or ask another crewman to move from his station to allow passage. Both were inconvenient, but the over-the-table method also often led to the 'foot in the food tray' result!

Upon leaving the main compartment through the hatch in the dome, the crewmans legs impacted the dome sufficiently enough to leave dents in the ceiling, bruising their legs in multiple hatch negotiations and frequent orientations about the spacecraft during a normal working day. It was also suggested that control consoles along an IVA traverse route be adequately protected in case the crew inadvertently bumped into them.

In the crew compartment, the lighting was on the ceiling, and most of the equipment was on the grid floor, and the crew tended to move perpendicular to the ceiling by the use of foot-restraints on the grid floor, following conventional orientation in the different 'rooms'. However, in the larger upper area, the crew tended to opt for head-first movement, indicating that compartments affected the selected method of movement – and in large numbers, walls and floors were not necessarily an inconvenience, nor an influence on modes of mobility.

It was suggested that in future designs, IVA architecture should incorporate translation routes which do not interfere with working, eating, sleeping or relaxing. In addition, at a critical point in translation in which an astronaut changes direction or moves through a hatch or opening, adequate protection should be provided to prevent injury or impact as he translates.

In general, the crews related quite well to the gravity-designed orientation of the Skylab architecture. This proved easy in construction and training, but also produced the familiar experience of 'up–down' visual gravity vectors when working the wardroom and upper work area. None of the crews expressed a desire to repeat this throughout the station, although a recognisable reference axis was desirable at the work-stations; and while the MDA looked confusing, with no set layout, the crews found it fairly easy to adjust to it, and commented on its efficient use of all available areas for items of hardware or stowage. On the third mission, Carr experienced a feeling of a pending 'fall' if he entered the MDA from the CM feet first, as he could see all the way 'down' to the wardroom floor at the opposite end of the complex.

Mass handling No difficulty was experienced by the crews in moving large masses within the spacecraft, but the items needed provision for handling, such as handles. Individual techniques varied between the astronauts, but all worked well. The problem was found in moving multiple small items without a container to 'fence them in'. This was found in moving food from the lockers to the table/heaters, and a suggestion was made that in future, small transfer bags should be provided. A limiting factor was that large items tended to block the view of the astronaut as he 'pushed' the item forward. The energy to begin the movement was also more difficult to stop at the end of the trajectory, and care had to be taken.

Mission Control One element often overlooked in summarising lessons learned from any programme is that of Mission Control experience. Like the astronauts, the ground support teams gained considerable experience in preparing, training for and flying Skylab. Regular meetings of the management, flight, and centre teams were useful in updating programme requirements to real-time data. Understanding the needs of the crew, and both their difficulties and success, was an important learning curve during the mission. Over-tasking the crews, and expecting that each crew would operate the same equipment in the same way, was a lesson learned the hard way. However, the planned press conferences with the crew in orbit were useful, although the difficulties in using the private channels added an extra edge to the distribution of information to the public. Throughout Skylab, medical support was essential, and a certain reluctance to discuss personal medical issues over open channels did not help ease the flow of communication to and from the ground. What was found to be beneficial to Skylab was the use of a Programme Scientist (astronaut Bob Parker) as an interface between the experiment investigators, the flight control team and through the CapCom to the crew. Parker held a science planning meeting with the PIs twice weekly, to mutually determine experiment priorities and scheduling over the following seven days. Unlike previous missions – during which every minute of every day up to two weeks could be scheduled and reasonably followed – the Skylab mission could not operate a fixed timetable of daily operations extending too far into the future, and the flight plan required constant updating and revision to respond to crew, equipment and programme changes.

Personal hygiene This was one area that received the most criticism from the astronauts, in that the equipment was not up to their usual standards, and items such as shower soap, shampoo, toothpaste and razors needed to be avoided. The shower soap left some with irritation or a stinging sensation and so they avoided it – and the shampoo just smelt bad. With the toothpaste not being digestible, spitting it out caused inconvenience. It floated. Using safety razors in zero g also meant there was no way to 'slosh in the water' a clogged razor, as they would have in their own homes. Skylab was rather overloaded with soap at the end of the mission. It was planned that one bar per man for two weeks, plus five bars per month for housekeeping and cleaning tasks, would be sufficient, so Skylab therefore launched with 55 bars of soap onboard. However, when the Skylab 2 astronauts came home, they had used only one of the eleven allocated. The stocking of limited space would be evaluated for future programmes not having the luxury of resupply craft.

Restraints The crews considered that the restraints should be the same, and not different, across the station, and they should be attached to the structure and not the astronaut. Thinking about release from a restraint should not be required until the astronaut was ready to do so. If it took time and concentration, this could detract from the job in hand. The triangle grid-shoes were deemed to be excellent, while a universal foot restraint was required for EVA work. It was also suggested that in future, shoes should have zipper fittings, and not laces to tie and untie. The astronauts also found that some of the restraints at work-stations were positioned to 1-g posture and not to zero g. In training they used a seated arrangement that placed some of the controls out of immediate reach or beyond eye level. In flight, the crew opted for a more upright stance to make the controls more accessible.

Repair and maintenance While demonstrating a capacity for repair and maintenance above and beyond what had been expected of them, they suggested a programme of guidelines for such work in the future. These included EVA as a major element in repair, servicing and maintenance tasks, and hardware, training and provisions to support this should be included in future programmes. All equipment should also be

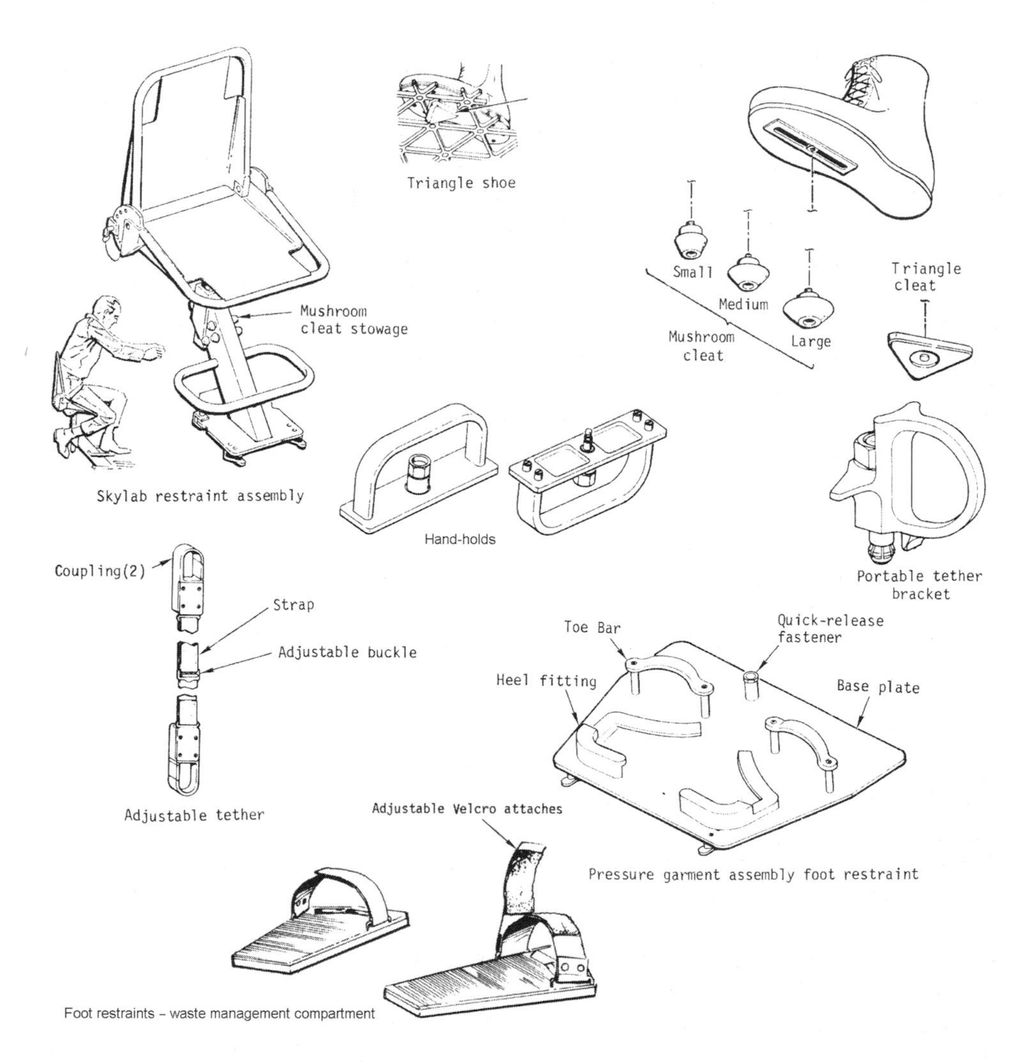

Grid foot restraints and portable restraints.

designed to facilitate repair and maintenance where possible, with a range of spares and tools. The containment of small item such as nuts, washers and bolts was also suggested as a useful addition, as well as a portable work-site or bench, which should be available for the repair effort; standardisation and familiarity training should help prepare the crews for any unforeseen repairs and routine tasks; and panels and work surfaces should be accessible from the front to access the equipment inside or behind.

The astronauts found that many of the restraint straps on the OWS were too large or too bulky for small items. They were ideal for launch restraint, but after crew occupation and unpacking, small items were found to be difficult to restrain. A small interim storage facility would be of help before relocating the used tools, equipment or hardware to the permanent storage place if they were to be used over a period of time. Bungee cords were suggested to restrain items on the walls and surfaces, and book restraints needed to allow a page to be displaced, or the ring binder would spring open, sending the pages across the spacecraft. A device for adequately spreading out the clothes to dry them while the crew slept was also suggested.

Trash It took a considerable amount of time to manage the trash in Skylab. The crew found that due to the busy, and sometimes rushed working day, several temporary collection sites were converted to stow waste, pending a routing 'trash collection' to dispose of it in the trash airlock. It was then collected and placed in the airlock, which towards the end of the third mission required a crewman to 'stand' on the lid to ensure that the bag was ejected into the almost full oxygen tank. Underfloor stowage of dry waste was also successfully used in the wardroom. Should the trash airlock have failed, this would have made a significant impact on the habitability in the spacecraft; and after Skylab, suggestions for trash compactors, alternative stowage areas (even outside!), back-up and contingency waste collection systems were investigated. In all, the trash airlock was used for about 660 cycles (four times a day) during the three missions.

Time management Working time in space would be improved if equipment was standardised so as not to induce errors or delays in the crew's efforts to complete a task. If equipment was standardised, then its applications across the spacecraft would be considerably increased allowing less spare to be flown, and reducing documentation, stowage and failure rates. It was also important to more accurately determine the exact amount of equipment, spares and materials required on orbit. Before Skylab flew, this was estimated with uncertainty, but with the Skylab data, a more accurate determination of equipment required and used reflected a better management of crew time in dealing with unnecessary procedures, equipment and stowage.

Skylab individual task times were equal to or slightly less than the 1-g baselines set up in training. There was an 'envelope of attack' available to complete the job, and in many cases the crews asked for additional tasks to fill voids in this envelope of conservative time-lining. However, where logistic management was required this tended to go the other way especially in housekeeping and maintenance, which could not be exactly detailed. The third crew found that by taking housekeeping out of the detailed flight plan and moving it to a shopping list, to be completed as and when possible, they could alleviate the pressure of completing experiments and observations. All three crews commented that they were being scheduled to work right up to the sleep period, making relaxation difficult. The Skylab 4 crew observed an unscheduled one-hour pre-sleep period that improved their rest quality and in turn their work productivity.

Stowage It was expected that the crew would utilise all areas of the station for stowage, including over the head. However, in practise it was found that the crews tended not to use spaces above tables, consoles or shoulder high when working on the crew quarters' lower deck, in spite of ease of access. When all three crew were eating, the astronaut located in the front of the pantry had to move out of the way while his colleague accessed the facility. It was suggested that the astronauts' food be stowed directly behind the person to ease access to his own menu. Spacecraft stowage and control panel numbers were confusing, as they used similar identifications numbering. It was suggested that a stowage system use a separate numbering system to prevent confusion. The crews found that a stronger friction on hinges of doors was needed to allow the door to remain open when used, and that they should have provision for attaching items on the open hatch door. Specialised storage restraints, filler material and separators were found to be unnecessary. It was far easier and simpler to stuff items of clothing, towels and soft goods into the lockers to restrain them, while Velcro stopped small items such as personal items, hygiene items, and so on, from floating out of the lockers.

Sleep The 'against the wall' sleep stations were adequate for all nine astronauts, and were only changed due to personal preferences of temperature and airflow. Throughout sleep they remained aware of noise (listening for alarm warnings). They objected to airflow 'up the nose' rather than head to toe, and suggested flexibility in blanket arrangements to allow for variations in thermal conditions. Firm restraint against the back was not sufficient, and more body coverage was suggested to give the impression of bed clothes. The compartments also needed to be longer than the length of the restraint.

Shower The first crew complained that the systems took too long to set up, but offered favourable comments about the stimulating and pleasant experience of a weekly shower. The SL-3 crew reverted to sponge baths because of the time it took to assemble, use, clean and disassemble the device – anywhere between 45–60 minutes, or even longer! The sponge bath was deemed 'adequate'. To use the shower, the astronaut stripped off and placed his feet in foot loops in the base plate, then extend the tube up to the top base plate and twist lock it by a few degrees to retain the water whilst showering. The outside portable container of just over a gallon of water was pressure fed and was filled from the hot water source in the hygiene facility. The liquid shampoo did not smell particularly pleasant – and it was stated that it was a reminder of pet shampoo, or something even worse! The water from the shower nozzle sprayed about 6–8 inches, then formed a globule of a gelatine consistency so that the astronaut learned very quickly to hold the nozzle close to the body. The nozzle was then stored, the water was moved across the skin to spread it, and the soap was added to produce a lather. To clean off the water, the astronauts used the vacuum cleaner attachment, the rest of the clean water as a rinse, and then vacuumed again and then they had to vacuum the inside of the shower before collapsing it to the floor. Too much use of the soap meant less water to rinse, and so it was a balancing act of water management. When the screen fell, the humidity in the OWS was quite cool and it was a quick dash to the waste management compartment for towels to dry the body to get warm again. So although the shower was a welcome relief, it was sometimes more trouble that it was worth, and needed further design if it was to be used on future stations.

Training The Skylab training programme proved very efficient in preparing both the flight crew and ground crew for the Skylab mission operation. The SMEAT was also useful in revealing shortcoming in operation hardware and procedures. Several simulations were halted early in the training programme due to hardware or data problems, and it was suggested that activities should be performed in check-out simulators and procedures before committing them for crew or flight control training. Astronaut participation in a training programme greatly improved the quality of training, and involvement of support astronauts in design, development, experiments and other activities allowed flight crews to concentrate on priority requirements. One interesting point raised by the evaluation of mission operations was that 'mission personnel and procedures were not always the same as those used for simulations. Longer-duration missions and diversified mission operations will dictate more personal turnover and therefore condensed training courses will be needed for rapid familiarisation of new personnel.' Having the crew's activity involved in the development of hardware and procedures greatly enhanced the success of Skylab.

Wardroom window Earlier programmes had demonstrated the value of having a 'window on the world'. When an observation window was proposed for the wardroom, which had no link to any specific experiment, there was objection that it was a weakness in the hull. However it was this window that offered the crews an important view of Earth and the cosmos, and it was the centre of their relaxation time using the binoculars to aid observations through the window. The value of the wardroom window was that it was in an area in which hardware would have prevented the crew from simply enjoying the view. The other windows that were not filled with equipment were also of great benefit including those in the CM and EREP work-station. During one rare quiet moment on the third mission, Jerry Carr was looking out of the window as Skylab passed over the western coastline of continental United States at orbital sunset. Below him he could see the illuminated kidney-bean shape of greater Los Angeles and both bridges of San Francisco twinkling as night covered the ground. Looking down the coast he could see most of the California peninsular where he had grown up and served in the Marine Corps. He could see as far down as San Diego and Mexico, and almost up to the Canadian border. This spectacular view, as the golden Sun was surrounded by the velvet black of night, would be one of his most treasured memories of the mission.

Worktable When the Lunar module was designed for Apollo, it had seats – until it was pointed out that in space, or even in the 1/6 g of the Moon, seats were irrelevant. So why did the Skylab astronaut miss a worktable? They had the wardroom table, but that took time to clear off before and after meal times, and it did not have adequate restraints for books and papers. Writing in space, and the management of multiple items of equipment in zero-g, were two of the more challenging tasks that the crews faced. Although writing was not the problem, it was difficult to hold books and papers. The crews suggested a table or workstation for books or to allow small maintenance tasks to be performed; or provision at workstations for books, papers and instruments. It need not be permanent or even table-shaped, as it would be against the wall, but it would be of great benefit. On Skylab, the crews found the ventilation screens useful as a makeshift table, due to the suction holding items to the filer screens!

Waste management The airflow system worked well in collecting solid waste, but a higher airflow was suggested as an improvement to the design. The seat was also found to be hard and too small, and widening it would affect a better airtight seal. The lap belt and hand-hold were an absolutely essential restraint aid. The urine collection system worked well, but was rather noisy, and because it was located near the sleep compartment, it disturbed sleeping crew-members. The foot loops in the area were not so useful, as they required curling of the toes to restrain the position, becoming loose or uncomfortable. Storage of urine caused a surprise one morning when Gibson opened his urine storage draw and saw a 2-inch diameter ball of urine float up towards him – but he could not find a leaking bag!

SUMMARY

The research programme conducted on Skylab was, in its day, ground breaking in its complexity and scope. Beyond the main scientific programmes in solar physics. Earth resources, space physics and astronomy, and biomedical experiments were the additional science and technology programmes. Perhaps just as important, but smaller, were the first investigations from high-school students that offered them the opportunity to perform their own experiments in space, and the series of space demonstrations that opened up space science to the general public.

Earth resources revealed new vegetation patterns, forestation, and geological land features of value to mineral and resources prospecting; and they also obtained oceanographic data for sea transportation and fisheries, and on the natural water cycle for the forecasting of floods and droughts.

Material sciences investigated new ways to heat, mix, cool and form alloys possessing unique electrical properties that would benefit the developing electronics industry, while crystals were larger and purer than Earth-grown samples for application in computer chip production

Life sciences included studies of humans, and recorded changes in body weight related to both diet and adequate load-bearing exercise. Physiologically significant body changes were recorded, and over a period of time, work and performance levels were monitored in relation to a range of time and motion studies; and there were also studies of how self-induced and self-imposed work-rates increased or decreased productivity.

Bright spots were discovered all over the Sun, and prominences and flares were recorded more frequently than expected.

At a time when future manned space operations were, if not in doubt, certainly several years in the future, Skylab clear demonstrated the potential of prolonged spaceflight and the volumes of material produced by living and working in that environment. But with the Space Shuttle in the planning stage for missions of initially a week or ten days, it was to be twenty years before that potential was realised by Americans once more flying long-duration missions.

However, as the benefits of Skylab were being reaped, and as the results were beginning to be analysed in the mid-1970s, the OWS was still in orbit, and plans were still being evaluated to revisit the station before it decayed.

Beyond Skylab

On 10 January 1974, with the Skylab 4 crew still onboard the station, studies were conducted to determine how to configure the OWS for unmanned operations and keep the option open for a revisit 'at some future date'. A series of special deactivation procedures were required for the ATM and MDA and OWS to ensure ground support monitoring, and control options were still available from mission control at JSC. In addition MSFC preferred that the OWS be left in a configuration that permitted a revisit in the future without the need to reactivate the whole station.

When the Skylab 4 crew departed the OWS in February 1974, they left, in the MDA a bag that contained food, clothing and a few other items that could be retrieved by a revisit crew, who could then determine the effects of orbital storage. Although it was unlikely that these items would be retrieved, it remained an option for NASA to go back to the station on a future mission, although it was realised that this would not be until the Space Shuttle was operational later in the decade. With the possibility in mind, the orbit of Skylab was raised by the engines on the Skylab 4 SM prior to undocking.

Early in Skylab flight planning there was a desire to keep it in orbit long enough for it to be used as the core of a larger station. In discussing this prospect during management meetings, it was realised that it would take a fair slice of the NASA budget to maintain the station in a suitable orbit and condition it for a later visit, and so the idea of using Skylab OWS as the core off a new and much larger station was not an idea that was long-lived. The two options that remained for disposal of the station after the final crew came home were boosting it into a higher orbit for a period of years allowing the possibility of a short revisit, or sending it to destruction into the atmosphere over a large expanse of ocean.

REVISIT, REBOOST AND RE-ENTRY, 1974–1979

In addition to the three crews that manned Skylab, there was a fourth Skylab crew that trained to fly at least two types of mission profiles – either a rescue attempt to bring back a stranded crew, or an OWS de-orbit manoeuvre. Vance Brand and Don Lind were the Skylab rescue crew for all three manned missions, and almost flew such a mission to recover the Skylab 3 astronauts in August 1973. These two astronauts were also back-up crew-members, with Bill Lenoir, for the second and third missions. They had undergone all the training for what the Bean and the Carr crews had planned to accomplish on their missions, and with their rescue training they were in some respects probably better trained than the back-up crew for Skylab 2, (Schweickart, Musgrave, McCandless).

A discussion into the possibility of a fourth crew visit to the station was held prior to the first launch in the spring of 1973. This would very much depend on the availability of consumables and status of the station after the third crew had come home, and apparently proceeded no further than coffee-time discussions. There was no firm duration agreed, no crew activities planned, and no crew training conducted. In reality, the idea never ever progressed beyond a desire to obtain as much from the programme as possible. Had such a mission flown, it would probably have been around 21 days to close out the station after the third crew completed their planned 56-day mission, supplementing the science programme and moving some of the deactivation and orbital storage chores to the hypothetical fourth mission. The three-person crew likely to have performed such a mission consisted of Brand (Commander), Lind (Pilot) and Bill Lenoir (Science Pilot). As a scientist as well as a pilot, Lind could have equally served as the Science Pilot, and the advantage of this would have become apparent on such a short mission.

When the gyros, coolant loops and power supply reserves were indicating any useful operational life beyond the SL-4 mission was not practical, there only remained the plan to turn off and abandon Skylab after the Carr crew returned home. The extension of SL-4 from 56 to 84 days added the 28 days that was suggested for SL-5, and thus rendered that mission unnecessary. Although it remained theoretically possible to dock to the OWS and briefly enter the MDA in a pressure suit, there was no clear reason why this should be attempted, and so a lengthy revisit to Skylab shortly after the departure of the Carr crew was not further explored.

Skylab 5 reboost mission planning

What had progressed somewhat further in the mission planning was a possible deorbit mission using a fourth Apollo CSM for the re-entry burn after the three main crews had returned home. In May 1971, following a simulated 144-day (1 + 28 + 1 + 56 + 1 + 56 + 1 man-supported) OWS mission, the station's reaction control system engines were test-fired at White Sands Test Facility in New Mexico, in a test of depletion of the onboard propellant supply, and as a test of the back-up (to the CSM) deorbit propulsion mode. However, with the excessive use of propellant early in the Skylab 1 mission, supplies were low. Therefore, the two options remained of using the CSM to either push the OWS upwards to a higher storage orbit, or downwards to destruction in the Pacific Ocean.

As part of their training, Brand and Lind practised these options in the CM simulators. During tests it was found that at full throttle, if the SM engine fired for longer than planned, this could result in 'jack-knifing' the combination and placing it in a less than desirable re-entry configuration. To prevent this, the full throttling of the SM engine was restricted so that a longer burn was needed, but if over-burn was encountered, it would not spin the spacecraft around and cause premature break up.

The second problem was of more concern to the astronauts. The burn to point Skylab into the atmosphere was fairly close to the point of entry, which meant that the undocking and CM re-entry had to follow this manoeuvre fairly rapidly. If the crew were unable to undock the CM for some reason, it would still remain attached to the station as the OWS re-entered. This, of course, was not a healthy situation for the crew, for even though they may have achieved their long awaited mission, they preferred to survive the flight to tell the tale!

In this mission scenario, even a normal undocking would have meant a fairly close re-entry path to the Skylab, and the thought of their CM following the flaming wreckage into the atmosphere and possibly sustaining structural damage or ripped parachutes was also not a desirable situation.

During their simulations, Brand and Lind determined how long they had to complete a manual unlatching in the CM tunnel if they could not undock before the inevitable re-entry began. Wearing a full pressure suit, and depressurising the spacecraft, Lind would need to move to the docking tunnel, undo the hatch and float inside to manually release the twelve latches in the docking ring. With Brand handling the control of the CM, the simulations and evaluations indicated that Lind had just fourteen minutes to achieve this, but proved that if required it could be done – just; and this information was passed up to management for a final decision.

The astronauts also reported that the second the last latch was released, Brand had to perform a separation burn to move away from Skylab to line up for a separate re-entry trajectory to miss the main debris from the station. If it took all of the fourteen minutes, Lind realised that there was a possibility that he would be still in the docking tunnel as the CM separated from the SM. The possibility of how to close the hatch and return to the seat to strap in, with all the manoeuvring of the CM before they began the re-entry profile, 'still had to be worked out' in their training. As Lind recalled almost three decades later: 'There was a real possibility that I would be standing looking back up at the fire train with an open hatch; it [should have been] no problem, [and] would have been spectacular!'

The trajectory planners had stated that they could choose any 25-mile footprint in the Pacific and aim Skylab for that target, but the high risks on the manned deorbit mission outweighed the benefits of de-orbiting the station in 1974. It was therefore decided to place the station in orbital storage after the third crew had left, until natural decay pulled it back to Earth – which, hopefully, was after the Space Shuttle became operational and could fly to either reboost it again, or de-orbit it over an unpopulated area.

A second Orbital Workshop

In 1973, Skylab was called America's first space station, implying that other workshops were to follow. Although there was a desire to launch a second workshop, there were no longer the funds to achieve it.

Some of the earliest detailed planning for a second Skylab OWS, were outlined in a preliminary mission definition document dated 5 September 1969. The design of the OWS would feature the same physical appearance and capabilities as the first OWS and would use the back-up hardware assigned to the first station. A total operational lifetime would be between twelve and 24 months, with rotational crews of three astronauts, of whom at least one should be a scientist-astronaut.

The launch of the first station was then planned for late 1972, so the programme planning for a second OWS suggested a launch early in Fiscal Year 1974 (October 1973–October 1974). The second OWS was planned for an orbit that would be at 55° inclination, at 242 × 310 miles altitude, and use existing hardware, or that under development to keep launch and operational costs down, requiring minimal funding from the Fiscal Year 1970 or 1971 budgets.

Experiments were again to be devoted to solar, stellar and Earth resources, but this time reflect an experimental facility rather than individual experiments, using a learning curve from the first OWS missions. One of the suggested experiments would be in attempting the creation of artificial gravity. By 30 September an Ad Hoc group was formed at MSFC to look at the prospects of a second OWS programme. The justification for such a mission was investigated, and the payloads, constraints and mission goals were balanced against the launch dates, budget restrictions, and its long-term benefit in the larger manned spaceflight programme

By adding artificial gravity to the combination, it was found that it complicated the operations, and required relocation of certain elements on the cluster. A series of studies during the summer of 1969 indicated that the strength of the CSM and MDA docking interfaces would need to be improved, and the ATM would need to be strengthened and its position reviewed. There were additional issues in maintaining attitude, control and propellant usage in ensuring that the ATM still kept pointed at the Sun as the station rotated, which indicated that attempting to add artificial gravity to the station would require a significant redesign of the cluster.

On 8 August 1969, MSFC defined the existing contract with McDonnell Douglas for the fabrication and delivery of two OWSs by July 1972, in which the second workshop would serve as a back-up for the first, to be launched in mid- to late-1972, and then, if not required, could be launched as a second vehicle. Running concurrent with these plans were studies to define the costs, schedules and performance characteristics of a second telescope mount, designated ATM-B. By October, a study team at MSFC was tasked to determine whether such a programme was worth the investment, and if this second ATM could be used to support the development of even larger instruments. In November 1969 the objectives for a second station were identified as a year-long mission with four three-man crews (4 × 3-month rotating missions), possibly with artificial gravity experiments, with replacement of the solar telescope with an advanced stellar ATM, and an expanded EREP package.

By early January 1970 some of the hardware that could be assigned to a second Skylab was becoming available. The Apollo 20 mission had been deleted from the programme, and soon two more lunar missions would also be cancelled. Hardware from Saturn V SA-513 was to be assigned to the first OWS, and while the fate of SA-514 remained undecided, the use of SA-515 was under consideration as the launch vehicle for the second workshop. On 29 January 1970 a preliminary report produced to support the congressional hearings to propose the second OWS, to be followed by a work statement by July and a preliminary design review early in 1971.

By March, definition studies were well under way. They aimed to continue the ability to live and work effectively in orbit, continued observation of the Earth and its environment for global benefits, and use space for scientific research. However, by 7 March a decision had been made that there was no advantage in progressing with a stellar telescope any further, or for the foreseeable future, and that all efforts should be aimed towards artificial gravity and EREP experiments that could be seen by the general public as the tangible benefits of a second Skylab. There had been strong support in expanding the EREP studies from MSC, and only small funding would be required for a twelve-month mission of four visits, with the advanced solar ATM, no major design changes, or with an advanced EREP in place of the ATM. By adding artificial gravity experiments, it would also indicate to the public and Congress that this was not just a repeat of the first set of Skylab missions.

The following month, during a meeting of a Manned Spaceflight Management Council, there was further discussion on what was now being called Skylab II (also known as Skylab B) would look like, and what its primary objective should be. What experiments should be included, and how many crews should visit it? The report was due by May, but did not appear until September.

Houston's commitment also pointed towards a far larger station that was much more sophisticated than the plans for Skylab II/B. Huntsville disagreed, and considered that such a move would be a threat to the first Skylab missions; and went as far as stating that to support a major OWS twelve-month manned habitation period was impossible without serious admendments to the hardware, and that experiments with artificial gravity would double or triple costs. In its place, MSFC offered an alternative eight-month mission with solar astronomy/EREP objectives.

In a Skylab programme schedule review dated 6 April 1970, Ken Kleinknecht indicated that the impending decision of cancelling further Apollo missions would affect both the 1972 Skylab A launch and Skylab B planning. This included several staff, hardware, processing and budget implications that needed to be addressed in either overlapping Apollo and Skylab operations, moving Apollo forward in launch intervals to fly Skylab after Apollo, or cancelling two Apollo flights. In Kleinknecht's briefing, the proposed Skylab B would launch in late 1974, followed by three missions, the first at the end of 1974, then two more in the first and second quarter of 1975. There were plans to scrap Apollo 19 and use the booster to launch Skylab B, and hold Apollo 18 in storage, for launching a third station – a core module that could remain aloft for over 10 years.

Studies into a second Skylab continued all that summer in preparation for Fiscal Year 1972 requests, but the payload weight also increased, and became a serious problem that threatened the lift capability and structural strength of the second stage of the Saturn V, adding even more costs. Skylab II, it was estimated, would require $1.32 billion to $1.5 billion. After 31 July, discussions with Office of Management and Budget mission planners knew the money would not come easily. Following a further review on 31 August, it was nevertheless agreed to go ahead with the recommendation, and so on 4 September 1970 the delayed and final planning study report indicated that there were sufficient data to proceed with a second set of Skylab missions. Such a series would not only provide addition and complementary data to the first series of flight but also continue the development and expansion of US manned spaceflight operations, even though no new hardware would be flown. The study also determined that the costing would be a worthwhile investment that offered an 'economically feasible programme option if future funding for the Space Shuttle programme fell behind the anticipated growth rate'. In other words, it would keep Americans flying in space until the Space Shuttle was ready to take over later in the decade.

Later that same month, two more Apollo missions were cancelled from the lunar programme, releasing hardware for the space station programme, and talks were in hand with the Soviets about the possibility of a Soyuz–Skylab mission.

By the end of 1970, however, support for a second Skylab dwindled, due to the complicated and expensive modifications required to include artificial gravity. Not wanting to merely repeat the first Skylab missions, and with other more pressing requirements, Skylab II would require a larger overall NASA budget, or would need to divert funds from the Shuttle programme, resulting an even longer delay in making that programme operational. Therefore, despite strong support in House Space Committee to authorise a second Skylab, the Nixon administration was unwilling to underwrite costs, and NASA not wishing to jeopardise other programmes, examined other options from the already built hardware in Apollo and Skylab

On 17 March 1971 it was suggested that the uncommitted hardware from Apollo and Skylab programmes indicated that there was an opportunity to fly 'low Earth orbit manned missions' prior to commencing Space Shuttle operations. In addition to a joint mission with the Russians two additional Earth resources surveys were possible using the Apollo CSM/Saturn 1B, and an option for a fourth solo CSM mission remained a possibility, although again these proposed missions still required additional funds to mount them.

On 19 April 1971 – the day that the Soviets launched Salyut – a meeting held at McDonnell Douglas discussed plans for the back-up Skylab hardware. Representatives at this meeting were from NASA Headquarters, MSFC, MSC, KSC, Martin Marietta and McDonnell. Preliminary plans agreed on preparations for a Skylab B launch ten months after the launch of the first OWS and the three missions planned. A contractor's manufacturing and KSC payload testing profile would feature the format as on the first OWS.

Progress on the first OWS continued, and the option on the second remained open, although in reality it was unfunded. Over the next two years, the idea of a second Skylab became more attractive, but little progress was made other than in its role of supporting the first OWS in the event of a launch mishap. By May 1972 it was estimated that to meet the turnaround time of a ten-month processing schedule to launch the second set of hardware, the acceptance check-out of the back-up OWS/

AM/MDA would be required in May 1973, and in support of this, the option remained open.

During December 1972, concerns were raised that all activities with the second OWS were to be terminated after the first manned mission of Skylab A. The back-up rescue system was available, all the flight hardware was ready, and the cost of storage would be minimal, allowing the hardware and support operation to be retained until at least the spring of 1974. It was argued that in the event of a major failure prior to, during, or shortly after the first manned flight, with a back-up system already being wound down and with major ATM and EREP planned for the second and third missions, 'the possibility existed that the programme could end up with only 28 days of medical data, no science data, no EREP data, and just a $2.6 billion failure'. Ironically, with the launch problems that occurred during Skylab 1, this almost became true.

On 19 March 1973, consideration was still being given to flying Skylab B between ASTP in 1975 and the beginning of Space Shuttle operations in 1979. The worth of a second Skylab was recognised, but there was a reluctance to fully recommend it, as this would bring about a far more serious risk of further delaying the already restricted funds for the Space Shuttle and other programmes already in the budget.

Finally, on 13 August 1973, with the second crew half-way through their mission, NASA announced the decision to delete the second Skylab, effective 15 August, and that all Skylab launch processing work would be cancelled immediately, except that directly associated with the SL-3 and SL-4 missions and rescue capability.

Even after this decision, NASA was not prepared to let Skylab B slip away easily. On 3 January 1974 there remained a flexibility for a second Skylab OWS to be retained until all planning was completed for Fiscal Year 1976 budget. The launch umbilical tower number two was retained for a possible future Skylab launch until a decision was made to launch Skylab or convert to the Space Shuttle. The continued storage of existing hardware required for the Skylab B mission, at minimum storage costs, and including back-up and spares, continued through June 1974.

Skylab B died in the budget discussions of Fiscal Year 1972, but struggled on until Skylab 1 was successfully placed in orbit, and manned by the first two crews. The problems associated with the first launch did not help the case for a second mission of the same design. However, despite Skylab being saved from the brink of failure by the astronauts and controllers in a further demonstration of the usefulness of man in space, there was a decision to cancel all thoughts of a 1976–1977 second Skylab. Not wanting to threaten or delay Shuttle fundings with a repeat of Skylab A, the opportunity for launching a second Skylab was finally lost.

Instead of being launched into orbit in 1976, the second OWS was taken out of storage, cut up and displayed in the National Air and Space Museum to represent the first laboratory configuration. Skylab B remained forever grounded, like the three remaining Saturn V vehicles located in the visitor centres at KSC, MSFC and JSC – a clear reminder that the Apollo/Skylab hardware would fly no more.

Skylab II crewing

One of the attractions of following manned spaceflight operations is to re-examine cancelled programmes, and ask: 'What if? ...' In this case, 'What if Skylab B had flown; who would be the likely candidates to crew it?

With the case of four Skylab astronauts reassigned from two cancelled Apollo lunar missions (Pogue and Carr from Apollo 19, and Weitz and Lousma from Apollo 20) this also prompts the question 'If the flights had remained, and they had indeed gone to the Moon, who might have taken their place on Skylab A?'

If the three Apollo lunar flights had been retained, Pete Conrad might have tried to obtain a second lunar landing flight (on Apollo 20), but probably would have opted for the command of the first Skylab mission. Al Bean would still probably have flown a Skylab mission, and Cunningham was in line for a flight, but was faced with an uphill struggle to secure it. Certainly Kerwin, Garriott and Gibson were always prime contenders for the three Science Pilot seats, and possible Pilots could have been McCandless, Lind (if there was still no seat for him on Apollo) and the first Group 7 members with Bob Crippen (who commanded SMEAT) the leading contender.

The three Skylab missions with Apollos 18–20 still on the manifest could have looked like this:

Mission	*Commander*	*Science Pilot*	*Pilot*
SL-2:	Conrad	Kerwin	McCandless
SL-3	Bean	Garriott	Lind
SL-4	Cunningham	Gibson	Crippen

With these assignments, the back-up crews would probably have been made up from other reassigned astronauts from flown Apollo missions as Commanders, the unflown scientist astronauts selected in 1967, and Pilots from the MOL transfers from 1969. These could have gone on to fly Skylab B, had it been funded.

Under the Skylab B programme there were four crewed missions planned, but it was recognised that supporting a long-duration mission would be complicated, with on-orbit limit in the lifetime duration of the CSM. Therefore, Skylab B could have featured two long-duration and two 'visiting' missions to exchange the CSM – similar to the exchange and repositioning of Soyuz spacecraft on Salyut and Mir missions.

This also allowed a limited resupply of consumables, with a two-men 'visiting crew' Apollo crews for short overlapping missions with the main long-duration crew. In 1998 a series of communications posted on the Internet by space sleuths Mike Cassutt and Dave Portree offered suggestions for a Skylab B crewing profile – dependent, of course on sufficient hardware and funds, and a desire to accomplish it. These suggestions form the basis for the following:

1975 December	Skylab 5	Unmanned OWS Skylab B launched
1975 Dec–1976 Jun	Skylab 6	Schweickart, Musgrave, Henize (two scientist astronauts – long-duration and back-up to SL-2)
1976 Mar	Skylab 7	Roosa, Truly (mid-mission visit to SL-6 crew to exchange CSM, possibly SL Rescue crew 1)

1976 Jun–1977 Apr	Skylab 8	Worden, Lenoir, Thornton (long-duration flight) BU to SL-3 and SL-4)
1976 Nov	Skylab 9	Evans, Bobko, mid-mission visit to SL-8 crew CSM switch, a possible second SL Rescue crew

The assignments of Group 6 and Group 7 astronauts are suggested, as they were assigned to AAP/Skylab in 1969. The assignment of Roosa and Worden follows the sequence of former Apollo CM Pilots normally offered the commands on their next flight. From Apollo 12, Dick Gordon was in line to back up Apollo 15 and then command Apollo 18, but Slayton also wanted to include lunar landing experience on the last two Apollos, as they were to involve the most difficult landing areas. Therefore, former Apollo 13 LMP Fred Haise was offered Apollo 19, while CMP Swigert was appointed to the joint Russian mission and Apollo 14s Ed Mitchell was a leading contender for the command of Apollo 20. Next in line were Apollo 14 CMP Roosa, Apollo 15 CMP Worden and Apollo 17 CMP Evans, as suggested in this hypothetical crewing pattern.

In any event, no assignments were ever made, but when the Skylab group was in training during 1972/73, Brand, Lind and Lenoir were informed that they would have a prime seat on Skylab B if it flew. However, despite Lind's personal plea to convince James Fletcher to save Skylab B, the assignments progressed no further. If Skylab B had flown as planned after the cancellation of the three Apollo missions and after the Skylab A series, the second set of missions might have been as follows:

Skylab 5	Unmanned OWS B (followed by four long-duration crews)
Skylab 6	Schweickart (or Weitz)–Musgrave–McCandless (there remains some doubt that the back-up SL-2 crew would have made a flight as an intact crew)
Skylab 7	Brand–Lind–Lenoir (possibly flying as the first crew?)
Skylab 8	Lousma (or Crippen?)–Thornton–Bobko (with Crippen this would have been the highly compatible former SMEAT crew)
Skylab 9	Pogue–Henize–Hartsfield?

Jerry Carr was also eligible for a second flight, and each of the first three Science Pilots offered considerable experience if it was decided to fly two Science Pilots on each long-duration mission. Of course, suggested back-up is very difficult, but it could have come from the flown Skylab A crews in dead-end assignments (as seen on Gemini and Apollo final crews), Group 6/7 astronauts or from a new selection that would have been needed around 1972/73 to replace CB attrition, and certainly well before Group 8 was selected in 1978.

The combinations for these 'what-if' missions are endless, but reflect the long-term planning that would have been required to assign astronauts for Skylab in the early 1970s when more flights were planned and before three Apollo missions were lost.

Shuttle mission to Skylab

Between 1974 and 1977, Skylab continued to orbit unmanned while the Soviets launched and operated Salyut 3, Salyut 4 and Salyut 5, with mixed success but with five successful resident missions to these stations lasting 14, 30, 63, 49 and 18 days. By the end of 1977, the record duration of Skylab 4 remained – but not for long.

Meanwhile, the American flew the last Apollo as the ten-day Apollo 18 in the joint ASTP docking mission in the summer of 1975, and waited for the first of the fleet of Space Shuttles. *Enterprise* (Orbital Vehicle OV-101) would conduct atmospheric landing tests in the summer of 1977, followed by an orbital test-flight by the second orbiter *Columbia* (OV-102) beginning in 1978. Optimistically, at the beginning of 1977 NASA HQ had directed JSC and MSFC to define schedules to use an early Space Shuttle mission to reboost Skylab to a higher orbit.

A decision on the mission would be made by September 1977. At Houston, engineers looked at the condition of the station in orbit and the status of the shuttle programme, and concluded that it would not be feasible to attempt any type of rendezvous before the fifth flight-test, then set for late 1979, as there had been no studies of docking to an inert station, and the Space Shuttle was still an unflown and unproven vehicle, despite good results from the atmospheric tests then being conducted in California.

It also became apparent that the activity of the Sun was higher than any previous prediction had indicated since the Skylab astronauts left the station in 1974. The solar cycle was climbing to a new solar maximum in 1980–1981, and this would heat up the upper atmosphere, which would increase the density of the layers thought which Skylab was orbiting, and thus create drag that would slow the station, allowing it to 'fall' to Earth at a much faster rate.

By 1 September the decision to proceed with the mission was official, and allowed two years to prepare for the development and production of the hardware and systems necessary to effect a return to Skylab. Of course, all of this depended on the first four Space Shuttle missions flying on time without any serious delays. The whole project to reboost Skylab was a huge gamble and a race against time.

NASA administrator Robert Frosh had informed President Carter's science advisor Frank Press about the slow decline of the space station, warning that the space agency had little experience in understanding the effects of such a large and complex spacecraft performing an uncontrolled re-entry or in analysing surviving debris from space. It would be difficult for the agency to determine how much material might burn up, survive re-entry, or impact the ground.

By November, a letter contract was awarded by MSFC to Martin Marietta, to study the best way in which a mission could be accomplished, and to design and fabricate a system to be carried by the Space Shuttle and attached to the station in order to reboost it to a higher orbit. Wherever possible, existing and qualified hardware was to be used to save time and money, and this also reflected the origins of Skylab itself, in which most of the hardware had evolved from other programmes. The plans were set for a review in March 1978 – less than two years away from the estimated Skylab natural decay date.

It would clearly be a tight time-line, and a race against which NASA was beginning to lose. By the end of the year, the National Oceanic and Atmospheric Administration (NOAA) had indicated that the current solar cycle was the second most intense in records going back 100 years, which indicated that the station would be drawn back to Earth much quicker than NASA was estimating. Despite comments that there were inadequate resources available to predict accurate solar activity even in light of recent Skylab data that was only just beginning to provide information, NASA seemed to have ignored NOAA forecasts in 1976 that the trend for a major solar peak was imminent by the end of the decade.

In the official NASA history of Skylab, *Living and Working in Space* (NASA SP4208,1983) authors David Compton and Charles Benson indicted (p. 363) that there was a certain amount of interest at Marshall for ignoring the data, as the desire to still use the station was a strong influence and that 'it was not in their centre's interest to acknowledge that the space station might fall to Earth before it could be rescued'. They also surmise that an equally credible answer is that because there was a more pressing effort to get the Space Shuttle flying, in which Marshall was a leading field centre. Marshall was responsible for the management of the development of the solid rocket boosters, the main engines, the external tank and the Spacelab science module that was to take over from Skylab in the next decade. There were never any formal plans to use Skylab after 1974, and in effect the station was simply forgotten!

Meanwhile, other aspects of the Shuttle programme progressed. January 1978 saw the announcement of a new class of astronauts, the first since the scientist-astronauts of 1967 and the MOL transfers of 1969. On 16 January thirty-five astronaut candidates were named, including the first six women selected for astronaut training by NASA. A week later the space agency indicated that the Skylab reboost attempt would be assigned to the third and not the fifth Shuttle mission, then set for October 1979, improving the chances of reaching the station before it decayed. After boosting Skylab to a higher orbit, the propulsion system being developed by Martin Marietta would remain in orbit, powered from a solar panel to be retrieved by the fifth Shuttle flight.

On 24 January 1978 the Soviet satellite Cosmos 954 re-entered in the atmosphere above Canada after suffering a systems failure, and burned up in the descent, resulting in parts of its nuclear-fuelled electrical power module that held 100 lbs of high radioactive uranium 235 being spread over a wide area of the Canadian countryside, creating world headlines on the hazards of radioactive space debris and large objects falling from the sky. The Soviet satellite caused no injuries and little damage; but there was a large 'clean-up bill' and as a result, attention turned to Skylab and what consequences a 100-ton vehicle might have on entering the atmosphere. NASA responded, stating that there was no radioactivity material onboard, that there was a Shuttle mission being planned to reboost the station, and that when the Shuttle was launched, the station would not be lower than 170 miles in orbit.

Although appearing confident, it was a cut in expectations of four years in anything previously quoted about the lifetime of the station in orbit. This did not reassure the State Department, which was beginning to receive requests from governments around the world seeking assurance that when Skylab did fall to Earth it would not be on their citizens. In response, NASA set up a Skylab contingency working group to co-ordinate interagency planning that encompassed a range of activities from keeping up to date with the latest condition on Skylab to informing foreign governments on the latest developments and acting as a liaison group for the space agencies and the Departments of State, Defense and Justice.

NASA reviewed the conditions onboard the station, and determined that by controlling its, attitude, MCC could either increase or decrease the drag on the station. In the most favourable situation, it might have been possible to configure the attitude to reduce the drag and increase the orbital lifetime by five months, but not its altitude. Skylab was orbiting over 90% of the world, but studies had shown that risk to human injury from falling debris was very small – although not totally impossible. Indeed, the agency received updated calculations of risk – after presenting the figures publicly – that increased the percentage from first estimates, but was still very small. Even so, a small risk was large enough for the media to start worldwide doomsday story-lines.

Lousma's return

Since returning from Skylab, the nine astronauts had either left the agency or had moved on to support the development of the Space Shuttle. On 17 March the first four two-man Shuttle crews were announced by NASA, to complete the orbital flight test programme, including Fred Haise (Commander) and Jack Lousma (Pilot). A few weeks later Haise indicated that he and Lousma were in training as prime crew for the third Shuttle flight. Since the third mission was tentatively assigned to conduct the Skylab reboost mission, it seemed that Lousma was to make an unexpected rendezvous to the station which he had lived onboard for two months, although he would not have a chance to reboard it. Their back-up crew was former Skylab crewman Vance Brand and rookie Gordon Fullerton.

The plan was to launch the Shuttle on a four-day rendezvous approach to Skylab, and to then send a remotely controlled rocket stage – a 10,000-lb Teleoperator Retrieval Systems – (TRS) that would be deployed by the robot arm from the Shuttle cargo bay. The Shuttle crewman could control the box-like TRS from the aft flight deck, using two TV cameras mounted on the system to view the docking approach to the Skylab MDA. Link-up would be by an Apollo type docking system. With the TRS docked to Skylab, there were two options available. If the decision was made to keep the station in orbit then, two 13.5-minute burns of the 24 100-Newton rocket thrusters on the TRS would lift the complex to a higher orbit. If the decision was to deorbit the station, then a long burn of 27 minutes would point the spacecraft to a selected footprint in the Pacific Ocean well, away from inhabited areas. A final decision would await the latest orbital prediction early in 1979.

After completing the burn, the TRS would then undock from the station, to be left in orbit and then retrieved on the fifth mission for a return to Earth. The system was evolved from a concept from of the mid-1960s, and with rescue capability was expected to be a regular payload feature of Shuttle missions, in which payload boosting, stabilisation, retrieval and delivery missions were planned.

The completed unit was scheduled for delivery to the Cape in August 1979, just a month before the launch. It was a very tight schedule, but training commenced as NASA Administrator Robert Frosh tried to explain how much it was costing to keep

An artist's impression of the intended Space Shuttle mission to boost Skylab into a higher orbit, using TRS hardware.

Skylab in orbit. Even his best estimate indicated only a 50/50 chance that the mission would be ready in time, and that Skylab would still be in orbit, as re-entry prediction was still uncertain. NASA Associate Administrator for Spaceflight, John Yardley, stated that a proposed launch date of 28 September 1979 was only 'probable'.

By June, NASA had spent $750,000 on the project, and expected to extend this by another $3 million by the end of 1978. Funding the system was an additional burden on NASA in Washington. The Senate Appropriations Committee had restored $20.5 million to NASA's Fiscal Year 1979 $4.359 billion funding bill originally cut by an earlier request, so that the development of the TRS could proceed. However there was a catch. NASA had to obtain further approval for funds more than $10 million, and the balance of $10.5 million could only be used for Shuttle funding and nothing else without committee approval. There was still scepticism on Capital Hill that the proposed mission would be impossible due to delays in the Shuttle development programme and a deteriorating Skylab orbit.

Regaining control

As the mission was evolved and the crew trained, the control of the station was accomplished in an attempt to prolong the decay as long as possible. In February 1978, a joint team of engineers from the Marshall and Johnson space centres had travelled to Bermuda to establish contact with the station. The reason that the site at Kindley Naval Air Station, Bermuda, was selected, was that NASA had been updating its tracking network in preparation for the Shuttle, and the station at Bermuda was the only one that could still communicate with the now obsolete UHF equipment onboard Skylab!

Over the next few months the controllers gradually gained control of the station, one system at a time. It was a long, frustrating process when systems that were appraised to be back on line suddenly dropped off line again, and using only one ground station allowed only a brief contact once each orbit. It was determined that Skylab was spinning at about 10 rpm, and it was to be a long process to charge the batteries one at a time, initiate the control moment gyros, the thruster attitude control systems and attitude sensing gyros. Gradually the team extended their achievements, but by June it became apparent that the remote site could not handle the control of the station.

Back at JSC a two-shift flight control team began manning a control centre in ten-hour shifts to handle the task, and soon tracking stations in Madrid, Goldstone and Santiago were brought into the network.

During the summer months, the battle continued to keep Skylab stable and working, but there was a recurrence of the deterioration in the operation of the gyros. The refrigeration systems that cooled the batteries were loosing their cooling fluid, and propellants were reduced to lower limits which added to the problems.

Despite all the efforts to prevent Skylab from falling, each time the low point (perigee) of the orbit was reached, the atmospheric resistance pulled on the vehicle, which resulted in a lower high-point (apogee). Seventeen times a day, seven days a week, Skylab skirted the low point of its orbit, and moved ever closer to Earth. Close monitoring of the orbit was handled through mission control, so that by October there were five teams working three shifts a day. One member of this enlarged team was future astronaut Bonnie Dunbar, who would be selected to the astronaut programme in 1980.

In addition to the four veteran astronauts assigned to the reboost flight, some of the new astronaut candidates received their first CB technical assignments in supporting efforts to evaluate the Shuttle–Skylab reboost mission in simulators and the Shuttle Avionics and Integration Laboratory (SAIL) and Flight Simulation Laboratories (FSL) at JSC. Those who worked on the reboost concept included John Fabian, Robert Gibson, Terry Hart and Brewster Shaw.

What had started as a project with little confidence had grown into a challenging effort that was certainly difficult, but also very rewarding, for the flight controllers. The task was help by the discovery that Skylabs systems were, in effect, better than expected. With hours spent testing and rerouting systems to compensate for restrictions in batteries and the attitude control fuel, and by not using the thrusters but by refining in the attitude in which Skylab orbited, as despite all expectations the gyros continued to work until the very end.

Cancellation of the reboost mission

As feared, it was the Shuttle itself that was to deliver the final blow to all hopes of rescuing Skylab. The director of JSC, Chris Kraft had expressed his personal opinion that the effort to save Skylab was a waste of money and that the effort that was better directed to other efforts. Despite the progress on the TRS, there were several problems with getting the Shuttle ready to fly its first mission on time let alone its third mission.

Towards the end of 1978 it seemed evident that a first launch was not expected before the third quarter of 1979, which slipped the third mission to a tentative launch date of 9 November 1979, if there were no other surprises in store. In December 1978, duel test failures on two different Space Shuttle Main Engines again delayed an already troubled development programme further. In turn this seemed to indicate that no Shuttle would launch before 1980 and the earliest that the TRS could fly was April 1980. With NORAD predictions of a Skylab re-entry in July or August 1979, it became clear that the delays in the development of Shuttle hardware had placed the rescue of Skylab beyond reach.

On 19 December 1978 NASA announced that efforts to mount a rescue mission to reboost Skylab would be abandoned, as there was little chances of success. Production of the TRS would be terminated, and instead its funds would be directed to support Shuttle development to help put the first vehicle in orbit as soon as possible.

In the summer, during the success of regaining control of Skylab, Marshall had resurrected interest in Skylab by issuing in a press release that suggested that it offered large living quarters and crew accommodation that could be 'a welcome addition to Shuttle and Spacelab missions involving long duration'. It was surmised that useful experiments could be performed with Skylab – instruments supplemented by new experiments that could be realised 'with [the Shuttle] orbiter and Spacelab docked with Skylab'.

Speculation continued about the release which stating that the combined hardware could perform a base platform for larger space structures that could be termed 'public service facilities' Unfortunately, by the end of the year it was a dream not to be realised. Like it or not, Skylab was coming down in 1979 – but exactly where? (This saga was relived with Mir, 22 years later.)

The final months

In June 1978, Robert Frosh informed President Carter that NASA believed everything was under control after giving a great deal of thought to the situation, and that if everything went to plan, the Shuttle could deliver the TRS to reboost the station or initiate a controlled re-entry. It was the best decision available, and posed no unacceptable risk to the Shuttle after the first flight tests.

The President's Science Adviser, Frank Press, still had reservations that the pressure of time should not become the driving force that controlled the Shuttle development tests first. By November it was a different story. Frosh reported that despite NASA's best efforts the plan to the rescue Skylab, it would not work in time, and all efforts should be abandoned. NASA had evaluated an option to launch TRS on an expendable launch vehicle instead, but even this offered less than a 1% chance of success in boosting the station in time and a less than one in ten chance of performing a successful controlled deorbit The other alternative was to let Skylab re-entry at random – and hope for the best!.

Present Carter was offered two clear options: firstly, terminate TRS, relocate $30 million in the troubled Shuttle programme, and let Skylab fall out of the sky; or secondly, spend up to $30 million on continued TRS development to launch on an unmanned launch vehicle that offered probably less than a one in ten chance of success. It was also proposed to President Carter that in taking option 2, the TDS was a useful vehicle that could have other applications, and in the eyes of America and the world, the US Government could be seen as taking responsibility for the re-entry of Skylab to prevent as much as possible property damage or personal injury – and if such attempts failed then it would at least be seen to have been tried! At least in international eyes America would have attempted to do more with Skylab re-entry than had the Soviets with Cosmos 954.

Advisors suggested the second option, but Carter selected the first option, as he believed that there had been enough taxpayers money spent on the effort to save the station, and he ordered NASA to devise an Action Plan that would cover the imminent entry and the resultant clean-up. In February 1979, NASA's latest estimates pointed to re-entry as early as the middle of 1979, and on a memo that informed Carter, he wrote 'the sooner the better,' and underlined the predicted date.

For the first six weeks of 1979, NASA controllers maintained Skylab in a low-drag attitude, while the politics and protocol informing the management of the re-entry was determined. There would not be many choices open, as the station's orbit was not decaying rapidly, and in order to obtain the latest information there daily contact with NORAD, which established a radar tracking network on the station from where the information would be relayed to NASA field centres and then via a central co-ordination centre in Washington to the Office of Space Transportation Systems (successor to the Office of Manned Spaceflight), Associate Administrator, John Yardley, to direct the re-entry operation.

With the certainty that Skylab was on it way down, the world's media began running weird and wonderful doomsday stories of the pending disaster and debris falling from the sky. This author fondly remembers the BBC TV news reports of opportunists trying to sell construction worker hard hats to protect worried citizens, and how almost every country in the world was certain that Skylab was heading its way.

During the final months there little that could be done except to monitor the orbital parameters, update predictions and wait. In January 1979, Skylab had been placed in attitude which allowed its solar arrays to face the Sun. While keeping the batteries charged, allowing the controllers in Houston to maintain some control as long as possible, it increased the drag, but the controllers had realised that by changing the attitude of the station they could control the point of entry and avoid highly populated areas. By letting Skylab fly with the MDA forward and perpendicular to the ground track, the flight could be extended by several hours. Turning it sideways increased drag, which shortened the flight by several hours. This was termed 'drag modulation, and became a new challenge for the controllers so that by late spring the predicted re-entry of Skylab could be increased or decreased by one or more revolutions.

Between NORADs estimates and Houston controllers, Skylab was twisted and turned, so that by June it was placed in the high drag mode that was also the most stable so that the station entry could be more accurately predicted. Meanwhile, studies by the Batytelle Institute indicated that several items would probably survive entry – amounting to 25 tons, that would impact in around 500 or more pieces.

There was the 4,950-lb airlock shroud and the 39,600-lb lead-lined film vault, which was expected to fall in one pieces at 393.6 fps! Largest of all was the 14,960-lb bulkhead, and six 2,650-lb oxygen tanks.

The impact footprint predicted for between July 7 and 25 July was estimated to be 4,000 miles long and 100 miles wide, and though NASA press releases assured the public that they stood a greater risk from being hit by a meteorite, it did not prevent the increase in sales of Skylab T-shirts and protection hats with bull's-eye targets. There may not have been a panic mood, but there was certainly an incredible 'silly mood' growing. In Congress, Frosh was asked why Skylab had been launched with no means of controlling re-entry. It was too expensive, they were told.

Each orbit had been analysed by NASA, and calculated population figures indicated high- and low-risk orbits for the estimated re-entry. Obviously, the lowest risk was over smaller land mass and more ocean, but this still took Skylab over Canada, which seeked assurances from America that the lowest-risk orbit would not pass over its territory, as it might see a repeat of the Cosmos 954 incident.

By 4 July, entry was predicted for 10–11 July. Controllers had decided to set an end-over-end tumbling motion when the 87-mile altitude was passed by the station. This would allow a more accurate prediction of re-entry. On 10 July that re-entry was predicted for the next day, during an orbit that crossed southern Canada, the east coast of the United States, and then over a long stretch of open ocean before flying over Australia and then the Pacific Ocean. There were indications that the footprint would overlap the US if the end-over-end tumbling was initiated as planned. It was decided to tumble the station sooner, to head it further downrange to about 800 miles south-east of Cape Town, South Africa, outside of shipping lanes, and half way between the US and Australia.

NASA relied totally on the USAF NORAD tracking data that operated at a factor of 10 (10 days allows 1-day notice, 10 hours equate to 1 hour notice and so on.) Updated forecasts were being received for each orbit as the entry time approached. The tumble, initiated as planned, moved the foot-print off Canada and the US, but then NORAD came back with the latest figures that indicated that the station was not breaking up and therefore shifted its re-entry further east, and then latest predications indicated a landing footprint possibly over Australia! It was now too late to change the inevitable.

Skylab's rendezvous with Australia

In the Houston Control Centre, Chris Kraft was monitoring the final minutes of Skylab, received the information that the station would not, as predicted, fall south of Africa, but instead was heading for a probable impact in Australia. Kraft suddenly realised the implications, and shouted for more data.

NORAD predicted impact at 00.37 EDT on 12 July, and shortly afterwards the first report was received in Houston that an airline pilot had reported that something had passed him at 42,000 feet, and looked like an aircraft with his locked brakes, as sparklers were coming out of it 140 miles out from Perth Airport, Skylab was coming back to Earth. The controllers waited, and waited. A small community reported small fragments, and each of the controllers breathed a sign of relief that of all the land masses to hit, if there had to be one, then western Australia was probably the best, as there were fewer people living there. After five hours, and no reports of injury or death, the mood in the control centre relaxed a little.

There had been spectacular visual effects – whizzing sounds and roof impacts – but by a stroke of luck and good fortune, what was left of Skylab was back on Earth without hurting anyone or causing serious damage. Most of it had burned up, hit the ocean or fell in desolate areas. Charlie Harlan, Flight Director in charge of the team in Houston, stated: 'We assume that Skylab is on the planet Earth, somewhere'.

The *San Francisco Examiner* had offered a reward of $10,000 for the first authenticated piece of Skylab to arrive in its office within 48 hours of the re-entry. And on 13 July a 17-year-old Australian beer-truck driver, Stanley Thornton Jr, arrived at the paper's San Francisco office from the small coastal town of Esperance, Australia. He had found 24 charred items in his back yard, put them in a bag, and caught the next plane to California. When he arrived with his parents and girl-friend, courtesy of a Philadelphia businessman, he did not have a passport – only a razor. The items were identified as insulation from Skylab, and he won his prize.

Post-entry data analysis indicated that Skylab re-entered during orbit 34,981, and began breaking up until it was ten miles above the Earth. Its debris spread over a 40 × 2,400-mile footprint. In the ensuing weeks the largest item found was a 180-lb piece of aluminium that was thought to be the door of one of the film vaults, two

Skylab B – the second flight hardware that could have been launched into orbit in 1976, but instead was cut up and installed in the National Air and Space Museum in Washington DC, and configured as the Skylab 1 OWS. (Courtesy Rex Hall.)

oxygen tanks, and a couple of titanium tanks containing nitrogen, which were found 275 miles east of Perth, near the town of Rawlina. The spread of debris extended to about 500 miles north-west of the town.

The pieces were returned to NASA, and were examined before being mounted for museum display. Also relegated to museums and public viewing sites was the Skylab B OWS, at the National Air and Space Museum, the Skylab trainer at JSC, the three Saturn Vs that should have carried astronauts to the Moon, or other Skylabs. Just over six years after leaving Earth, Skylab – or what was left of it – had returned home. It was over.

THE LEGACY

In October 2000, the first resident crew of one American astronaut and two Russian cosmonauts left the Baikonur cosmodrome for a four month mission on the International Space Station. It had been more than 25 years since the last Skylab crew had come home, and five Shuttles had completed almost 100 missions but none had lasted more then 18 days. During the same time period, cosmonauts had occupied six different space stations, and fifteen of them had each accumulated more than a year in orbit.

Salyut, 1974–1986

During the period 1974–1986, while America was trying to find the best way to stop Skylab falling back to Earth and get the Shuttle off the planet, the Soviets made steady progress with their space station Salyut. After the early tragedy and setbacks of 1971–1973 the first successful Salyut mission was the 1974 fourteen-day Soyuz 14 mission to the military orientated Salyut 3. In 1975, two cosmonaut crews occupied the Salyut 4 space station for 30 and 63 days, and in 1976 Salyut 5, the final military orientated Soviet station, played hosts to a crew for 49 days; and then, in 1977, to a second crew for 18 days.

These stations were much smaller than Skylab, and did not have the range of scientific investigations or capacity for resupply, but nevertheless afforded the Soviets considerable experience in handling orbiting stations. This experience was built upon between 1977 and 1986 with the next-generation stations Salyut 6 and Salyut 7. These two much improved Salyut's featured two docking ports, supported EVAs, and incorporated facilities to receive unmanned re-supply and refuelling ferry craft for prolonging the mission durations of the cosmonauts – a capability that was never available to Skylab.

Improved versions of the Soyuz spacecraft, designated Soyuz T (for transport) ferried rotating resident crews to the station for missions of up to several month in duration, while other visiting crews arrived, often with guest cosmonauts from the Eastern block countries, Cuba, Vietnam, Mongolia, France, and India, for a week's visit, and to exchange the ageing Soyuz with a fresh vehicle.

One of the advantages of the Salyut 6 and Salyut 7 programmes was the series of Progress unmanned resupply craft, which were based on the Soyuz design, and were used to ferry cargo and fuel to the station and act as a trash can for disposing of unwanted rubbish and equipment in eventual burn up in the atmosphere at the end of its mission.

The lessons that the Soviets had learned very early in their station operations were that to secure successful station operations, resupply was essential, visiting crews were helpful, and disposing of rubbish was highly desirable. The welcoming relief of opening the hatches for the Progress to deliver fresh fruits, new equipment, mail and messages from home, and even musical instruments, was a bonus for the crews on long missions. Years later one of the American astronauts who journeyed on Mir for several months commented on the delight of opening a bar of fresh chocolate, and that unpacking Progress was given priority so that they could find the goodies from home as soon as possible.

Once empty, the Progress could then be stuffed with dirty linen, broken equipment, empty food containers, packaging, and all the other unwanted and disposable material that had accumulated since the previous cargo ship had departed. The problem with Progress was the limitation in just how much could be pushed into the compartments. Often there was more rubbish than room. The other disadvantage found on Salyut, and later on Mir, was that sometimes visitors, though welcoming, also disrupted daily life on board, and if they were only short-term visitors, the residents had to clean up after they had left.

Nevertheless, with Salyut 6 the Soviets began to significantly increase their experience over the Americans in long-duration spaceflight, as well as conducting more extensive programmes of research and observations on the Earth, in space and in life sciences that provided a baseline programme of data leading to the next generation stations.

A luxury not available on Skylab was the Progress resupply craft, here seen approaching the ISS in 2000.

In March 1978, Salyut 6 cosmonauts finally surpassed the Skylab 4 endurance record with a mission of 96 days, only to be followed later that year with a crew duration of 140 days over four months. But this was just the beginning. In 1979 a crew remained in orbit for 175 days, and the following year the crew duration was increased to 185 days, and included Valeri Ryumin, who had flown the mission the previous year, becoming the first man to accumulate almost a year in space on three flights. In 1981 the final resident crew on Salyut 6 logged just 75 days – nine days short of the longest American flight.

Salyut 7 missions began in 1982 with a record-breaking 211-day resident crew mission, followed by a 149-day mission in 1983, and a new endurance of 237 days set in 1984. The Salyut 7 station also played host to the first woman to fly in space since Valentina Tereshkova in 1963. Svetlana Savitskaya flew two visiting missions to Salyut 7 – the first in 1982, and the second in 1984, on which she also became the first woman to walk in space. Salyut 7 experienced several operational and hardware setback that threatened the programme, including a serious power failure, but in 1985 a rescue crew restored the station in a 'Soviet Skylab' scenario, that allowed for 112-day mission and the first partial main crew exchange for a further 65 days. In 1986 Salyut 7 was occupied for the final time for 51 days, by the first crew to occupy Mir in the first twin station mission.

A Shuttle view of the Mir complex in 1997, showing a docked Soyuz TM crew vehicle surrounded by the four technology modules and a Progress resupply craft docked at the rear port.

Between 1961 and 1971, the longest Soviet missions had logged 18 days on Soyuz 9; but the crew returned in poor health – in part because they did not complete their exercise programme. Three Soyuz 11 cosmonauts also spent 23 days on Salyut. During the same period, American crews flew up to 14 days, and then Skylab increased this to 28, 59 and 84 days, totalling 171 days by the spring of 1974. Between 1974 and 1977 the Soviets added to their total with five long-duration missions totalling 174 days.

From 1977 to 1986, Salyut 6 and Salyut 7 cosmonauts added almost 1,500 crew days on the long-duration missions alone. As impressive as this was, it would pale to what they were to achieve on Mir in the fifteen years from 1986.

Mir, 1986–2000

The core module for Mir was launched in February 1986. It was based on the Salyut design, and over the years it was expanded to support several research modules and a wide range of investigations and experiments. The station still relied on Soyuz crew transport and Progress resupply craft, and this restricted the amount of logistics that could be carried to and from the station, which meant that the volume of unwanted and deteriorating requirement accumulated onboard Mir over the years, and it became very cluttered, humid and smelly.

The programme survived the delays and cancellation of other modules and the larger Soviet Shuttle Buran, which would have helped in resupplying the station instead of the smaller Progress craft. The break-up of the Soviet Union in 1991, constant shortage of operating budgets, hardware shortages and equipment failures all plagued the Mir programme throughout the 1990s, and despite a desire to keep

Cosmonaut Yuri Onufriyenko inside the cluttered base block of Mir in 1996.

Mir flying until 2000 it became clear that Mir would need to be deorbited, as it was becoming too expensive to keep it flying, due to the creation of the ISS programme and significant Russian participation. By the spring of 2001, the memories of Skylab re-entry resurfaced as controllers evaluated the best way to dispose of Mir in the atmosphere over the Pacific Ocean (as they had done with each of the Salyut's at the end of their operational lives) at the same time of keeping control as long as possible to ensure it landed in the ocean and not on populated land masses. The lessons of Skylab would be once more in the minds of both Russian and American space officials – and those under the flight path, as Mir re-entered on 24 March 2001.

On the plus side, 29 Soviet/Russian expedition crews occupied the station many for over 6 months. Crew rotation was a standard operation, and from February 1987 to April 1989 there were cosmonauts onboard, including one crew who stayed for twelve months. After a five-month period of unmanned activity, on 5 September 1989 a new team of cosmonauts reoccupied the Mir complex, and until 28 August 1999 there were always cosmonauts on Mir. In almost a decade, successive crews lived and worked on Mir, pushing the endurance to an average of six months. For Valeri Polyakov, a stay of 437 days (14 months!) to add to his earlier 240-day visit in 1988/89, made him the undisputed holder of the world spaceflight endurance record of 677 days – almost four times the duration of the three Skylab crews combined! It is a record which is unlikely to be surpassed in the near future.

Shuttle/Spacelab, 1981–1999

Commencing on the twentieth anniversary of the first manned spaceflight, the first Space Shuttle was launched on 12 April 1981. Until 1992, the average Shuttle flight was about a week to ten days in duration, and as such could not compete with the Soviet space station programme. From 1992 a mission extension kit was available for missions up to 17 days in the Extended Duration Orbiter programme, and most of these supported scientific research missions using the European Spacelab pressurised module and unpressurised pallets in the cargo bay. The Spacelab programme was intended to expand the research carried out on Skylab, and was managed by the Marshall Spaceflight Center on shorter but more frequent trips into space.

From 1983 to 1998, sixteen flights of the Shuttle carried the Spacelab module into orbit on a range of missions investigating life and material sciences, and a further twelve missions carried experiments and payloads on unpressurised modules in the cargo bay that included solar physics, astrophysics, astronomy, Earth observations, and engineering experiments as well as supporting the Hubble Space Telescope.

One of the first crew-members to fly a Spacelab mission was Owen Garriott, Science Pilot on the second Skylab, and with him was Bob Parker, Skylab Mission Scientist. Several of the astronauts who had worked on Skylab, later worked and flew on Shuttle and Spacelab missions (see Appendix – astronaut biographies).

The short of Shuttle missions (seventeen days maximum) meant that not all lessons learned from Skylab could be applied to working and living in the orbiter or Spacelab. For example, crews worked in rotating shifts 24 hours a day to maximise the return from each short flight. This added to the complexity of crew scheduling, rest and disturbance, and in addition there was only so much room in the Shuttle;

An artist's impression of the Shuttle/Spacelab configuration.

Ten years after flying on Skylab, Owen Garriott (*left*) is a crew-member of the first Shuttle/Spacelab mission – STS9/Spacelab 1. At left is ESA astronaut Ulf Merbold.

and even 50-odd lockers could only hold so much equipment, and with no trash disposal airlock available, bags of rubbish had to be stored in the airlock or transfer tunnel to the Spacelab module.

The benefit of flying specialist research scientists was once again clearly demonstrated early in the programme during Spacelab 3, when the Drop Dynamic Module failed shortly after it was turned on. Fortunately its designer, Taylor Wang, was on aboard, and literally rewired the experiment and brought it back on line. Bill Thornton found that he learned more on his two Shuttle flights on space adaptation than he had learned after years of ground-based study and post-flight evaluation of other astronauts. Don Lind finally had his flight into space, and was able to conduct some of his own experiments. Both Thornton and Lind recalled that although neither had flown to Skylab, their work in supporting the programme greatly benefited their preparation for flying on Spacelab just over a decade later.

Spacelab was an outstanding success, and complemented the Shuttle well, but it was no space station. Despite plans to add further mission extension kits to increase the orbital duration up to several weeks, using extra fuel supplies and solar arrays for power, the Shuttle/Spacelab combined mission would not exceed eighteen days. What was needed was an American space station programme, and in 1984, after years of planning, one was finally authorised by President Reagan. The successor to Skylab was to be named Freedom.

Freedom, 1984–1992

Freedom was to be a big and impressive space station. It would take several years to assemble, using the Shuttle, and it included about 1,500 hours of EVA to construct it – and that was one of the problems. Freedom was very complex, and despite the

Full-size Freedom habitation modules at JSC in 1991. (Astro Info Service collection.)

Freedom research racks. (Astro Info Service collection.)

An individual crew compartment on Freedom, with a sliding privacy door, a large observation window, and personal IT facilities. The sleeping restraint was on the wall to the left (out of frame). (Astro Info Service collection.)

participation of European, Canadian and Japanese partners, it was extremely expensive.

For almost a decade Freedom became a programme of design studies, conceptual models, redesign studies, budget cutbacks, more design studies, compromise and delays; and all the time the costs escalated and the launch date slipped further into the future. While the Shuttle and Spacelab flew, and while cosmonauts lived on Salyut and Mir, Freedom remained on drawing boards. In the late 1980s a full-size mock-up was created at JSC in Houston, and featured a network of proposed habitation modules (as illustrated in the accompanying photographs).

Several former Skylab astronauts worked as consultants on the space station programme, offering their assistance to primary contractors and NASA, but it was a new space programme, and not all of their suggestions, based on their experience, were accepted or even agreed upon.

One of the basic problems on Skylab was in maintaining position in one place for a long time. Experiences on Spacelab and on the Shuttle indicated that simple foot-loops were sufficient for this, and these were incorporated into Freedom. The Skylab astronauts pointed out that while these were satisfactory for short-duration restraint in confined locations, you would not want to curl your toes up for hours on end while monitoring a materials science process or manning a telescope. Far better were the triangular foot cleats that allowed you to relax the leg muscles and work more comfortably. Freedom officials disagreed, and kept the foot-loops.

A further problem was in keeping track of items on Skylab. A bar code system would be ideal, the Skylab astronauts suggested, because the bigger the station became, the larger the logistics problem would become. If this was adapted for Freedom, everything could be stored in purpose-built lockers as an when required, and the manifest would be constantly updated. 'Sure', answered some of the Skylab veterans.

Room on Freedom for up to eight crew-members. The 'tables' could tilt, and the pantry was overhead. (Astro Info Service collection.)

Incorporated on Freedom, however, were individual crew habitation cubicles with direct contact with home, an entertainment centre, a shower, a improved toilet, and a central crew galley with state-of-the-art facilities. Space was at a premium, however, and to help alleviate the build-up of waste, recycling was also a feature, with Shuttle visits being available to return a significant amount of waste back to Earth after delivering new cargo. Unfortunately, the Shuttle was not unmanned and not indispensable, and so atmospheric destruction was not an advisable option for Freedom's waste disposal!

By 1992, Freedom's costs had escalated to 'astronomical' proportions, forcing once more a redesign, of the redesign and a major redirection towards what has developed into the International Space Station. The redirect of Freedom to ISS also brought in the Russians, and the chance for the Americans to gain much-needed experience in long-duration spaceflight on Mir between 1995–1998.

Shuttle–Mir, 1992–1998

Seven American astronauts live onboard Mir between 1995 and 1998, and from March 1996 an American was constantly in orbit until June 1998. In addition, ten missions (nine docking) of the Shuttle to Mir were the first American docking activities since ASTP in 1975 – a valuable experience before attempting the ISS construction from 1998.

The seven astronauts who exceeded the Skylab 4 endurance record held for 21 years. (*Left–right*) Thagard (115 days), Thomas (140 days), Blaha (128 days), Lucid (188 days), Linenger (132 days), Foale (144 days) and Wolf (127 days).

The opportunity to place astronauts on Mir gave NASA more than just long-duration flight experience. For all of the astronauts who were to live onboard Mir, the first hurdles to overcome were the language barrier, and the separation from home for many months to train in Russia before even flying to the station.

The opportunity to fly to Mir was not received enthusiastically by many of the astronauts as assignment to Mir resident crew training was as out of step with ISS construction scheduling as the AAP was to Apollo crews. It almost appeared that NASA felt obliged to resupply crews and logistics to Mir in response to Russian offers to provide the first elements for the ISS to begin construction.

Comments from some of the astronauts assigned to Russia suggest a lack of NASA support which they had expected for such a demanding assignment. The cultural shock began almost immediately, as they lived in the cosmonaut training centre, trained under Russian instructors, spoke only Russian, and had little contact with home. Even some of the cosmonauts with whom they trained were exchanged and reassigned, so that the crew they flew in space with was not necessarily the same they trained with.

Onboard Mir, several astronauts felt even more isolated, while others – notably Mike Foale – accepted the situation, and actually took positive actions to ensure that he was accepted not only by the trainers and cosmonauts, but also by the flight controllers.

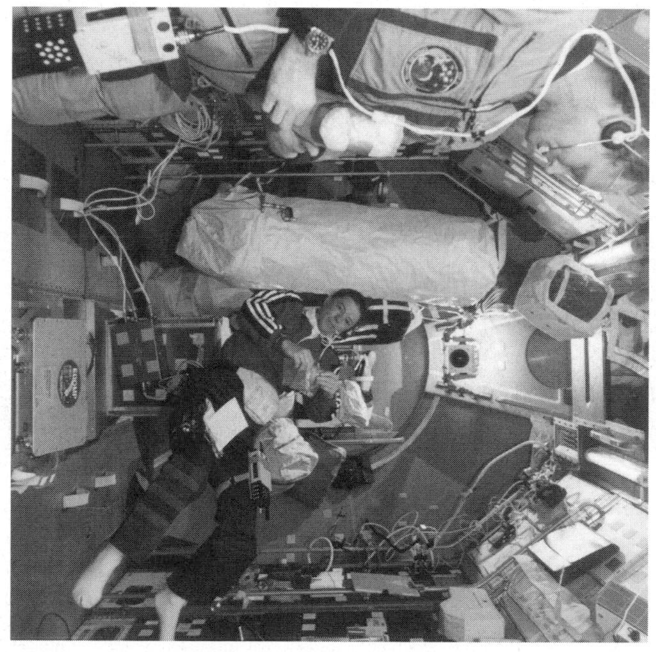

After spending four months on Mir, Norman Thagard is shown in the Spacelab module undergoing medical tests and data collection before return to Earth. With him is Bonnie Dunbar, who sixteen years earlier was part of the Skylab re-entry control team.

Showing the importance of keeping track of items on aboard the Shuttle before transfer to Mir, loadmaster Tom Akers reviews the inventory by hand. A bar-code automated system is under development for the ISS.

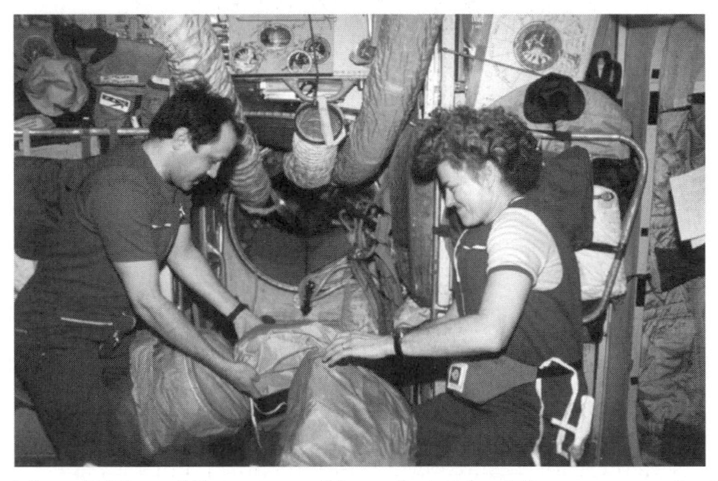

As on Skylab, rubbish on Mir was a problem – but at least there were regular Progress supply craft to collect it. Here, cosmonaut Usachov and astronaut Lucid prepare to move the supplies off-loaded from Progress to another part of Mir.

Norman Thagard was the only astronaut launched to Mir on a Soyuz, although he returned on the Shuttle. All the others launched and returned on the Shuttle, and one of the emotional moments occurred when watching the American craft depart, leaving them behind, and counting down the days until they were picked up again. The events on Mir are documented and referenced in the Bibliography, but some of the things learned on Mir, that had not been learned or even remembered from Skylab, were adequate communications and support from Mission Control in Houston. Despite control of the mission being from Moscow, the astronauts at times

Andy Thomas uses a Russian treadmill on the Kristall module onboard Mir. An improvement over 'Thornton's Revenge', the principle is nevertheless the same – to place loads on the leg muscles to maintain condition.

felt isolated and at times depressed in the lack of recognition from NASA or the US media in what they were doing. After the fire and collision onboard Mir in 1997, the last two missions were almost ignored, in the West, as nothing went wrong!

Another important lesson that was learned from Shuttle/Mir was in pacing crew time and in working effectively as a team. At the height of the crisis on Mir in 1997, the blame was all too soon pointed at the cosmonauts, and this tended to create even more of a 'them and us' situation. In some reports, the Russians stated this was an intended ploy to make the crews work together as a team, and to take out the frustration on the ground. If this was the case, it was certainly not the method used on Skylab.

Lessons learned on Mir should be adopted on the ISS: scheduling crew time, communications, and the amount of housekeeping, repair and maintenance and scheduling new and unplanned activities in the crew day, as well as trying to balance research periods with housekeeping, meal-times, exercise and sleep. Mir crews experienced these difficulties over a decade, as did the seven American astronauts; and they were exactly the same problems faced by the Skylab astronauts.

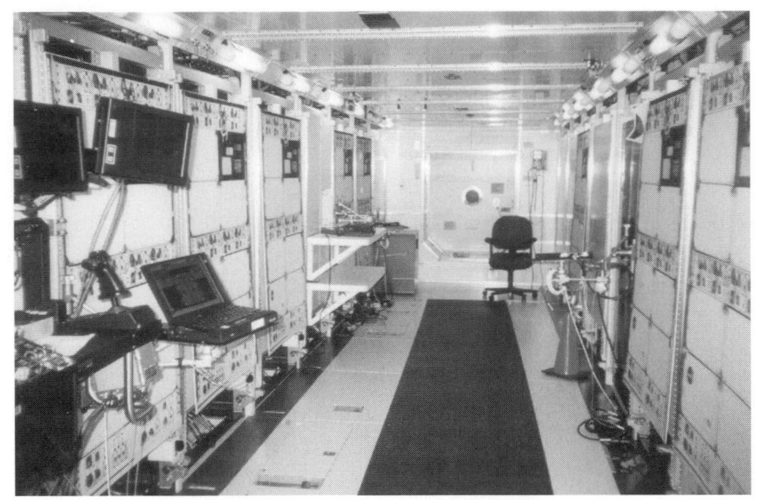

US laboratory Destiny training simulator configuration at JSC. (Astro Info Service collection.)

Towards the future: the International Space Station

When ISS operations began it was also the time of the 25 anniversary celebrations of Skylab. In an interview for NASA's oral history programme, Skylab 2 astronaut Joe Kerwin reflected on any lessons that could be learned from Skylab: 'Well, it's a lot that is now deeply ingrained that it would be almost hard to point them out; but the habitability, the diet and exercise, the workday structure, a lot of these things . . . one thing I find good about the long gap is that it's kept me usefully employed for 25 years [laughter]. People still come back and ask for data.'

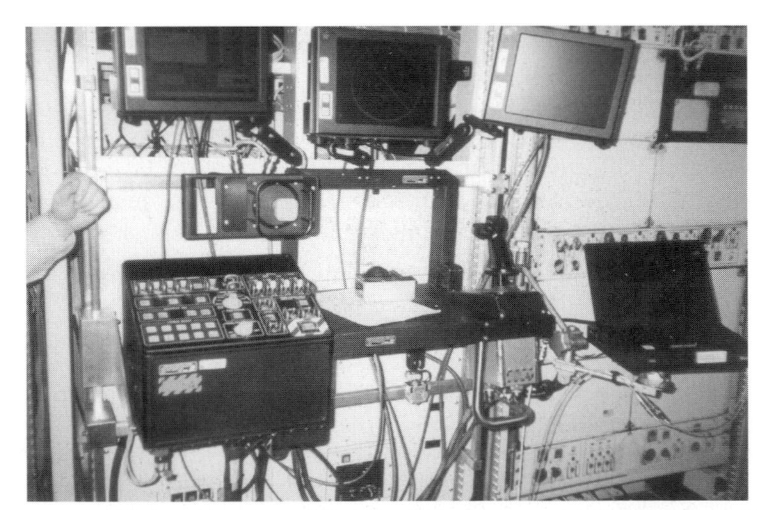

The command and control centre of Destiny, linked to portable computers and laptops in the training mock-up at JSC in June 2000. (Astro Info Service collection.)

This was certainly true when the plans for astronauts to live on Mir were first proposed. The archive files of Skylab had been at Rice University for many years, and they were often re-examined to discover just how things were done, how things were managed onboard, what went wrong, and what was found to be a success. As Kerwin recalled: 'We had a lot of things we did wrong that should not be done that way again. So I figure once the ISS has been in operation for about two years, Skylab will be as interesting as Columbus' voyage [laughter]. All those lessons will have been learned over again, and new ones will be being learned. But now we're still popular, so it's kind of fun'.

And maybe they are learning from Skylab, there appears to be an item tracking system for ISS to maintain up-to-date records of consumables and locations on the station as it grows. The only problem with this is that the astronaut tends to become a storekeeper. The Skylab astronauts felt that this was not a task for the crew.

Exercise is recognised by the Russians as important, and conditioning machines are already on the station. Progress freighters are to be a frequent visitor to deliver fuel, cargo and take away the rubbish, if Russia can afford to supply them. With the arrival of the US laboratory Destiny in February 2001, the station is expanding, allowing storage issues to be addressed further, although by just looking at photographs from the first mission STS-88 in 1998, to more recent flights, shows just how cargos logistics tend to clog up the passageways! At the end of construction, however, the living quarters for NASA astronauts should have improved with the docking of the US Habitation Module. In the 2002 budget, the station was expected to be $4 billion over budget, and President Bush settled into the White House the verbal support and encouragement for pioneering a new frontier and heading to Mars seemed as strong as ever; but just as before, the cheque remains unsigned.

In the ISS Press Information Book, the US Habitation Module was described as the last element to be added to the ISS, replacing the living quarters in the Russian

STS-88 begins space DIY. Members of the first mission to dock to the ISS are inside a pristine Zarya in December 1988.

Twenty-two months later, and Zarya is becoming crowded as an STS-92 astronaut is just visible in the 'standing room only' Zarya module. As with Skylab, space is at a premium, but the volume of the upper deck of Skylab was a luxury that was not incorporated into the ISS.

Following Skylab, the first ISS resident crew Cosmonaut Gidzenko (*left*) astronaut Shepherd (*centre*) and cosmonaut Krikalev (*right*) open a new era, and celebrate with fresh fruit delivered by Progress and the Shuttle. Note the tool cupboard and restraint straps behind the crew.

modules. A sleeping area would allow the crew to rest away from the main part of the station. The galley would have an oven/microwave combo and a non-alcoholic drink dispenser, and there will be a food freezer and refrigerators, offering menus 'astronauts could only dream about': broiled lobster tails, stir-fired chicken, fruits, salads, ice cream and on demand. A wardroom for two, three or four would serve as a crew meeting place, a conference facility or a viewing area. There would be a toilet, a sink and a shower, and the crew size would increase from three to four resident crew-members.

That was the plan, but it appears that it will be either seriously scaled down or cancelled altogether, and that only a permanent crew of three, and not seven, will be able to live on the station. This, of course, affects the international crewing of the station and the science programmes. The Crew Return Vehicles are also under threat, and Soyuz and Shuttle must be relied upon for the foreseeable future.

So, although lessons have been learned from Skylab and Mir, as Kerwin suggested, new lessons will be learned as the ISS develops.

Skylab: a place in history

The Skylab missions were developed on a small budget, from developed hardware and the skills of a small group of talented people. They presented a wealth of information that perhaps represents the best set of medical data on nine people in space (with a control data of the three SMEAT astronauts), studies of the Sun, stellar astronomy and Earth, that rewrote the text books and still provides basic resource science data. It presented the first opportunities to really understand what it was like to live and work in space, and how to interact with mission control and ground support over many months. It provided opportunities for science and discovery by the scientists and engineers, challenges and hurdles to overcome by the families, and above all the chance for the astronauts to experience spaceflight. It also offered them an experience of a lifetime, and an opportunity to have great fun!

In their words

In closing this account of a pioneering programme, I turn to the astronauts themselves.

When John Glenn was assigned a second flight at the age of 77, the medical objectives of the flight were often promoted. There were also suggestions, at the time, for reflying some of the Skylab astronauts, who offered a much better comparative set of data based on their first missions. Without doubt, the flight of Glenn in 1998 was an enormous public relations programme, but which also offered the opportunity to secure important data on an older test subject. It would be advantageous to assign several Skylab astronauts to Shuttle flights to continue this trend, and if this was the case, then many of them could form a queue in Washington to obtain the chance to once more experience the thrill of spaceflight.

The benefit of Skylab was best summed up by Bill Pogue, that who thought the National Geographic said it best, in calling Skylab 'an ugly duckling of space, but the stately swan of science', and if given the chance of a second Skylab flight he thought this would be an attractive proposition, but this time ensuring that the work pace was more relaxed, with the flight activities not so hectic. In fact, many of the Skylab astronauts said that they would fly a second Skylab mission, but the long wait and the cancellation of Skylab B, meant that it would be many years before they flew again.

Don Lind, who almost made it to Skylab, stated that it was the first real demonstration of the usefulness of a scientific research laboratory in space, and that the demonstration was eminently successful. Bill Thornton added that without the SMEAT ground test, there would not have been a Skylab, as it was during SMEAT that serious design errors were discovered that would have posed major problems if they were experienced during the first time in space, and that comparative flight data from different individuals provided clear medical results that allowed counteracting measures to be provided in time for the third crew to show significant improvements in their condition.

Paul Weitz, who on his second flight on STS-6 a decade after Skylab, was struck by the deterioration in the environment since he first flew, said: 'Skylab demonstrated that humans can live and function for long periods of time in a microgravity environment. It also demonstrated the ability of the crew and the ground to work together to enhance the quality of work done or data gathered. I would like to think that it helped make people in advanced countries aware that addressing our environment problems requires consideration on global, not just local scale.'

Bob Crippen, who participated in SMEAT ground control but not flight operations on Skylab, and became a four time Shuttle flyer and Director of KSC, recalled: 'It has been almost thirty years since we ran Skylab. It was a programme

2001: a new space odyssey, as the International Space Station sails into the future.

that had many challenges, which we learned to overcome. It gave us our initial data on the effects of long-term spaceflight on the human body, and provided some great astronomy data. It was our initial space station. However, the long time between that programme and the current ISS resulted in the loss of a lot of corporate memory. Consequently we've had to learn many of the same lessons again. From my perspective it was a great success and was rewarding to me personally for having had an opportunity to participate.'

Lousma likened Skylab to an application of experience gained from the earlier programme. It offered the opportunity to evolve from simply operating in space to applying the skills of living there.

This book opened with the comments of Pete Conrad, the Commander of the first crew. Al Bean, Commander of the second crew, told me that he would have loved to contribute to the memory of Skylab by painting scenes that he remembered, but was so busy with Apollo commissions that he felt that he would unfortunately be unable to do so.

The Commander of the third crew, Jerry Carr, has recollected his conversation with the man who evolved the Skylab concept, Wernher von Braun: 'I've thought a lot about what I might say in hindsight regarding my Skylab flight. It turns out that I could have been the sixteenth man to step on the Moon (Apollo 19, if I calculate correctly) as opposed to being the commander of the longest flight in history at that time. Looking back on Skylab, I feel very fortunate that I had that opportunity. We did really good science, and established that humans can live in weightlessness for a very long time without enduring physical effect. Wernher von Braun told me once that if he had a choice he would have preferred to fly Skylab 4 to any other mission that had been flown. He felt that human destiny in space depended upon the ability to adapt physiologically and psychology to that environment, and he would have liked to have played an intimate role in determining that ability.'

The words of Wernher von Braun, published in *Collier's* in March 1952, now seem particularly poignant at the beginning of ISS operations: 'Scientists and engineers know how to build a station in space ... The job would take ten years [and] not only preserve the peace, but we can take a long step towards uniting mankind. From this platform a trip to the Moon would be just a step.'

Skylab was one of von Braun's steps in returning to the Moon and beyond.

Appendix

THE ASTRONAUTS

The careers of the 22 astronauts of Skylab (from Conrad's selection to NASA in September 1962, to the retirement of Musgrave from the CB in September 1997) span 35 years of American manned spaceflight from the end of the Mercury project to the threshold of the International Space Station. They represent a significant contribution to the human exploration of space.

The following are biographical sketches of every astronaut assigned to Skylab in either a prime, back-up, support or CapCom position. Rank is correct at time of their Skylab flight. Assignment and details of their career is current to December 2000.

Bean, Alan Laverne (Captain USN), *Commander Skylab 3,* was born on 15 March 1932 in Wheeler, Texas. He grew up in Fort Worth, graduating from high school there in 1950. Following a year as an electronics technician with the Naval Reserve in Dallas, he attended the University of Texas at Austin, graduating with a Bachelor of Science (BS) degree in aeronautical engineering in 1955, and receiving his naval commission. After completing flight training in 1956 he was designated a naval aviator and subsequently completed a four-year assignment with Attack Squadron 44 at Naval Air Station (NAS) Jacksonville, Florida. A 1960 graduate of the USN Test Pilot School, Patuxent River, Maryland, he remained as a test pilot at Patuxent River until 1962. He was with Attack Squadron 172 at Cecil Field, Florida, when selected by NASA as one of fourteen astronauts on 14 October 1963 (Group 3). He received technical assignments in spacecraft recovery systems, and during the Gemini programme he served as a CapCom for Gemini 7/6 and 11, and as back-up Command Pilot for Gemini 10. He became Chief of the APP Branch of the Astronaut Office (Code CB) in September 1966, but when astronaut C.C. Williams was killed in a flying accident in October 1967, Bean was reassigned to Pete Conrad's Apollo crew, to replace him as Lunar Module Pilot (LMP). Bean served as back-up

LMP to the Apollo 9 crew before flying as LMP on Apollo 12 to the Moon. On 19 November 1969 he became the fourth man to walk on the Moon. He completed two EVAs on the lunar surface, totalling 6 hours 24 minutes, during a surface stay of 31 hours 31 minutes. Following Skylab 3 he served as the back-up American Apollo Commander for the 1975 Apollo–Soyuz Test Project with the Soviets, and was then assigned to the Shuttle programme in August 1975, before retiring from the USN in October 1975 with the rank of Captain. Bean subsequently served as Acting Chief of the CB, supervising the training of the 1978 and 1980 classes of astronaut candidates. Although widely expected to take the Commander's seat on STS-9 carrying the first Spacelab mission, he resigned from NASA on 26 June 1981 to pursue his hobby of painting. Since leaving NASA, Bean has developed a highly successful career as a space artist, focusing on the Apollo lunar missions, with his work being exhibited around the world and featured in several books. He logged 1,671 hours on two spaceflights, and 10 hours 30 minutes during three EVAs.

Bobko, Karol Joseph (Major USAF), *Pilot on SMEAT,* was born on 23 December 1937 in New York City, graduating in 1955 from Brooklyn Technical High School. He received a BS degree in 1959 as a member of the first graduating class of the USAF Academy. After completing pilot training the next year, he served with the 523rd Tactical Fighter Squadron (TFS) at Cannon AFB, New Mexico, and the 336th TFS at Seymore Johnson AFB, North Carolina. Bobko is a 1965 (Class 65B) graduate of the USAF Test Pilot School, Edwards AFB, California, and was selected to the USAF Manned Orbiting Laboratory Programme on 30 June 1966 (Group 2). After MOL was cancelled in June 1969, he transferred to NASA on 14 August 1969 (Group 7). Before assuming NASA assignments, he obtained a Masters degree in aerospace engineering from the University of Southern California in 1970. His only major Skylab assignment was as a crewman on the 56-day altitude test conducted in 1972. After SMEAT he was assigned to the support crew of the American ASTP mission (1973–1975), and to the Shuttle Approach and Landing Test (ALT) programme (1976–1977), in which he served alternately as a T-38 chase pilot and as CapCom. Bobko completed three flights into space onboard the Shuttle – once in 1983 (Pilot STS-6, *Challenger*), and twice in 1985 (Commander STS 51-D, *Discovery*; and Commander STS 51-J *Atlantis*), before resigning from NASA and retiring from the USAF with the rank of Colonel on 1 January 1989. He joined the Houston Space Systems Division of Booz, Allen and Hamilton as a consultant for the Freedom space station programme, before working for Grumman Corporation on the ISS. He logged 386 hours 04 minutes on his three spaceflights.

Brand, Vance DeVoe (civilian), *back-up Commander Skylab 3 and Skylab 4, and Commander of Skylab Rescue,* was born on 9 May 1931 in Longmont, Colorado, where he graduated from high school in 1949. After graduating from the University of Colorado in 1953 with a BS degree in business, he served four years as a Marine Corps pilot before returning to the University of Colorado. After completing work for a second BS (this time in aeronautical engineering) in 1960, he worked for the Lockheed Aircraft Company initially as a flight test engineer and, after graduating from the USN test pilot school, as a test pilot. He continued flying with the Marine Corps Reserve and Air National Guard from 1957 until 1964, and in the same year received a Masters in business administration from the University of California at Los Angeles. At the time of his selection to NASA, Brand was working as a Lockheed test pilot at the flight test centre at Istres, France. His application for the 1963 astronaut intake was not successful, but he was one of nineteen astronauts selected on 4 April 1966 (Group 5). He trained as an Apollo Command and Service Module (CSM) specialist, and completed thermal vacuum chamber tests of the CSM with Joe Kerwin and Joe Engle in 1968. He was a support crew-member and shift CapCom for Apollo 8 and Apollo 13, served as back-up CMP for Apollo 15, and would have been assigned to fly around the Moon on Apollo 18 as CMP, had the flight not been cancelled. He was reassigned to Skylab, during training for which he was assigned as CMP for the American ASTP crew. Brand made his first spaceflight in July 1975 in the historic joint docking mission with the Soviet Soyuz 19. He also completed three Shuttle missions as Commander, in 1982 (STS-5), 1984 (STS 41-B) and 1990 (STS-35), and trained for two others that were cancelled. He retired from the astronaut office in 1981 to work on the National Aerospace Plane programme at Wright Patterson AFB, Ohio. When that programme was cancelled, Brand transferred to NASA's Dryden Research Center, located at Edwards AFB, California. He logged 746 hours 04 minutes on four spaceflights.

Carr, Gerald ('Jerry') Paul (Lt-Colonel, USMC), *Commander Skylab 4,* was born on 22 August 1932 in Denver, Colorado. He grew up in Santa Anna, California, graduating from high school in 1952. He received a BS degree in mechanical engineering in 1954 from the University of Southern California, and was commissioned in the USMC, completing pilot training in 1955. He was then assigned to Marine All-Weather Fighter Squadron (VMF) 114 until entering the USN post-graduate school, at which he earned a second BS degree in aeronautical engineering in 1961. Carr then achieved an MS degree in aeronautical engineering the following year, from Princeton University. He was assigned to VMF-122 in the US and Far East until he joined Marine Air Control Squadron 3 in 1965 as a test director. Carr became a NASA astronaut on 4 April 1966 (Group 5) and became an LM specialist, working on the support crews (and as a CapCom) for Apollo 8 and Apollo 12 before being assigned to the development of the Apollo Lunar Roving Vehicle in 1970. It was planned that he would train as the original back up LMP for Apollo 16, leading to assignment as LMP for Apollo 19 and a chance of becoming the sixteenth man to walk on the Moon. But budget cuts cancelled the latter flight. After Skylab, Carr was assigned as the head of the CB Shuttle design support group in 1975, working on a number of technical issues including evaluating methods of post-landing crew escape. Carr retired from the USMC, as a Colonel, in September 1975 and from NASA in June 1977, to become Vice-President of Bovey Engineers Inc, based in Houston, Texas. He also worked for Applied Research Inc, and as a consultant on Shuttle payload integration and crew training procedure development before creating the family business Camus, which focuses on the development of zero-g human factors in engineering procedures. In partnership with former Skylab astronaut Bill Pogue, Jack Lousma, and other former astronauts, cosmonauts and engineers from NASA and the space industry, Carr completed a number of space consultant projects. These included the Space Station, space habitability, EVA, lunar and Mars manned exploration. He retired from Camus in 1998 to devote his full attention to assisting his artist wife Pat Musick with her career and highly successful art business in rural Arkansas. Carr logged 2,017 hours 16 minutes on one spaceflight, including 15 hours 51 minutes on three EVAs.

Conrad, Charles ('Pete'), Jr (Captain, USN), *Chief of the Skylab Branch of the Astronaut Office, and Commander Skylab 2,* was born on 2 June 1930 in Philadelphia, Pennsylvania. He grew up in the suburb of Haverford and graduated from the Darrow School in New Lebanon, New York, before attending Princeton University, where he earned a BS degree in aeronautical engineering in 1953. He then entered active duty with the USN, and after completing pilot training, he attended the USN Test Pilot School at Patuxent River. Following graduation in 1957, he remained there as a test pilot, flight instructor, and performance engineer. Prior to selection to NASA, he was assigned to Fighter Squadron 96 at Miramar, California. Conrad was one of nine astronauts selected by NASA on 17 September 1962 (Group 2). He had been short-listed for the Mercury astronaut selection in 1959, but was not selected. One of his early assignments was in the development of the Apollo Lunar Module, but it was in the Gemini programme that Conrad made his first flights into space. He flew two Gemini missions; as Pilot on Gemini 5 in 1965 (which set a new world endurance record of 8 days) and as Commander of Gemini 11 in 1966 (which set a new altitude record of 850 miles). He also completed two back-up Commander crew assignments on Gemini 8 and Apollo 9 before flying as Commander of Apollo 12 in 1969. On 19 November 1969 he became the third man to walk on the Moon, completing two EVAs (with Al Bean), logging 6 hours 24 minutes during a 31 hour 31 minute surface stay. Conrad became Chief of the Skylab branch of the Astronaut Office in the summer of 1970 (replacing Cunningham), and following Skylab 2 he retired from NASA and the USN with the rank of Captain in December 1973. He became Vice-President (operations) and Chief Operating Officer for American Television Communications (ATC) corporation, a Denver-based cable TV company, until March 1976. He then joined McDonnell Douglas Corporation as Vice-President, working in various marketing and business development roles until 1993, when he joined the companies Delta Clipper launch vehicle technology development programme. He had also continued his association with the space programme, as a consultant and test subject for underwater EVA simulations for the ISS during his years as an executive at McDonnell. From 1993, he worked on McDonnell's revolutionary space launch system for three years, including serving as a remote pilot and CapCom for a series of flight tests at the White Sands Missile Test Range in New Mexico. Conrad retired from McDonnell on 31 March 1996 to head a private space launch firm called Universal Space Lines, based in Irvine California. He died, aged 69, from injuries sustained in a motorcycle accident in California on 8 July 1999. Conrad logged 1,179 hours 39 minutes in space in four spaceflights, including 11 hours 33 minutes on four EVAs. A larger-than-life character, Conrad was the epitome of the astronaut image: hard-working and fun-loving, always with a happy grin showing off the wide gap in his front teeth.

Crippen, Robert Laurel ('Crip') (Lt-Commander, USN), *Commander SMEAT, Support and CapCom on all three manned missions,* was born on 11 September 1937 in Beaumont Texas, and is a graduate of Caney High School, Caney, Texas. He attended the University of Austin, Texas, and earned a BS degree in aeronautical engineering in 1960. Entering the USN following graduation, he completed flight training in 1962 and was detailed as a carrier pilot on *USS Independence* in the Pacific. Crip is a 1965 (Class 65A) graduate of the USAF Test Pilot School and remained there as an instructor until selection to the USAF MOL programme on 30 June 1966 (Group 2). Following the cancellation of MOL in June 1969, he was one of seven former MOL astronauts transferred to NASA on 14 August 1969 (Group 7). He received an early assignment in Skylab development, before serving as SMEAT Crew Commander and as a support crew-member and shift CapCom for all three manned missions. In a dual assignment, Crippen also trained as a support crew-member for the American ASTP astronaut crew, and following Skylab he continued serving a support crew-member and as a CapCom for ASTP. Following the ASTP assignment, he was assigned to the Orbital Flight Test training group for early Shuttle flights. In 1981, Crippen was Pilot on the very first Shuttle flight, STS-1 (*Columbia*), with Commander John Young. He subsequently flew on *Challenger* as Commander for three other missions, once in 1983 (STS-7) and twice in 1984 (STS 41-C and 41-G). He also trained to command the first launch from Vandenberg AFB California (STS 62-A, *Discovery*), before it was cancelled in the wake of the 1986 *Challenger* accident. In 1987 he left the astronaut office to became Deputy Director for Shuttle Operations at the Kennedy Space Center (KSC), Florida until October 1989. He then moved to NASA Headquarters in Washington DC as Associate Administrator for Shuttle operations until January 1992, when he returned to KSC as the fifth Director of the Center. Crippen retired from the USN with the rank of Captain, and in 1995 left NASA to join the Lockheed Martin Corporation in Orlando, Florida, as Vice-President of simulations and training systems and subsequently as Vice President for automation systems. Since October 1996, Crippen has held the position of President for space operations with the Aerospace Group part of the Thiokol Corporation. Crippen logged 565 hours 48 minutes in four spaceflights.

Garriott, Owen Kay (Dr), *Science Pilot Skylab 3,* was born on 22 November 1930 in Enid, Oklahoma, and graduated from the University of Oklahoma in 1953 with a BS degree in electrical engineering. He completed a tour in the US Navy (1953–1956) as an electronics officer on the destroyers *USS Cowell* and *USS Allen M. Sumner*. He attended Stanford University, receiving his MS (1957) and his PhD (1960), both in electrical engineering. For the next five years he was a teaching associate professor in the Department of Electrical Engineering at Stanford. Garriott was selected by NASA on 28 June 1965 (Group 4) as one of six scientist-astronauts, and completed a 53-week jet pilot course at Williams AFB, Arizona, before receiving his first technical assignment in 1966 in design and development issues of the Apollo Applications Programme (AAP). In 1968 he served, briefly, as Chief CB representative for the AAP, and the following year was a CapCom during Apollo 11. Following the Skylab 3 mission he served as Deputy Director and later Acting Director for science and applications at Johnson Space Center (JSC) Houston. After a year's sabbatical in 1976 to teach at Stanford, Garriott returned to NASA and in 1978 was assigned to Spacelab 1 to be launched by the Shuttle. After training for more than five years, the mission was finally flown in 1983 (STS-9, *Columbia*) and following the flight, Garriott served as programme scientist in the Space Station Freedom Office at JSC while training for two Earth Observation Missions (EOM-1 and EOM-2). These were ultimately cancelled due to manifest problems and the *Challenger* accident. Garriott resigned from NASA in June 1986, before the missions had been remanifested under the Atlas series of flights, to become Vice-President of space programmes for Teledyne Brown Engineering, Huntsville, Alabama. One of his projects evaluated design studies that proposed to extend the normal seven-day duration of Shuttle missions up to eleven weeks. In 1995 he resigned from Teledyne Brown and became co-founder and President of Immunotherapeutics, also based in Huntsville. On 5 September 2000 he assumed the temporary role of Interim Director of the Huntsville Space Science and Technology Center. On two spaceflights, Garriott logged 1674 hours 56 minutes, with 13 hours 44 minutes on three EVAs.

Gibson, Edward George (Dr), *Science Pilot Skylab 4,* was born on 8 November 1936 in Buffalo, New York, graduating from the Kenmore High School. He gained his BS degree from the University of Rochester, New York, in 1959 before attending the California Institute of Technology, where he received his Masters in engineering in 1960 and his PhD in 1964. He also studied for his doctorate at Cal Tech Gibson and was a research associate in jet propulsion and atmospheric physics. In 1964 he joined Philco Corporations Applied Research Laboratory, Newport Beach, California, where he remained until joining the astronaut programme. One of six scientist astronaut selected by NASA on 28 June 1965 (Group 4), Gibson completed the required 53-week jet pilot training course at Williams AFB, Arizona, and in July 1966 he began work on the Apollo Applications programme. In 1969 he served as CapCom and support crew-member for Apollo 12, and after Skylab 4 he resigned from NASA in November 1974, to join the Aerospace Corporation in Los Angeles, working in solar physics and analysing the results from Skylab. From March 1976 he was a consultant at ERNO Raumfahrttechnik GmBH, Germany, for a year, working on the Spacelab science module design. He returned to NASA in 1977 to become Chief of selection and training of the new Group 8 mission specialist candidates. At the time of his second resignation from NASA in October 1980, he was in training as a CapCom for STS-1, but joined TRW Inc, Redondo Beach, California, as an Advanced Systems Manager for the energy development group. After working for the Bethesda-based aerospace and consultant firm Booz, Allen, and Hamilton, he set up his own aerospace consulting firm based in Carlsbad, California. More recently, he has become a successful author and consultant. On his only spaceflight, he logged 2017 hours 16 minutes, with 15 hours 20 minutes on three EVAs.

Hartsfield, Henry ('Hank') Warren, Jr (Lt-Colonel, USAF), *Support Crew and CapCom for all three manned missions,* was born on 21 November 1933 in Birmingham, Alabama. He graduated from West End High School there in 1950, and in 1954 he gained a BS degree in physics from Auburn University. Hartsfield has performed graduate work in physics at Duke University and in astronautics at the USAF Institute of Technology. He entered the USAF in 1955, and trained as a pilot before serving in Bitburg, Germany. He is a 1964 graduate (Class 64C) of the USAF Test Pilot School, Edwards AFB, California. Following graduation, he served at the school as an instructor until selection to the USAF MOL programme on 30 June 1966 (Group 2). When MOL was cancelled in June 1969, he was one of the seven MOL astronauts who transferred to NASA on 14 August 1969 (Group 7), where he served as support crew member and CapCom for Apollo 16 and then the three Skylab manned missions. Following his Skylab assignments, Hartsfield worked on Shuttle development for the next six years. In August 1977 he resigned from the USAF with the rank of Colonel. Hartsfield was back-up pilot for STS-2 and STS-3 before flying into space as Pilot for STS-4 (*Columbia*) in 1982, the final Orbital Flight Test. He made two further spaceflights as a mission Commander (STS 41-D in 1984, *Discovery*, and STS 61-A in 1985, *Challenger*), before completing a number of administrative roles. In 1987, after leaving the astronaut office, he became the Deputy Director for Flight Crew Operations Directorate (FCOD) at NASA JSC, and in June 1989 he received a temporary assignment as Director of Space Flight/Space Station, Office of Space Flight. He then became Deputy Director of Operations, Space Station Office at NASA Marshall Spaceflight Center in September 1990 for two years. Hartsfield returned to JSC in 1992, as manager of the man-tended capability for Space Station Freedom, then as an assistant director within the Office of Safety, Reliability and Quality Assurance. More recently, he worked as a training manger for the Raytheon Training Division at JSC. He logged 482 hours and 51 minutes on three spaceflights.

Henize, Karl Gordon (Dr), *Support Crew and CapCom for all three manned missions, and Principal Investigator S019*, was born on 17 October 1926 in Cincinnati, Ohio, and graduated from the Mariemont High School nearby. Henize earned both his BS in mathematics (1947) and his MA in astronomy in (1948) from the University of Virginia. This was the beginning of his career as a professional astronomer, which took him around the world. He was an observer for the University of Michigan at the Lamont–Hussey Observatory, Bleomfontein, Union of South Africa until 1951, when he returned to the US to complete his doctorate in astronomy. This was awarded by the University of Michigan in 1954. He was a Carnegie post-doctoral fellow at Mount Wilson Observatory in Pasadena, California, for two years until 1956, when he joined the Smithsonian Astrophysical Observatory. Henize joined the faculty of Northwestern University in 1959 as a teaching and research professor and was a guest observer at Mount Stromlo Observatory, Canberra, Australia, during 1961 and 1962. He developed the UV stellar spectra experiment cameras flown on Gemini 10–12 in 1966, and was selected as one of eleven scientist-astronauts by NASA on 4 August 1967 (Group 6). Following a 53-week jet pilot course at Vance

AFB, Oklahoma, Henize worked as a support crew-member (CM) and CapCom for Apollo 15 and all three Skylab missions, with one of his experiments being flown on the OWS. After Skylab, he worked on the development of ultraviolet telescopes for Spacelab missions, and participated as a mission specialist for ASSESS II airborne Spacelab simulation during 1977. He flew in space in 1985, aged 58, as Mission Specialist on STS 51-F (Spacelab 2, *Challenger*), an astronomy research mission. Henize retired from the astronaut office in April 1986 to work as a senior scientist in the JSC Space Science Branch. One of the things he worked on was orbital debris issues affecting the space station. At the age of 65 he took up mountaineering, and during an ascent of Mount Everest in October 1993 he suffered respiratory failure. He died on 5 October at Base Camp and was buried on the mountain. Henize logged 190 hours 46 minutes on his one spaceflight.

Kerwin, Joseph ('Joe') Peter (Commander, USN Medical Corps), *Science Pilot Skylab 2,* was born on 19 February 1932, in Oak Park, Illinois, graduating from the local Fenwick High School in 1949. After gaining his BA in philosophy from the College of the Holy Cross in Worcester, Massachusetts, he attended medical school at Northwestern University, receiving his MD in 1957. He then completed his one-year internship at the District of Columbia General Hospital in Washington DC. In December 1958, Kerwin graduated from the Naval School of Aviation Medicine at Pensacola, Florida, receiving the designation Naval Flight Surgeon. He served as a medical officer at the Cherry Point Marine Corps Air Station, North Carolina, from 1959 to 1961, and followed this by attending navy aviator training. After completing pilot training in Florida and Texas, he was designated a naval aviator in 1962. His last naval assignment prior to joining the astronaut programme was on the medical staff of Attack Carrier Wing 4, Jacksonville, Florida. Kerwin was one of six scientist astronauts selected by NASA on 28 June 1965 (Group 4), and since he was already a qualified jet pilot, he did not have to complete the 53-week pilot course. Instead, he was assigned to the AAP development group, and also worked on the biological isolation garment designed for the first lunar returning astronauts and the lunar receiving laboratory at the Manned Spacecraft Center (now JSC) Houston. Kerwin was assigned support duties in the thermal vacuum testing of the Apollo spacecraft and was commander of the eight-day altitude chamber simulation of the CSM, to qualify it for manned spaceflight, in June 1968, and was later CapCom for Apollo 13. Following Skylab, he held the position of Director of Life Sciences for the CB for three years, and was assigned to Space Shuttle development issues. Kerwin was also a member of the selection board for the first group of Mission Specialist astronauts in 1978, and was their first training chief. He worked on operational Shuttle mission planning, involving rendezvous, satellite deployment and retrieval, and Remote Manipulator system (RMS) operations. He was a leading candidate for assignment on the flight crew for the Solar Max rescue mission (STS 41-C), but instead assumed the role of NASA Senior Science Representative in Australia before the flight crew was chosen. After the two-year assignment ended in 1984, Kerwin returned to JSC as Director of Space and Life Sciences, retiring in April 1987 from both NASA and the USN (with the rank of Captain). He then joined the Lockheed Missile and Space

Company as Chief Scientist for the space station at their Houston office. More recently, he has worked for United Space Alliance, also in Houston. Kerwin logged 672 hours 50 minutes in one spaceflight and 3 hours 25 minutes on one EVA.

Lenoir, William Benjamin (Dr), *back-up Science Pilot Skylab 3 and Skylab 4, CapCom Skylab 4,* was born on 14 March 1939 in Miami, Florida, and attended MIT to study electrical engineering, receiving his BS in 1961, his MS in 1962 and his PhD in 1965. During studies for his doctorate, Lenoir also worked as an instructor at MIT, and in 1965 joined the school faculty as Assistant Professor of Electrical Engineering, working on the development of scientific experiments for satellites. He was one of eleven scientist-astronauts selected by NASA on 4 August 1967 (Group 6) and completed a 53-week jet pilot training course at Laughlin AFB, Texas, before assignment to the design and development of Skylab. Following his Skylab assignments, in 1974, Lenoir worked for two years with the NASA Satellite Power Team, investigating the potential of adapting space power for use on Earth. He subsequently worked on shuttle development issues, including EVA, and payload deployment and retrieval. He completed one spaceflight in 1982, as mission specialist on STS-5 (*Columbia)*, but his scheduled EVA was cancelled because of mechanical failures in the EVA suit. Lenoir left NASA in 1984 and for the next five years worked for Booz, Allen and Hamilton as a consultant on the Freedom space station programme. He returned to the space agency to become Associate Administrator for the space station programme in 1989, and before being promoted to Associate Administrator for Spaceflight, in charge of all of NASA manned spaceflight programmes. After resigning from NASA a second time in May 1992, Lenoir rejoined Booz, Allen Hamilton, as Manager, Applied Sciences Division. He logged 122 hours 14 minutes on one spaceflight.

Lind, Don Leslie (Dr), *back-up Pilot Skylab 3 andSkylab 4, Skylab Rescue Pilot, Co-Principle Investigator S230,* was born on 18 May 1930 in Midvale, Utah, and is a graduate of Jordan High School, Sandy, Utah. After attaining his BS (with high honours) in physics in 1953, Lind served in the USN from 1954 to 1957, qualifying as a jet pilot in 1955, and subsequently serving as a carrier pilot on *USS Hancock*. In 1957, he began work on his doctorate in high-energy nuclear physics at the University of California at Berkley – which was presented in 1964. At the same time as pursuing his doctorate, Lind was employed by the Lawrence Radiation Laboratory in Berkeley. In 1964, he began work as a space physicist at NASA's Goddard Spaceflight Center, Maryland. Selected to NASA in April 1966 (Group 5) as one of 19 astronauts, Lind specialised in the Lunar Module. He was instrumental in developing lunar surface techniques for the Apollo 11 lunar EVA and the deployment of the Apollo Lunar Surface Experiment Package (ALSEP). A leading candidate to become the second scientist to walk on the Moon, budget cuts cost him his chance of a lunar flight and he was reassigned to Skylab. After his assignments on the Skylab programme, Lind took a leave of absence for a year to perform post-doctoral work at the University of Alaska's Geophysical Institute. He returned to JSC and worked on the first four shuttle missions under the Orbital Flight Test

Programme, before assignment to Spacelab 3. He flew his only spaceflight in 1985 (after a 19 year wait since selection) as a mission specialist on STS 51-B (Spacelab 3 *Challenger)*, operating one of his own experiments on the mission to take three-dimensional recordings of Earth's aurora. He also had an interstellar gas experiment flown on the Long Duration Exposure Facility (deployed from on STS 41-C in 1984 and retrieved on STS-32 in 1990). Lind retired from NASA in April 1986 to become Professor of Physics at the University of Utah until retirement in 1995. Since then he has performed missions for his Church, the Mormons, giving inspirational presentations on his experiences in space and on Earth. Lind logged 168 hours 9 minutes on one spaceflight.

Lousma, Jack Robert (Major, USMC), *Pilot Skylab 3,* was born on 29 February 1936 in Grand Rapids, Michigan, but grew up in Ann Arbor, Michigan. He received a BS in aeronautical engineering from the University of Michigan in 1959 and entered the USMC, completing pilot training the following year. Over the next six years he served as an attack pilot with the 22nd Marine Air Wing, and with the 1st Marine Air Wing at Iwakuni, Japan. Prior to his selection as an astronaut, Lousma was a reconnaissance pilot with the 2nd Marine Air Wing, Cherry Point, North Carolina. He earned a degree in aeronautical engineering from the USN Postgraduate School in 1965, and was one of 19 astronauts selected by NASA on 4 April 1966 (Group 5), becoming a Lunar Module specialist. He worked on the support crews for Apollo 9, 10, and 13, and was to be assigned as LM Pilot for Apollo 20 – the last of the series – before it was cancelled by budget cuts. The cancellation of Apollo 20 also cost Lousma the chance of becoming the eighteenth, and last, man of Apollo to set foot on the Moon. After completing his Skylab assignments, Lousma served as back-up DMP for the American ASTP crew in 1975, followed by assignment to Space Shuttle development. He was originally Pilot on STS-3, the third OFT, on which one of the early objectives was to re-boost the unmanned Skylab workshop to a higher orbit. This was abandoned when the station re-entered the atmosphere before the first Shuttle had flown. When the original Commander of the OFT-3 crew, Fred Haise, resigned in 1979, Lousma was 'promoted' to Commander STS-3, which he flew in 1983 (*Columbia*). Lousma resigned from NASA on 1 October 1983 and retired with the rank of Colonel from the USMC on 1 November the same year. After an unsuccessful 1984 campaign to seek election to the US Senate, representing Michigan, he formed his own high-technology consulting firm based in Ann Arbor, Michigan, and has partnered former Skylab colleagues Jerry Carr and Bill Pogue on various projects with Carr's consultancy company Camus. Lousma logged 1619 hours 14 minutes on two spaceflights and 11 hours and 01 minutes on two EVAs.

McCandless, Bruce II (Lt-Commander, USN), *back-up Pilot Skylab 2, co-investigator M-509, CapCom Skylab 3,* was born on 8 June 1937 in Boston, Massachusetts, and is a graduate of Woodrow Wilson Senior High School in Long Beach, California. In 1958, McCandless graduated second in a class of 899 from the USN Academy at Annapolis, receiving a BS in science. He was awarded his wings in March 1960 following pilot training at Pensacola, Florida, and Kingsville, Texas, and spent the next four years as a carrier pilot aboard the *USS Forrestal* and *USS Enterprise*, participating in the 1962 Cuban blockade. McCandless was a flight instructor at the Naval Air Station (NAS) Oceana, Virginia prior to entering Stanford University graduate school, earning an MS degree in electrical engineering in 1965 and completing studies towards a PhD. One of nineteen astronauts selected on 4 April 1966 (Group 5), McCandless became a LM specialist and worked as a CapCom for Apollo 10, 11 and 14. He was the Apollo 11 lunar surface EVA CapCom on 20 July 1969 during the historic first steps on the Moon. After assignments on the Apollo 14 support crew he began work on Skylab and was a co-investigator for the M-509 astronaut manoeuvring unit, which was test flown inside the Skylab OWS by the second and third crews. Following Skylab, McCandless was assigned to various projects relating to Space Shuttle, including the Inertial Upper Stage (IUS), the Space Telescope (Hubble), and the Solar Maximum repair mission. He was also instrumental in developing the Manned Manoeuvring Unit for the Shuttle. His first spaceflight came in 1984, as mission specialist on STS 41-B (*Challenger*), after seventeen years as an astronaut. On 8 February 1984 McCandless became the first person to make an untethered spacewalk, flying the MMU 320 feet from *Challenger*'s payload bay. After developing EVA procedures for the Freedom space station and earning an MBA from the University of Houston, Clear Lake in 1987, his second spaceflight occurred in 1990,during which he participated in the deployment of the Hubble Space Telescope (STS-31, *Discovery*). McCandless resigned from NASA and retired from the USN with the rank of Captain on 31 August 1990 to become an aerospace consultant based in Houston. He has participated in developing EVA procedures for the HST service and repair missions, and more recently has joined the staff of Lockheed-Martin in Denver, Colorado. McCandless logged 312 hours 32 minutes on two spaceflights, including 12 hours 12 minutes on two EVAs.

Musgrave, Franklin Story (Dr), *back-up Science Pilot Skylab 2, CapCom Skylab 3 and Skylab 4,* was born on 19 August 1935 in Boston, Massachusetts, but considers Lexington, Kentucky, to be his home town. Musgrave has one of the most remarkable careers of any space explorer. Following graduation from St Marks School in Southborough, Massachusetts, in 1953, he entered the USMC, serving as an aviation electrician and aircraft crew chief onboard the carrier *USS Wasp* in the Far East. In 1958 he gained a BS in mathematics and statistics from Syracuse University, and was employed by the Eastman Kodak Company in Rochester, New York, as a mathematician and operations analyst. Musgrave also earned an MBA in operations analysis and computer programming from the University of California in Los Angeles (1959), and a BA in chemistry from Marietta College (1960). He then pursued a medical career, gaining a MD from Columbia University (1964) and completing his surgical internship at the University of Kentucky Medical Center. He also earned an MS in physiology and biophysics (1966), and worked as a USAF post-doctoral fellow, working on aerospace medicine and physiology. Prior to selection to NASA, Musgrave was working as a researcher and teacher in cardiovascular and exercise physiology. Selected to NASA on 4 August 1967

(Group 11), he completed a 53-week jet pilot training course and went on to log more than 16,800 flying hours, including 7,100 hours in jets (more than many military astronauts) in more than 160 different types of aircraft. He is the holder of several pilot and instructor ratings, has completed over 460 parachute jumps, and up to 1977 had logged more time in the T-38 than any other pilot. He added to his academic credentials, with an MA in literature from the University of Houston in 1987, and worked towards a PhD in physiology in addition to publishing more than 44 scientific papers and completing various CB technical assignments. Musgrave was assigned to the design and development of the OWS and after his Skylab assignments he worked on developing EVA equipment for the Shuttle. He also participated in two simulated Spacelab missions, and continued to be a part-time surgeon and teaching professor during his thirty-year career at NASA. Musgrave is the only astronaut to have flown on all five Shuttle orbiters, and on his last flight he became the first person over 60 to fly in space. Having spent sixteen years as an astronaut, he then flew six missions in thirteen years as a mission specialist. These were flown in 1983 (STS-6, *Challenger,* performed first Shuttle EVA), 1985 (STS 51-F, Spacelab 2, *Challenger*), 1989 (STS-33 *Discovery,* DoD), 1991 (STS-44, *Atlantis,* DoD), 1993 (STS-61, *Endeavour,* Hubble Service Mission, three EVAs) and 1996 (STS-80, *Columbia,* USMP). In September 1997, having completed six missions and been told that, at the age of 62, he would no longer be assigned to flight crews, Musgrave left NASA to become a consultant and public speaker. He logged 1,280 hours 50 minutes in six missions, and 26 hours 19 minutes on four EVAs.

Parker, Robert Allen Ridley (Dr), *Programme Scientist and CapCom for all three manned missions,* was born on 14 December 1936 in New York City, but grew up in Shrewsbury, Massachusetts, where he graduated from high school. He received a BA in astronomy and physics from Amherst College in 1959 and his PhD in astronomy from the California Institute of Technology in 1962, after which he worked as an associate professor of astronomy at the University of Wisconsin until selection to the astronaut programme. Parker was one of eleven scientist astronauts select on 4 August 1967 (Group 6) and completed a 53-week jet pilot training programme at Williams AFB, Arizona, before assignment to the support crews for Apollo 15 and Apollo 17. He served as CapCom for both flights and mission scientist for Apollo 17. After his Skylab assignments, Parker worked on Spacelab development issues, including the ASSESS airborne Spacelab simulation programme, and was assigned as a mission specialist for the first Spacelab mission in 1978. He flew on that mission, after sixteen years as an astronaut, in 1983 (STS-9, *Columbia).* His second spaceflight was as a mission specialist on the Astro-1 observatory mission in 1990 (STS-35, *Columbia).* He has also served as Director of the spaceflight/space station integration office at NASA HQ between his two spaceflights, before returning to NASA HQ in 1991 as Director of the division of policy and plans, Office of Spaceflight. He later served as Deputy Associate Administrator for operations and Director for Spacelab Operations. In June 1997, Parker became Director for program requirements at the NASA Jet Propulsion laboratory, Pasadena, California. He logged 462 hours 52 minutes on two spaceflights.

Pogue, William Reid (Lt-Colonel USAF), *Pilot Skylab 4,* was born on 23 January 1930 in Okemah, Oklahoma, and received a BS in secondary education from Oklahoma Baptist University in 1951, prior to entering the USAF. After pilot training, Pogue served as a member of the 5th Air Force and flew 43 combat missions in Korea, before assignment to the *Thunderbirds* air demonstration team from 1955 to 1957. He gained an MS in mathematics from Oklahoma State University in 1960 and then became a maths instructor at the USAF Academy. Pogue is also a graduate of the Empire Test Pilot School (Course Class Empire 22) in Farnborough, England. He was serving as an instructor at the USAF Test Pilot School, Edwards AFB, California when selected by NASA on 4 April 1966, as one of 19 Group 5 astronauts. Trained as a CM specialist, Pogue worked on AAP assignments, then served on the support crews for Apollo 7, 11, 13 and 14, and was preparing to be the CMP for the Apollo 19 lunar mission when it was cancelled – along with his chance of flying in lunar orbit. After his Skylab assignments, Pogue resigned from NASA and retired from the USAF with the rank of Colonel on 1 September 1975 to join former Apollo 15 astronaut Jim Irwin's, High Flight Foundation, a religious organisation. Pogue returned to NASA as an astronaut in 1976 and 1977 to work on a range of programmes to study the Earth from space. He left the agency a second time in 1977 to become a private consultant in aerospace and energy corporations, often working with former Skylab astronaut Jerry Carr's company Camus. He has also become a successful author. Pogue logged 2,017 hours 16 minutes and 13 hours 34 minutes on two EVAs.

Schweickart, Russell Louis ('Rusty'), *back-up Commander Skylab 2 and CapCom for all three missions,* was born on 25 October 1935 in Neptune, New Jersey, and is a 1952 graduate of Manasquan High School. He gained a BS in aeronautical engineering from MIT in 1956 and served as a pilot in the USAF for the next four years, until returning to MIT to study for his MS in aeronautics and astronautics (which he gained in 1963). He had been recalled to active duty for a year in 1961 and subsequently served with the Air National Guard. At the time of his selection to NASA on 17 October 1963 as one of fourteen Group 3 astronauts, Schweickart was a scientist at the Experimental Astronomy Laboratory at MIT, researching into the physics of the upper atmosphere and star tracking. He was assigned to technical assignments, working on scientific experiments for Gemini and Apollo missions, prior to assignment as LMP to the three-man back-up crew for Apollo 1 in 1966. After several crew reassignments, the same crew eventually flew as the prime crew on Apollo 9, where Schweickart tested the Apollo EVA suit that would be used on the Moon, during an EVA in Earth orbit. Following his Skylab assignments, he transferred to NASA HQ, as Director of User Affairs in the Office of Applications. Schweickart returned to JSC in November 1975 to work on Shuttle payload policies, and the following year, took a leave of absence to serve on the Californian Governor's committee staff as assistant for science and technology. He resigned from NASA in July 1979 and has since served on the Californian Commission for Energy, the US Energy Commission, and the US Antarctic Program Safety Review Panel. A founder member of the Association of Space Explorers, he has also served as President of NSR Communications and as Executive Vice President of CTA incorporated. He is a frequent lecturer on space and the environment. Schweickart logged 241 hours 1 minute in space and 1 hour 7 minutes during one EVA.

Thornton, William Edgar (Dr), *Science Pilot SMEAT, support crew-member and CapCom for all three manned missions, Co-principle Investigator M074 and Principle Investigator M172,* was born on 14 April 1929 in Faison, North Carolina. He graduated from the University of North Carolina in 1952 with a BS in physics, and served in the USAF as Chief of an instrumentation laboratory at the Flight Air Test Proving Ground, Eglin AFB, Florida. After leaving the USAF in 1956, he worked as an electronics engineer for Del Mar Engineering Laboratories in Los Angeles, rising to Chief Engineer of the electronics division and later heading the avionics research division. Thornton was a medical student between 1959 and 1963 at the University of North Carolina, completing his internship at Wilford Hall AF Hospital, Lackland AFB, Texas in 1964, before returning to active duty. While at the Aerospace Medical Division, Brooks AFB, Texas, he became involved in human adaptation to spaceflight and worked on the medical exercise and experiment procedures for the USAF MOL programme. Selected to NASA on 14 August 1967 (Group 6) as one of eleven scientist-astronauts, he completed a 53-week jet pilot training course before being assigned to the Skylab programme in 1969. After his Skylab assignments, Thornton worked on developing the Spacelab modules, completing a simulated Spacelab mission, and worked on Space Adaptation experiments and exercise regimes that were evaluated on STS 4–7. He flew in space at the age of 54 (after a wait of sixteen years) as Mission Specialist in 1983 (STS-8, *Challenger*), where he was able to experience the sensation of spaceflight adaptation for himself. His second flight was as MS on a Spacelab mission in 1985 (STS 51-B, *Challenger,* Spacelab 3). For the next nine years, Thornton continued his research into space adaptation and exercise for Shuttle and space station programmes. He retired from NASA on 31 May 1994, to become a clinical assistant professor at the Department of Medicine, University of Texas at Galveston, and adjunct professor at the University of Houston–Clear Lake. He logged 313 hours 18 minutes on two spaceflights.

Truly, Richard Harrison, (Lt-Commander USN), *support crew-member and CapCom for all three manned missions,* was born on 12 November 1937 in Fayette, Mississippi. He is a 1959 graduate of the Georgia Institute of Technology with a BS in aeronautical engineering. Truly received flight training in Beeville Texas in 1960 and was assigned to Fighter Squadron 33 aboard the carriers *USS Intrepid* and *USS Enterprise*. Truly is a 1964 graduate of the USAF Test Pilot School (Class 64A) at Edwards AFB, and remained there as an instructor until selected for the USA MOL programme on 12 November 1965 (Group 1, and his 28th birthday). When MOL was cancelled in June 1969, Truly was one of seven transfers to NASA on 14 August 1969 (Group 7), working on Skylab and then the support crew, and as CapCom in Moscow for ASTP in 1975. He then teamed with Joe Engle for the next six years, working on the ALT programme in 1977, participating in two of the five free flights of *Enterprise* at Edwards AFB, and then as back-up Pilot for the first Shuttle mission and prime crew for STS-2. As pilot STS-2 (*Columbia)*, he was launched into space on his 44th Birthday (12 November 1981), and less than two years later completed a second Shuttle flight as Commander (STS-8 *Challenger)* before leaving NASA in October 1983 to head the USN Space Command at Dahlgren, Virginia. Truly returned to NASA in February 1986 as Associate Administrator for the Shuttle programme, and in January 1989 he became the first former astronaut to be appointed Administrator of NASA, until February 1992 when he became a professor at Georgia Tech and director of the Georgia Tech Institute. In 1997, Truly accepted a position as head of the National Renewable Energy Laboratory in Golden, Colorado. He logged 19 hours 22 minutes on two spaceflights.

Weitz, Paul J. (Commander, USN), *Pilot Skylab 2,* was born on 25 July 1932 in Erie, Pennsylvania, graduating from the nearby Harborcreek High School in 1950. A 1954 graduate of the Pennsylvania State University with a BS in aeronautical engineering, he then served on destroyers in the 'black shoe' navy. He completed flight training in 1956 and was assigned as an instructor in tactics at Jacksonville, Florida. Weitz then transferred to the Naval Weapons Center, China Lake, California, in 1960, as a project officer on a variety of air-to-ground delivery system tests. After earning an MS in aeronautical engineering from the USN postgraduate school in 1964, he flew combat missions in the Vietnam conflict before assignment as detachment officer-in-charge at Whidby Island, Washington. Weitz was one of nineteen Group 5 astronauts selected on 4 April 1966, and became a CM specialist, serving on the support crew and as CapCom for Apollo 12. He was to have been selected to serve as back-up CMP on Apollo 17, leading to an assignment to fly around the Moon as CMP on Apollo 20, but the 1970 budget cuts eliminated his mission and he transferred to Skylab. Following those assignments he worked on Shuttle development, and retired from the navy with the rank of Captain on 1 June 1976, to continue as a civilian astronaut. Weitz flew as Commander of STS-6 (*Challenger*) in 1983, before assuming a series of administrative and management roles at NASA JSC, Houston, including Deputy Chief of the CB, Deputy Director of JSC in 1988, and Acting Director of JSC in 1993, before retiring from NASA in April 1995. He logged 793 hours 14 minutes in space and 2 hours 21 minutes on two EVAs during his two spaceflights.

Career experience of the Skylab astronaut group, 1965–1997

Astronaut	NASA Group	Year of Selection	Total Flights	Duration hrs:min	Total EVAs	Duration hrs:min
Bean	3	1963	2	1671:45	3	10:30
Bobko	7	1969	3	386:04	0	
Brand	5	1966	4	746:04	0	
Carr	5	1966	1	2017:16	3	15:51
Conrad	2	1962	4	1179:39	4	11:33
Crippen	7	1969	4	565:48	0	
Garriott	4	1965	2	1674:56	3	13:44
Gibson	4	1965	1	2017:16	3	15:20
Hartsfield	7	1969	3	482:51	0	
Henize	6	1967	1	190:46	0	
Kerwin	4	1965	1	672:50	1	03:25
Lenoir	6	1967	1	122:14	0	
Lind	5	1966	1	168:09	0	
Lousma	5	1966	2	1619:14	2	11:01
McCandless	5	1966	2	312:32	2	12:12
Musgrave	6	1967	6	1280:50	4	26:19
Parker	6	1967	2	462:52	0	
Pogue	5	1966	1	2017:16	2	13:34
Schweickart	3	1963	1	241:01	1	01:07
Thornton	6	1967	2	313:18	0	
Truly	7	1969	2	190:22	0	
Weitz	6	1966	2	793:14	2	02:21
Totals for all 22 astronauts			48	19,126:17	30	136:57

796 days 22 hours (2.18 years) cumulative experience in space (5.70 days on EVA)

In order of most experience

Career spaceflights			Career EVAs		
Pos.	Astronaut	hrs:min	Pos	Astronaut	hrs:min
=1	Carr	2017:16	1	Musgrave	26:19
=1	Gibson	2017:16	2	Carr	15:51
=1	Pogue	2017:16	3	Gibson	15:20
4	Garriott	1674:56	4	Garriott	13:44
5	Bean	1671:45	5	Pogue	13:34
6	Lousma	1619:14	6	McCandless	12:12
7	Musgrave	1280:50	7	Conrad	11:33
8	Conrad	1179:39	8	Lousma	11:01
9	Weitz	793:14	9	Bean	10:30
10	Brand	746:04	10	Kerwin	03:25
11	Kerwin	672:50	11	Weitz	02:21
12	Crippen	565:48	12	Schweickart	01:07

13	Hartsfield	482:51
14	Parker	462:52
15	Bobko	386:04
16	Thornton	313:18
17	McCandless	312:32
18	Schweickart	241:01
19	Henize	190:46
20	Truly	190:22
21	Lind	168:09
22	Lenoir	122:14

MISSION DATA

	SL-1 Cluster	SL-2 First crew	SL-3 Second crew	SL-4 Third crew
Internat No.	1973-27A	1973-32A	1973-50A	1973-90A
Launch date	1973 May 14	1973 May 25	1973 July 28	1973 November 16
Launch time	13.30 EDT	09.00 EDT	07.11 EDT	09.01 EDT
Launch vehicle	Saturn V (SA-513, two-stage)	Saturn 1B (SA-206)	Saturn 1B (SA-207)	Saturn 1B (SA-208)
CSM	n/a	116	117	118
Rescue vehicle	(BUp SA-515, two-stage)	SA-207/CSM 117 (SL-3)	SA-208/CSM 118 (SL-4)	SA-209/CSM 119 (back-up)
Recovery	(Re-entered 1979 July 11)	1973 June 22	1973 September 25	1974 February 8
Recovery time	n/a	09.49 EDT	18.19 EDT	11.17 EDT
Orbital parameters (SL-1 only)				
Perigee	268.1 miles	Distance travelled		930 million miles
Apogee	269.5 miles	Active life (1973 May 14 through 1974 February 8)		8 months 24 days
Inclination	50° 04′	Orbital life (through deorbit, 1979 July 11)		6 years 1 month 27 days
Period	93.18 minutes	Number of orbits		34,981
		SL-2	SL-3	SL-4
Prime crew	Commander	Charles Conrad Jr	Alan L. Bean	Gerald P. Carr
	Pilot	Paul J. Weitz	Jack R. Lousma	William R. Pogue
	Science Pilot	Joseph P. Kerwin	Owen K. Garriott	Edward G. Gibson
Back-up crew	Commander	Russell L. Schweickart	Vance D. Brand*	Vance D. Brand*
	Pilot	Bruce McCandless II	Don L. Lind*	Don L. Lind*
	Science Pilot	Story F. Musgrave	William B. Lenoir	William B. Lenoir

*Brand and Lind also served as Skylab Rescue crew for all three missions

	SL-2	SL-3	SL-4	Total
Distance travelled (millions of miles)	24.5	34.5	70.5	129.5
Mission duration (d:h:m)	28:00:49	59:11:09	84:01:16	171:13:14
Manned orbits	404	858	1,214	2,476

Man-hour utilisation	SL-2		SL-3		SL-4		Total	
	Hours	%	Hours	%	Hours	%	Hours	%
Medical activities	145.3	7.5	312.5	8.0	366.7	6.1	824.5	6.9
Solar observations	117.2	6.0	305.1	7.8	519.0	8.5	941.3	7.9
Earth resources	71.4	3.7	223.5	5.7	274.5	4.5	569.4	4.8
Other experiments	65.4	3.4	243.6	6.2	403.0	6.7	712.0	6.0
Sleep/rest and off duty	675.6	34.7	1,224.5	31.2	1,846.5	30.5	3,746.6	31.5
Pre/post sleep and eating	477.1	24.5	975.7	24.8	1,384.0	23.0	2,836.8	23.8
Housekeeping	103.6	5.3	158.4	4.0	298.9	4.9	560.9	4.7
Physical training/personal hygiene	56.2	2.9	202.2	5.2	384.5	6.4	642.9	5.4
Other (EVA etc.)	232.5	12.0	279.7	7.1	571.4	9.4	1,083.6	9.0
Totals	1,944.3	100.0	3,925.2	100.0	6,048.5	100.0	11,918.0	100.0

Experiment performance	SL-2		SL-3		SL-4		Total	
	Hours	%	Hours	%	Hours	%	Hours	%
Solar astronomy	117.2	29.9	305.1	28.2	519.0	33.2	941.3	31.0
Earth observations	071.4	18.2	223.5	20.6	274.5	17.6	569.4	18.8
Student	3.7	0.9	10.8	1.0	14.8	0.9	29.3	0.9
Astrophysics	36.6	9.4	103.8	9.6	133.8	8.5	274.2	9.0
Man/systems	12.1	3.1	117.4	10.8	83.0	5.3	212.5	7.0
Material science	5.9	1.5	8.4	0.8	15.4	1.0	29.7	1.0
Life science	145.3	37.0	312.5	29.0	366.7	23.5	824.5	27.2
Comet Kohoutek	–	–	–	–	156.0	10.0	156.0	5.1
Totals	392.2	100.0	1,081.2	100.0	1,563.2	100.0	3,036.9	100.0

Data returned	SL-2	SL-3	SL-4	Total
Solar observations (frames)	28,739	24,942	73,366	127,047
Earth observations: film (frames)	9,846	16,800	19,400	46,046
Earth observations: magnetic tape (feet)	45,000	93,600	100,000	238,600

Experiment summary	Planned	Actual	Deviation (%)
Earth observation passes	62	99	+60
Solar viewing time (unmanned hours)	565	724.7	+27.5
Manned solar viewing time (hours)	879.5	941.3	+7.1
Biomedical investigations	701	922	+32
Engineering/technical investigations	264	245	–3.4
Material/space manufacturing investigations	10	32	+220
Astrophysical investigations	168	345	+105
Student experiments	44	52	+18
Science demonstrations (optional, SL-4 only)	26	11	–42

Consumables	At launch (1973 May 14)	At end of manned mission (1974 February 8)	Used
Water (lbs)	6,000	1,710	4,290
Oxygen (lbs)	6,100	2,764	3,336
Nitrogen (lbs)	1,540	607	933
TACS* (lbs)	80,000	12,488	67,512

*More than 32% of the TACS was used during the first ten days of the mission

Extravehicular activity		SL-2	SL-3	SL-4	Total
Stand-up EVA	Date	1973 May 25			
	Duration	00 hr 37 min			
	Crew	Weitz*			
EVA 1	Date	1973 June 7	1973 August 6	1973 November 22	
	Duration	03 hr 30 min	06 hr 29 min	06 hr 33 min	
	Crew	Conrad and Kerwin	Garriott and Lousma	Pogue and Gibson	
EVA 2	Date	1973 June 19	1973 August 24	1973 December 25	
	Duration	01 hr 44 min	04 hr 30 min	07 hr 01 min	
	Crew	Conrad and Weitz	Garriott and Lousma	Carr and Pogue	
EVA 3	Date		1973 September 22	1973 December 29	
	Duration		02 hr 45 min	03 hr 28 min	
	Crew		Bean and Garriott	Carr and Gibson	
EVA 4	Date			1974 February	
	Duration			05 hr 19 min	
	Crew			Carr and Gibson	
Totals		5 hrs 41 min	13 hrs 44 min	22 hrs 21 min	41 hrs 46 min

*Supported by Kerwin on Stand-up EVA, while Conrad flew CSM

Mission achievements

SL-1 Skylab orbital workshop
- First American space station
- Only domestic American space station
- First 100% mission successful space station (all manned missions launched, docked and transferred crews who resided on station and successfully recovered); record held until Mir, 1986–2000
- Successful demonstration of utilisation of former Apollo lunar mission hardware

SL-2 First manned mission (28 days)
- Installed solar shield 'parasol' from scientific airlock
- Released jammed solar wing on EVA

	Restored OWS to operational space station
	Doubled previous length of time in space (Gemini 7 – 14 days)
	Set world endurance record for mission (28 days) and accumulated individual (Conrad)
SL-3	Second manned mission (59 days)
	Installed twinpole solar shield on EVA
	Performed major in-flight maintenance
	Doubled previous length of time in space (Skylab 2 – 28 days)
	Set world endurance record for mission and individual time (Bean)
SL-4	Third manned mission (84 days)
	Observed and photographed Comet Kohoutek
	Increased previous length of time in space by about 50% (Skylab 3 – 59 days)
	Set world, mission and individual crew accumulated time in space
	Held world endurance record for spaceflight for more than three years (Soyuz 26/Salyut 6, 96 days, 1977–78)
	Held US space endurance /space station record for more than 21 years (Norman Thagard, Mir, 21 March–July 1995)

Adapted from MSFC Skylab Mission Report – Saturn Workshop (NASA TM X-64814), October 1974, pp.3–39.

EXPERIMENT DATA

No.	Designation	Location on Skylab	Principle Investigator	Field Center	Manned mission 1	2	3
Solar Studies							
S020	Ultraviolet and X-ray solar photography[1]	OWS/SAL	Dr R. Tousey, USN Research Laboratory	JSC		x	x
S052	White-light coronagraph	ATM	Dr R. MacQueen, High Altitude Observatory	MSFC	x	x	x
S054	X-ray spectrographic telescope	ATM	Dr R. Glacconi, American Science and Engineering Corporation	MSFC	x	x	x
S055	UV scanning polychromator spectroheliometer	ATM	Dr L. Goldberg, Kitt Peak National Observatory	MSFC	x	x	x
S056	X-ray telescope	ATM	J.E. Milligan, MSFC	MSFC	x	x	x
S082A	Extreme ultraviolet spectroheliograph	ATM	Dr R. Tousey, USN Research Laboratory	MSFC	x	x	x
S082B	Ultraviolet spectrograph	ATM	Dr R. Tousey, USN Research Laboratory	MSFC	x	x	x
Stellar Astronomy							
S019	Ultraviolet stellar astronomy	OWS/SAL	Dr K.G. Henize, NASA astronaut, JSC	JSC	x	x	
S150	Galactic X-ray mapping[2]	IU	Dr W.L. Kraushaar, University of Wisconsin	MSFC			x
S183	Ultraviolet panorama telescope	OWS/SAL	Dr G. Couttè, Laboratoire d'Astronomie Spatiale, France	MSFC	x	x	
Space Physics							
S009	Nuclear emission package	MDA	Dr M.M. Shapiro, USN Research Laboratory	MSFC	x		
S063	Ultraviolet airglow horizon photography	OWS/SAL	Dr D.M. Packer, USN Research Laboratory	JSC	x		
S073	Gegenschein and zodiacal light	OWS/SAL	Dr J.J. Weinberg, Dudley Observatory	MSFC	x	x	x
S149	Micrometeorite collector[3]	OWS/EVA	Dr C.L. Hemenway, Dudley Observatory	JSC	x	x	x
S228	Transuranic cosmic rays	OWS/EVA	Dr P.B. Price University of California at Berkeley	MSFC	x	x	x
S230	Magnetospheric particle composition	ATM/EVA	Dr D.L. Lind, NASA astronaut, JSC, and Dr Johannes Geiss, University of Berne, Switzerland	MSFC	x	x	x

No.	Designation	Location on Skylab	Principle Investigator	Field Center	Manned mission 1	2	3
Earth Resources Experiments							
S190A	Multispectral photographic cameras	MDA	Project scientist: K. Demel, JSC	JSC	x	x	x
S190B	Earth terrain camera	OWS/SAL	Project scientist: K. Demel, JSC	JSC	x	x	x
S191	Infrared spectrometer	MDA	Dr T.L. Barnett, JSC	JSC	x	x	x
S192	Multispectral scanner	MDA	Dr C.K. Korb, JSC	JSC	x	x	x
S193	Microwave radiometer/scattermeter and altimeter[4]	MDA	D.E. Evans, JSC	JSC	x	x	x
S194	L-band radiometer	MDA	D.E. Evans, JSC	JSC	x	x	x
Life Science Projects							
M071	Mineral balance	OWS	G.D. Whedon MD, National Institute of Health, and L. Lutwak, MD, Cornell University	JSC	x	x	x
M073	Bioassay of body fluids	OWS	Dr C.S. Leach, JSC	JSC	x	x	x
M074	Specimen mass measurement	OWS	W.E. Thornton, MD, NASA astronaut JSC, and Col. J.W. Ord, Medical Corps, Clark AFB	JSC	x	x	x
M078	Bone mineral measurement	–	J.M. Vogel, MD, US Public Health Service Hospital, San Francisco, and Dr J.R. Cameron, University of Wisconsin Medical Center	JSC	(pre-flight and post-flight, all three manned flights)		
M092	Lower-body negative-pressure device	OWS	R.L. Johnson MD, JSC, and Col J.W. Ord, Medical Corps, Clark AFB	JSC	x	x	x
M093	Vectrocardiogram	OWS	N.W. Allenbach, MD, USN Aerospace Medical Institute, and R.F. Smith MD, School of Medicine, Vanderbilt University	JSC	x	x	x
M111	Cytogenic studies of blood	–	L.H. Lockhart MD., University of Texas Medical Branch, Galveston, and P.C. Gooch, Brown and Root-Northrop	JSC	(pre-flight and post-flight, all three manned flights)		

M112	Man's immunity, *in vitro* aspects	OWS	S.R.E. Ritzmann MD and W.C. Levin MD, University of Texas Medical Branch, Galveston	JSC	x	x	x
M113	Blood volume and red cell life span	OWS	P.C. Johnson MD, Baylor University Medical School	JSC	x	x	x
M114	Red blood-cell metabolism	OWS	C.E. Mengel MD, University of Missouri School of Medicine	JSC	x	x	x
M115	Special haematological effects	OWS	Dr S.L. Kemsey and C.L. Fisher MD, JSC	JSC	x	x	x
M131	Human vestibular function	OWS	A. Graybeil MD and Dr E.F. Miller, USN Aerospace Medical Institute	JSC	x	x	
M133	Sleep monitoring	OWS	J.D. Frost Jr MD, Baylor University College of Medicine	JSC	x	x	
M151	Time and motion study	OWS	Dr J.F. Kibis, Fordham University, and Dr E.J. McLaughlin, NASA Hq. OMSF	JSC	x	x	x
M171	Body mass measurement	OWS	W.E. Thornton MD, NASA astronaut, JSC	JSC	x	x	x
S015	Effects of zero gravity on single human cells	CM	P.O. Montgomery MD and Dr J. Paul, University of Texas Southwestern Medical School, Dallas	JSC	x		
S071	Circadian rhythm, pocket mice[5]	CM	Dr R.G. Lindberg, Northrop Corporation Laboratory	ARC		x	
S072	Circadian rhythm, vinegar gnats[6]	CM	Dr C.S. Pittendrigh, Stanford University	ARC		x	
Materials Science and Manufacturing in Space							
M479	Zero-gravity flammability	MDA	J.H. Kimzey, JSC	MSFC			x
M512	Materials processing facility[7]	MDA	P.G. Parks, MSFC	MSFC	x		x
(M551)	Metals melting	MDA	R.M. Poorman, MSFC	MSFC	x		
(M552)	Exothermic brazing	MDA	J. Williams, MSFC	MSFC	x		
(M553)	Sphere forming	MDA	E.A. Hasemeyer, MSFC	MSFC	x		
(M555)	Gallium arsenide crystal growth	MDA	Dr M. Rubenstein, Westinghouse Electric Corporation	MSFC	x		
M518	Multipurpose electric furnace system[8]	MDA	A. Boese, MSFC (Project Engineer)	MSFC			x
(M556)	Vapour growth of II-VI compounds	MDA	Dr H. Wiedemeir, Rensselaer Polytechnic Institute	MSFC			x

No.	Designation	Location on Skylab	Principle Investigator	Field Center	Manned mission 1	2	3
(M557)	Immiscible alloy compositions	MDA	J. Reger, Thompson Ramo Wooldridge	MSFC			x
(M558)	Radioactive tracer diffusion	MDA	Dr T. Ukanwa, MSFC	MSFC			x
(M559)	Microsegregation in germanium	MDA	Dr F. Padovani, Texas Instruments	MSFC			x
(M560)	Growth of spherical crystals	MDA	Dr H. Walter, University of Alabama, Huntsville	MSFC			x
(M561)	Whisker-reinforced composites	MDA	Dr T. Kawada, National Research Institute for Metals, Japan	MSFC			x
(M562)	Indium antimonide crystals	MDA	Dr H. Gatos, Massachusetts Institute of Technology	MSFC			x
(M563)	Mixed III-V crystal growth	MDA	Dr W. Wilcox, University of Southern California at Los Angeles	MSFC			x
(M564)	Halide eutectic	MDA	Dr A. Yue, University of California at Los Angeles	MSFC			x
(M565)	Silver grids melted in space	MDA	Dr A. Deruythere, Catholic University of Leuven, Belgium	MSFC			x
(M566)	Copper–aluminium eutectic	MDA	E. Hasemeyer, MSFC	MSFC			x
Zero-Gravity Systems Studies							
M487	Habitability/crew quarters	OWS	C.C. Johnson, JSC	MSFC	x	x	x
M509	Astronaut manoeuvring equipment	OWS	Maj C.E. Whitsett Jr. USAF Space and Missile Systems Org., and co-investigator B. McCandless, NASA astronaut, JSC	JSC		x	x
M518	Crew activity and maintenance study	OWS	R.L. Bond, JSC	JSC	x	x	x
T002	Manual navigation sightings	OWS	R.J. Randle, ARC	ARC		x	x
T013	Crew/vehicle disturbance	OWS	B.A. Conway, LaRC	LaRC		x	
T020	Foot-controlled manoeuvring unit	OWS	D.E. Hewes, LaRC	LaRC		x	x

Spacecraft Environment							
D008	Radiation in spacecraft	CM	Capt A.D. Grimm, USAF Kirtland AFB	AF, JSC	x		
D024	Thermal control coating	AM	Dr W. Lehn, Wright-Patterson AFB	AF, JSC	x		
M415	Thermal control coatings	IU	E.C. McKannan, MSFC	MSFC	x		
T003	In-flight aerosol analysis	OWS	Dr W.Z. Leavitt, Dept of Transportation	MSFC	x	x	x
T025	Coronagraph contamination measurements	OWS	Dr M. Greenberg, Dudley Observatory	JSC	x	x	x
T027	ATM contamination measurements	OWS	Dr J.A. Muscari, Martin-Marietta Corporation	MSFC	x	x	x
Skylab Student Project			*Secondary School Student Winners*				
ED11	Absorption of radiant heat in the Earth's atmosphere[9]	–	J.B. Zmolek, Oshkosh, Wisconsin	MSFC	x	x	x
ED12	Space observe/prediction of volcanic eruption[9]	–	T.A. Crites, Kent, Washington	MSFC	x	x	x
ED21	Photography of liberation clouds[9]	–	A. Hopfield, Princeton, New Jersey	MSFC		x	
ED22	Possible confirmation of objects within Mercury's orbit[9]	–	D.C. Bochsler, Silverton, Oregon	MSFC	x	x	x
ED23	Spectrography of selected quasars[9]	–	J.C. Hamilton, Alea, Hawaii	MSFC	x		
ED24	X-ray Content in association with[9] stellar spectral classes[9,10]	–	J.W. Reihs, Baton Rouge, Louisiana	MSFC			x
ED25	X-ray emission from the planet Jupiter[9,11]	–	J. Leventhal, Berkeley, California	MSFC		x	
ED26	A search for pulsars in UV wavelength[9]	–	N.W. Shannon, Atlanta, Georgia	MSFC	x		
ED31	Behaviour of bacteria and bacterial spores in the Skylab space environment	OWS	R.L. Staehle, Rochester, New York	MSFC	x		
ED32	An *in vitro* study of selected isolated immune phenomena	OWS	T.A. Meister, Jackson Heights, New York	MSFC		x	
ED41	A quantitative measure of motor sensory performance during prolonged flight in zero gravity	OWS	K.L. Jackson, Houston, Texas	MSFC			x
ED52	Web formation in zero gravity	OWS	J.S. Miles, Lexington, Massachusetts	MSFC		x	

No.	Designation	Location on Skylab	Principle Investigator	Field Center	Manned mission 1	2	3
ED61	Plant growth in zero gravity	OWS	J.G. Wordekemper, West Point, Nebraska	MSFC			x
ED62	Phototropic orientation of an embryo plant in zero-gravity	OWS	D.W. Schlack, Downey, California	MSFC			x
ED63	Cytoplasmic streaming in zero gravity[12]	OWS	C.A. Peltz, Littleton, Colorado	MSFC		x	
ED72	Capillary action studies in a state of free fall[13]	OWS	R.G. Johnson, St Paul, Minnesota	MSFC			x
ED74	Zero-gravity mass measurement	OWS	V.W. Converse, Rockford, Illinois	MSFC		x	
ED76	Earth orbital neutron analysis	OWS	T.C. Quist, San Antonio, Texas	MSFC	x	x	x
ED78	Wave motion through a liquid in zero gravity[14]	OWS	W.B. Dunlap, Youngstown, Ohio	MSFC			x

Abbreviations

AFB	Air Force Base	AM	Airlock Module	OWS	Orbital Workshop
ARC	Ames Research Center, Moffett Field, California	ATM	Apollo Telescope Mount	SAL	Scientific Airlock
LaRC	Langley Research Center, Hampton, Virginia	CM	Command Module		
JSC	Johnson Space Center, Houston, Texas	EVA	Extravehicular Activity		
MSFC	Marshall Spaceflight Center, Huntsville, Alabama	IU	Instrument Unit		
USN	United States Navy	MDA	Multiple Docking Adapter		

Notes

1 The solar airlock was blocked by the parasol sunshade, and the experiment could not be operated as planned by second crew, although the third crew operated it by EVA.
2 Component failure caused the instrument to shut off after operating 110 minutes of a planned 265 minutes.
3 Deployed through antisolar airlock and left between first and second manned mission.
4 Fore and aft scanning failed. Following repair by the third crew, the fault was locked out and cross-track scanning restored. 80% of data were recovered.
5 Short-circuit in equipment prevented acquisition of telemetric data.
6 See note 5.
7 M512 was a multipurpose vacuum chamber with an electron bream generator, used for conducting the next four experiments.
8 M518 was an electric furnace attached to M512, used in performing the next eleven experiments.

9 No special equipment required. Experiments used data from other Skylab sensors.
10 Skylab's X-ray detectors were not sufficiently sensitive to collect the data that this experiment required.
11 Could not be performed. When Jupiter was in the best observing position, the power crisis prevented manoeuvring to point at the target. An alternative target was below the detection limit of the Skylab sensors.
12 Only partially completed. The water plants used in this experiment did not live long enough for the planned observations to be carried out – except for one successful observation.
13 Leakage of fluids from the experiment hardware led to inconclusive results.
14 Hardware failure negated this experiment.

Adapted from

Skylab: A Guidebook (NASA EP-107, 1973), Chapter VIII – Listing of Skylab Experiments, pp.228–232
Living and Working in Space: A History of Skylab (NASA SP-4208, 1983), Appendix D, Experiments, pp.381–386

Bibliography

In addition to the references listed below, considerable assistance was afforded by the staff of the NASA History Office at JSC and at Rice University, Houston, Texas (also for the Curt Michel collection) in accessing the Skylab archives. Not all references are listed here, but I have included the most important references pertaining to the data in this book.

In addition, over many years former NASA astronauts Bill Thornton and Jerry Carr have repeatedly granted access to their personal archives or via personal interviews recalling events and experiences from almost three decades ago. Their direct participation in Skylab allowed a greater understanding of the 'bigger picture', and their assistance, as well as the interviews, comments and correspondence offered by the other former NASA astronauts listed in the acknowledgements, has greatly enhanced my research, although the recording of this detail in this book is my own personal interpretation of the material.

Taped interviews including Skylab discussions

Jerry Carr: August 1988, August 1989, September 1994, December 1997, December 1998, and July 2000
Bill Pogue: September 1999, and July 2000
Vance Brand: August 1989
Don Lind: May 2000
Story Musgrave: August 1988
Bill Thornton: August 1988, August 1989, December 1997, and June 2000
Bruce McCandless: August 1989
Karl Henize: August 1988, August 1989, and July 1991
Curt Michel: June 1992
Walt Cunningham: July 1989, and August 1989

Reference sources and further reading

These books offer not only further information on the history, development and operations of Skylab, but also on the evolution of the Apollo hardware, both

American and Soviet/Russian approaches to extended-duration spaceflight, and the problems that are associated with such efforts.

1963-1979 Aeronautics and Astronautics, NASA SP-4000 annual reference series

1966 Medical Aspects of an Orbiting Research Laboratory, Space Medicine Advisory Group Study (Jan–Aug 1964), ed S.P. Vinograd, NASA SP-86

1969 *Living in Space: the astronaut and his environment*, Mitchell R. Sharpe, Aldus Books

1970 *The Making of an Ex-Astronaut*, Brian O'Leary, Michael Joseph

1971 *Pioneering in Outer Space*, Hermann Bondi *et al*, Heinemann Educational Books

1973 Skylab News Reference, March, NASA PAO, Washington DC

1973 Skylab Preliminary Chronology, Roland Newkirk, NASA HHN-130, May 1973

1973 Skylab 1/2 Technical Crew Debriefing, June 30, 1973, JSC-08053, NASA JSC

1973 Skylab 1/3 Technical Crew Debriefing, 4 October1973, JSC-08478, NASA JSC

1973 Skylab Astronyms, Philco-Ford Corporation, Houston, PHO-BR2-73-1500

1973 Skylab: a Guidebook, Leland F. Belew & Ernst Stuhlinger, NASA EP-107

1973 Skylab: Diary of a Rescue Mission, David Baker, *Spaceflight,* **15**, Nos. 9, 10 and 11, British Interplanetary Society (BIS)

1974 Skylab 1/4 Technical Crew Debriefing, February 22 1974 JSC-08809, NASA JSC

MSFC Skylab Mission Report – Saturn Workshop, NASA Technical Memorandum NASA TM-X-64814 (from the files of Gerald P. Carr)

1974 Proceedings of the Skylab Life Sciences Symposium (August 27–29,1974), LBJ Space Center, Volume I and Volume II, JSC – 09275, NASA TMX – 58154 (November)

1974 Skylab: 59 Days in Space, David Baker, *Spaceflight,* **16**, nos. 2, 5, 6 and 8, BIS

1974 Skylab: The Three-Month Vigil, David Baker, *Spaceflight,* **16**, nos. 11 and 12, **17** No. 1, BIS (1975)

1975 Lessons learned on the Skylab programme, NASA; Skylab Program, MSFC, 22 February; Saturn Program Office, MSFC, 22 February; Johnson Space Center, 6 March; Skylab Program Office, Engineering Directorate, NASA HQ, 11 March; Kennedy Space Center, 1 April

1977 Skylab: a Chronology, Roland Newkirk, Ivan Ertel with Courtney Brooks, NASA SP-4001

1977 *A House in Space*, Henry S.F. Cooper Jr, Angus & Robertson (UK)

1977 Biomedical Results from Skylab, ed. Richard S. Johnston & Lawrence F. Dietlein, NASA SP-377

1977 Skylab Explores the Earth, NASA SP-380

1977 Skylab: Our First Space Station, ed. Leland Belew, NASA SP-400

1977 *The All-American Boys*, Walter Cunningham, Macmillan

1978 Skylab EREP Investigations Summary, NASA SP-399

1978 The Partnership: A History of the Apollo–Soyuz Test Project, Edward Clinton Ezell and Linda Newman Ezell, NASA SP-4209

1979 A New Sun: the Solar results from Skylab, John A. Eddy, NASA SP-402

1979 Skylab's Astronomy and Space Sciences, ed. Charles A. Lundquist, NASA SP-404 (from the Gerald P. Carr collection)

1979 Chariots for Apollo: a History of Manned Lunar Spacecraft, Courtney Brooks, James Grimwood and Lloyd Swenson, NASA SP-4205

1979 Spaceflight research relevant to health, physical education and recreation with particular reference to Skylab life science experiments, NASA EP-148, (June)

1980 *History of Manned Spaceflight*, David Baker, New Cavendish

1980 The Manned Orbiting Laboratory, Curtis Peebles, Part 1, *Spaceflight*. **22**, 4 April 1980; Part 2, **22**, 6 June 1980, Part 3, **24**, 6 June 1982

1981 Stages to Saturn: a technological history of Apollo–Saturn, Roger E. Bilsein, NASA SP-4206

1982 Space Physiology and Medicine, Arnauld Nicogossian and James F. Parker, NASA SP-447

1983 Living and Working In Space: a History of Skylab, W. David Compton and Charles D. Brown, NASA SP-4208

1984 *Spacelab – Research in Earth Orbit*, David Shapland and Michael Rycroft, Cambridge

1984 *The Voyages of Columbia*, Richard S. Lewis, Columbia University Press

1985 Civilian Space Stations and the US future in Space, Office of Technical Assessment, US Congress, (November)

1986 Living Aloft: Human Requirements for Extended Spaceflight, Mary Connors, Albert Harrison and Faren Akins, NASA SP-483

1986 The Human Factor: Biomedicine in the Manned Space Program to 1980, John A. Pits, NASA SP-4213

1986 *Space Patches*, Judith Kaplan and Robert Muniz, Sterling Publishing Inc.

1987 Spacelab: an International Success Story, Douglas R. Lord, NASA SP-487

1988 NASA Historical Data Book Volume II, Programs and Projects 1958—1968; and Volume III, Programmes and Projects 1969—1978, Linda Newman Ezell, NASA SP-4012

1988 *Lift-off: the Story of America's Adventure in Space*, Michael Collins, Grove Press

1989 *Survival In Space*, Richard Harding (with Foreword by Joe Kerwin) , Routledge

1991 *How do you go to the bathroom in space?* William R. Pogue, TOR Books

1992 *Space Shuttle: the Beginning through STS-50*, Dennis R. Jenkins
1993 Suddenly, Tomorrow Came A History of the Johnson Space Center, Henry C. Dethloff, NASA SP-4307
1994 *The US Manned Space Program from Mercury to the Shuttle*, Donald K. Slayton with Michael Cassutt, Forge Books
1994 *US Space Gear: Outfitting the Astronaut*, Lillian D. Kozloski, Smithsonian
1995 Exploring the Unknown, Selected Documents in the History of the US Civil Space Program, ed. John M. Logsdon, NASA SP-4407 (3 vols.)
1995 Mir Hardware Heritage, David S.F. Portree, NASA RP-1357, (March)
1996 *Living and Working in Space*, Philip Robert Harris, Second edition, Wiley–Praxis
1996 *The New Russian Space Programme,* Brian Harvey, Wiley–Praxis
1997 Walking to Olympus: an EVA Chronology, David S.F. Portree and Robert C. Treviño NASA Monograph #7
1997 *The Mir Space Station*, David M. Harland, Wiley–Praxis
1997 *Living in Space,* G. Harry Stine, Evans Co, New York,
1998 *Who's Who In Space: the International Space Station Edition,* Michael Cassutt, Macmillan
1998 *The Space Shuttle: Roles, Missions and Accomplishments*, David M. Harland, Wiley–Praxis
1998 *Dragonfly: NASA and the Crisis aboard Mir*, Brian Burrough, Harper Collins
1999 *Power to Explore: a History of Marshall Space Flight Center, 1960–1990* Andrew J. Dunar and Stephen P. Waring, NASA SP-4313
1999 *Waystation to the Stars: the Story of Mir, Michael and Me*, Colin Foale, Headline
1999 *Off The Planet*, Jerry M. Linenger, McGraw-Hill
2000 *Accidents and Disasters in Manned Spaceflight*, David J. Shayler, Springer–Praxis
2000 *Challenges of Human Space Exploration*, Marsha Freeman, Springer–Praxis
2000 Challenge to Apollo: The Soviet Union and the Space Race, 1945–1974, Asif A. Siddiqi, NASA SP-2000-4408
2000 Skylab: the fall and demise of America's first space station, M. Damohn, *JBIS*, **53**
2000 *The History of Mir, 1986–2000*, ed. Rex Hall, British Interplanetary Society
2001 Flight of the Falcons: the 18-day Space Marathon of Soyuz 9, David J. Shayler, *JBIS* **54**, 1/2 (January/February)

Index